Math

ADVANTAGE

HARCOURT
BRACE

Orlando • Atlanta • Austin • Boston • San Francisco • Chicago • Dallas • New York • Toronto • London

http://www.hbschool.com

Printed in the United States of America

ISBN 0-15-310692-1

5 6 7 8 9 10 030 03 02 01

Senior Authors

Grace M. Burton
Chair, Department of Curricular Studies
Professor, School of Education
University of North Carolina at Wilmington
Wilmington, North Carolina

Evan M. Maletsky
Professor of Mathematics
Montclair State University
Upper Montclair, New Jersey

Authors

George W. Bright
Professor of Mathematics Education
The University of North Carolina at Greensboro
Greensboro, North Carolina

Sonia M. Helton
Professor of Childhood Education
Coordinator, College of Education
University of South Florida
St. Petersburg, Florida

Loye Y. (Mickey) Hollis
Professor of Mathematics Education
Director of Teacher Education and Undergraduate Programs
University of Houston
Houston, Texas

Howard C. Johnson
Dean of the Graduate School
Associate Vice Chancellor for Academic Affairs
Professor, Mathematics and Mathematics Education
Syracuse University
Syracuse, New York

Joyce C. McLeod
Visiting Professor
Rollins College
Winter Park, Florida

Evelyn M. Neufeld
Professor, College of Education
San Jose State University
San Jose, California

Vicki Newman
Classroom Teacher
McGaugh Elementary School
Los Alamitos Unified School District
Seal Beach, California

Terence H. Perciante
Professor of Mathematics
Wheaton College
Wheaton, Illinois

Karen A. Schultz
Associate Dean and Director of Graduate Studies and Research
Research Professor, Mathematics Education
College of Education
Georgia State University
Atlanta, Georgia

Muriel Burger Thatcher
Independent Mathematics Consultant
Mathematical Encounters
Pine Knoll Shores, North Carolina

Advisors

Anne R. Biggins
Speech-Language Pathologist
Fairfax County Public Schools
Fairfax, Virginia

Carolyn Gambrel
Learning Disabilities Teacher
Fairfax County Public Schools
Fairfax, Virginia

Lois Harrison-Jones
Education Consultant
Dallas, Texas

Asa G. Hilliard, III
Fuller E. Callaway Professor of Urban Education
Georgia State University
Atlanta, Georgia

Marsha W. Lilly
Secondary Mathematics Coordinator
Alief Independent School District
Alief, Texas

Judith Mayne Wallis
Elementary Language Arts/ Social Studies/Gifted Coordinator
Alief Independent School District
Houston, Texas

CONTENTS

Getting Ready for Grade 1 1

Theme: All About Me

*** Algebra Readiness**

CHAPTER 1 Understanding Addition .23

Theme: Silly Stories

CHAPTER 2 Understanding Subtraction .37

Theme: Fun and Games

Chapters 1–2 ✓Checkpoint

* Algebra Readiness

ADDITION AND SUBTRACTION FACTS TO 10 — CHAPTERS 3–6

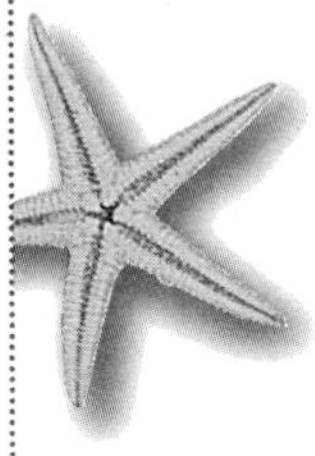

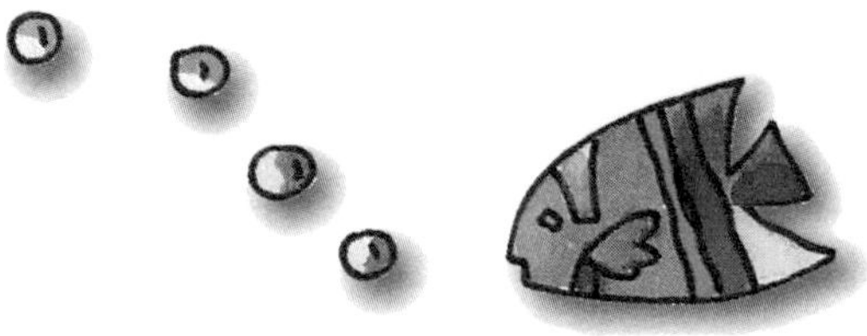

* **Algebra Readiness**

CHAPTER 5 Subtraction Combinations 85

Theme: Forest Creatures

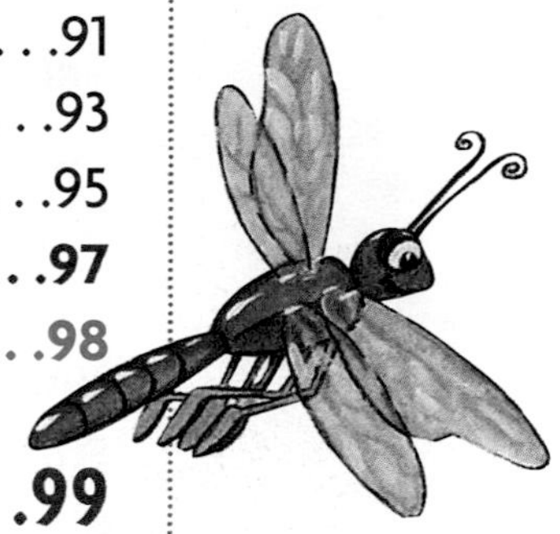

CHAPTER 6 Subtraction Facts to 10 99

Theme: In My Backyard

Chapters 3–6 ✓Checkpoint

* **Algebra Readiness**

GEOMETRY, SPATIAL SENSE, AND PATTERNS — CHAPTERS 7–10

*** Algebra Readiness**

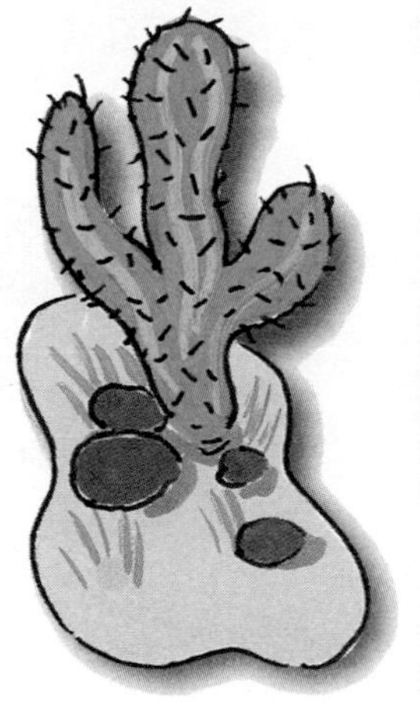

CHAPTER 9 Location and Movement 145

Theme: Westward Ho!

CHAPTER 10 Patterns 157

Theme: Patterns in My World

Chapters 7–10 ✓Checkpoint

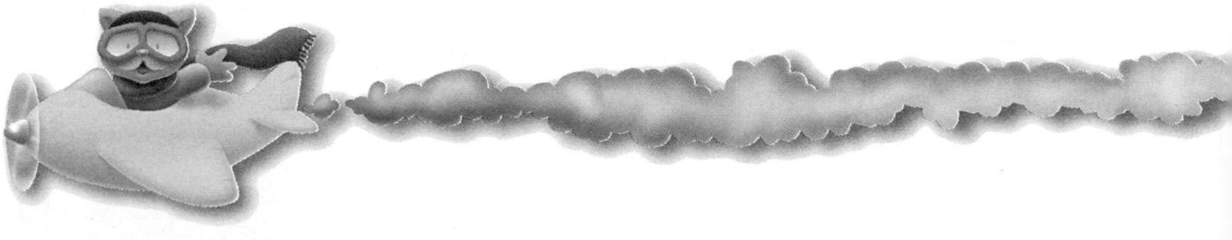

* **Algebra Readiness**

ADDITION AND SUBTRACTION FACTS TO 12 — CHAPTERS 11–12

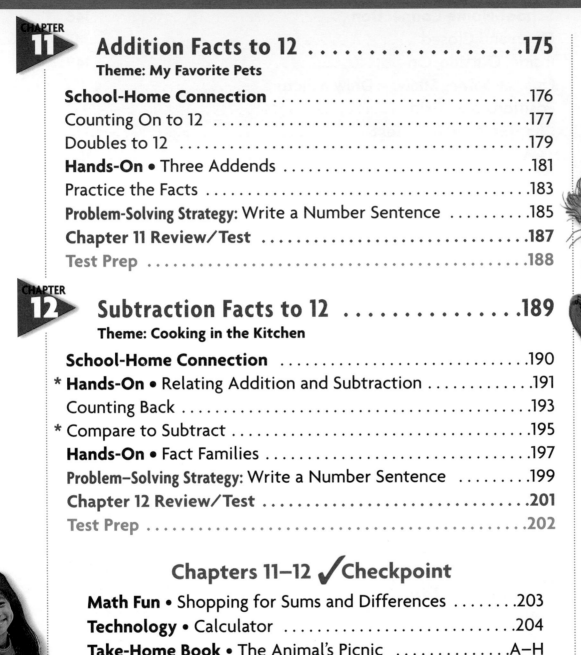

* **Algebra Readiness**

PLACE VALUE AND NUMBER PATTERNS — CHAPTERS 13–15

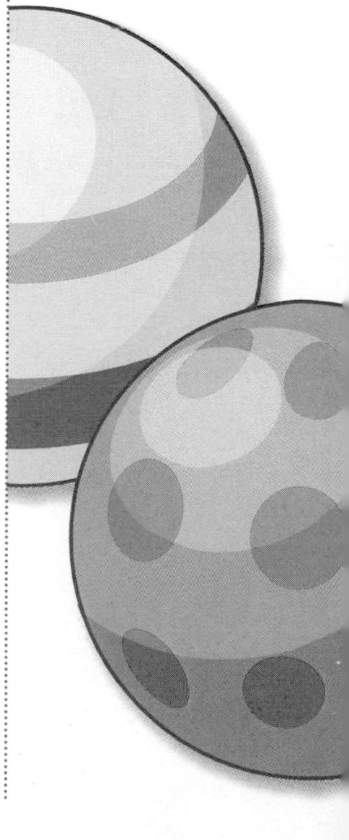

MONEY AND TIME — CHAPTERS 16–19

CHAPTER 16

Counting Pennies, Nickels, and Dimes . . .257

Theme: Animals! Animals!

CHAPTER 17

Using Pennies, Nickels, and Dimes 271

Theme: What's for Sale?

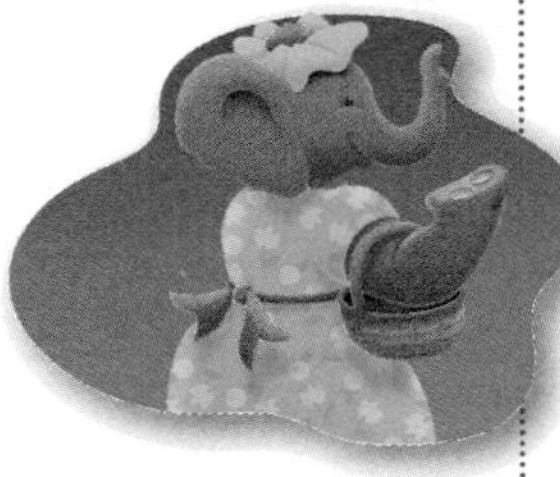

* **Algebra Readiness**

CHAPTER 18

Using a Calendar 285

Theme: Everyday Events

CHAPTER 19

Telling Time 297

Theme: My School Day

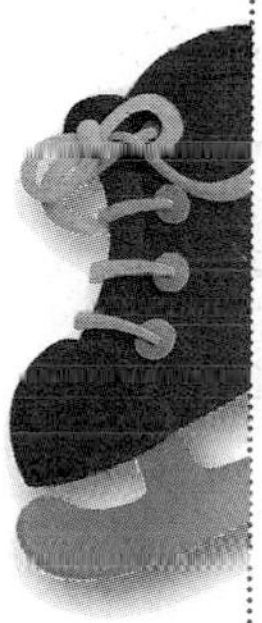

Chapters 16–19 ✓Checkpoint

* **Algebra Readiness**

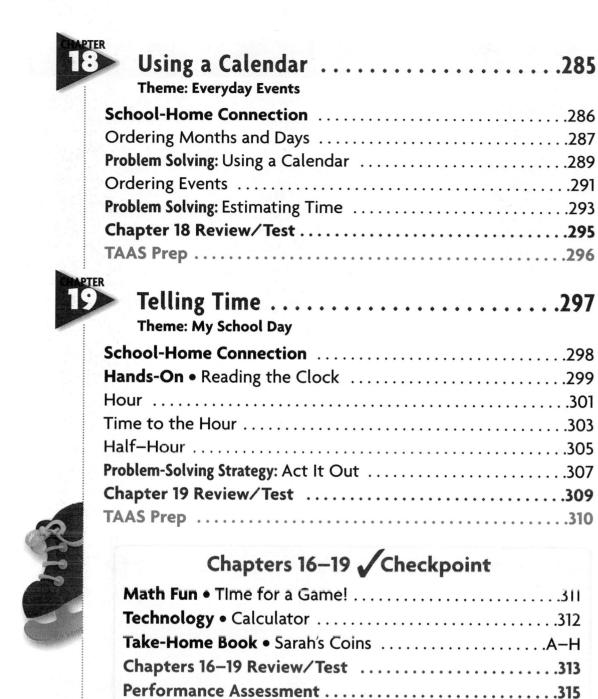

CHAPTER 22

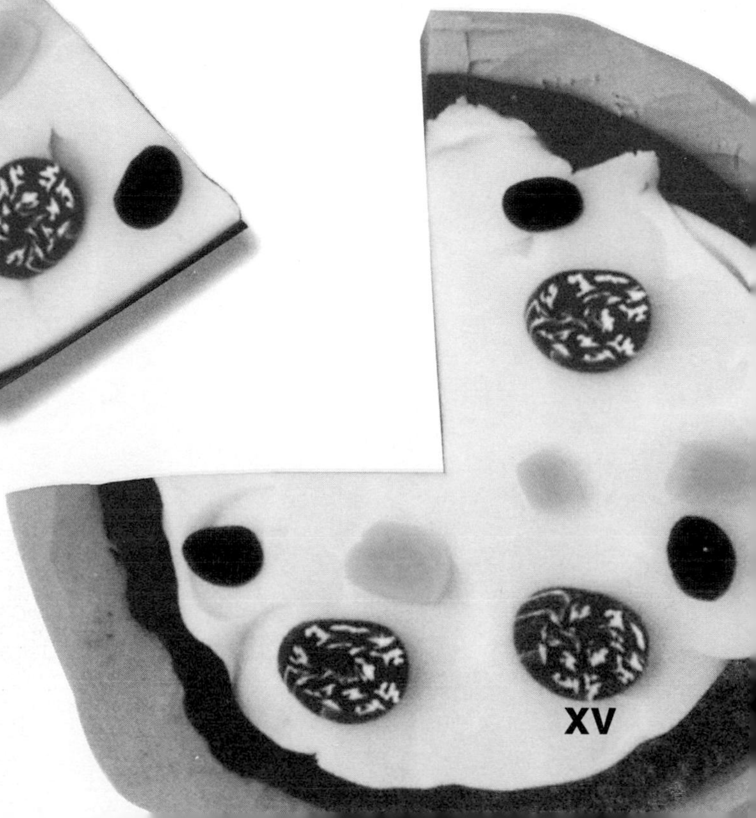

*** Algebra Readiness**

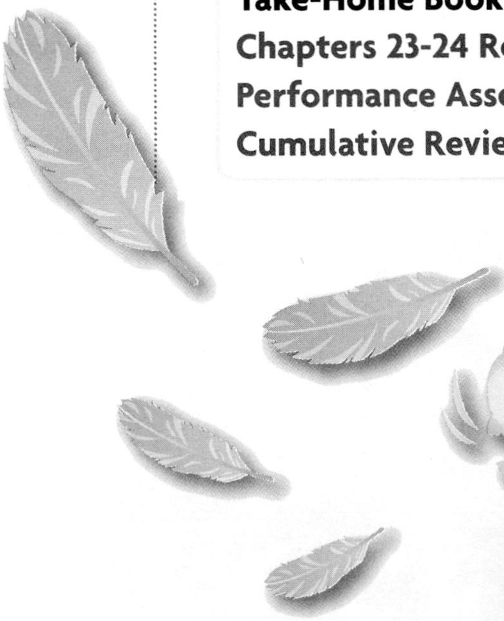

* **Algebra Readiness**

ADDITION AND SUBTRACTION FACTS TO 18

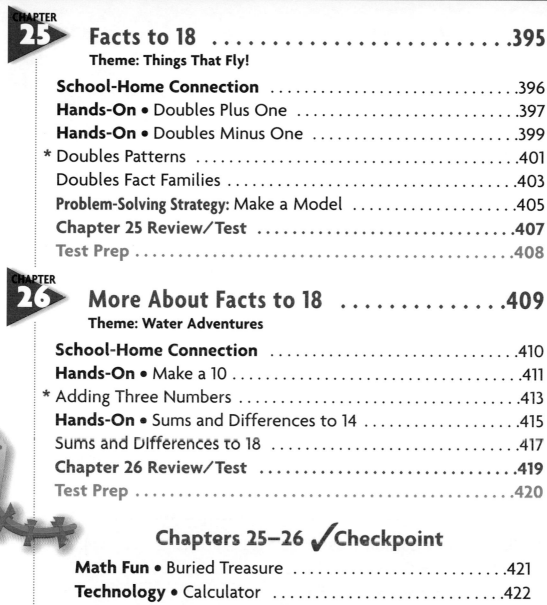

CHAPTERS 25–26

Chapters 25–26 ✓Checkpoint

*** Algebra Readiness**

EXPLORING MULTIPLICATION, DIVISION, AND TWO-DIGIT ADDITION AND SUBTRACTION CHAPTERS 27–28

CHAPTER 27

CHAPTER 28

Chapters 27–28 ✓Checkpoint

* **Algebra Readiness**

Getting Ready for Grade 1

Hi! My name is K.C.
I will be your math friend.
I am really excited about
starting first grade.

This book belongs to

____

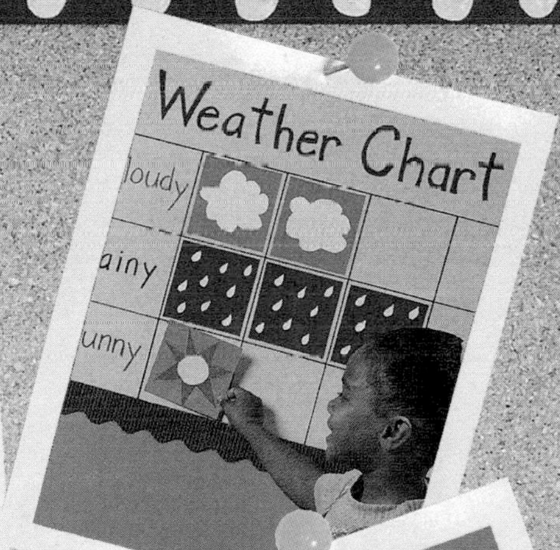

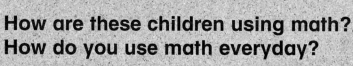

How are these children using math?
How do you use math everyday?

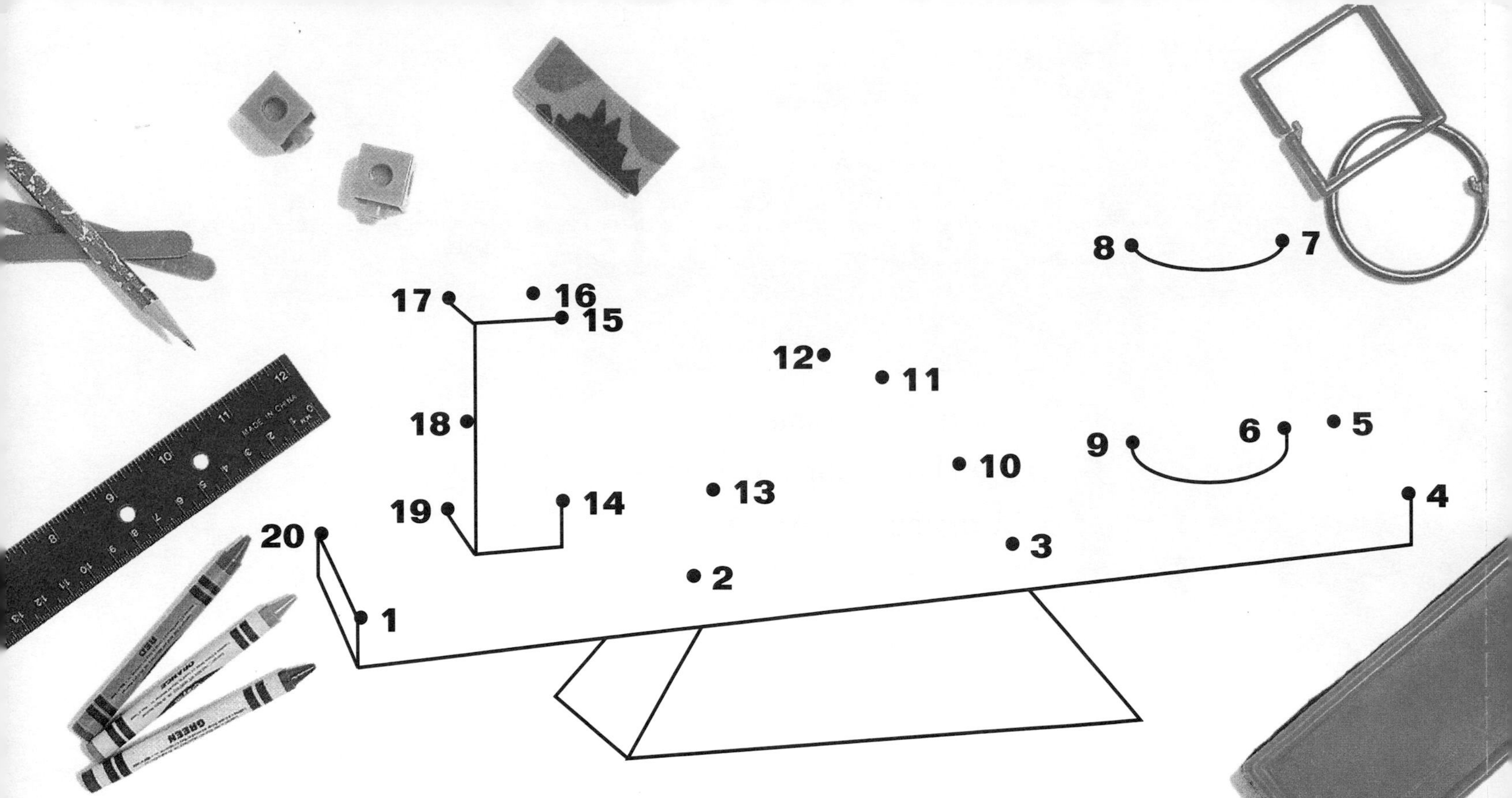

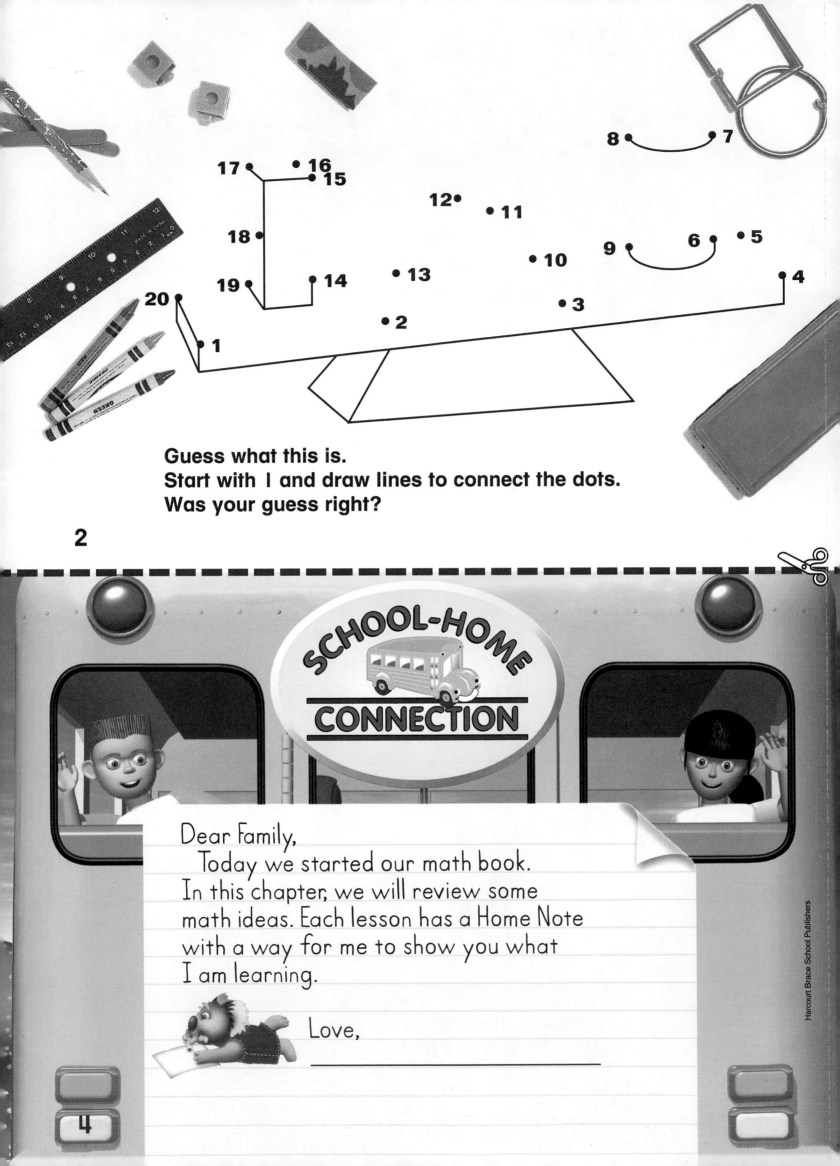

Guess what this is.
Start with 1 and draw lines to connect the dots.
Was your guess right?

SCHOOL-HOME CONNECTION

Dear Family,

Today we started our math book. In this chapter, we will review some math ideas. Each lesson has a Home Note with a way for me to show you what I am learning.

Love,

____

Name ____________________

One-to-One Correspondence

1

2

3

4

1. Draw one knife for each fork.
2. Draw one saucer for each cup.
3. Draw one straw for each glass.
4. Draw one spoon for each bowl.

REVIEW

REVIEW

1
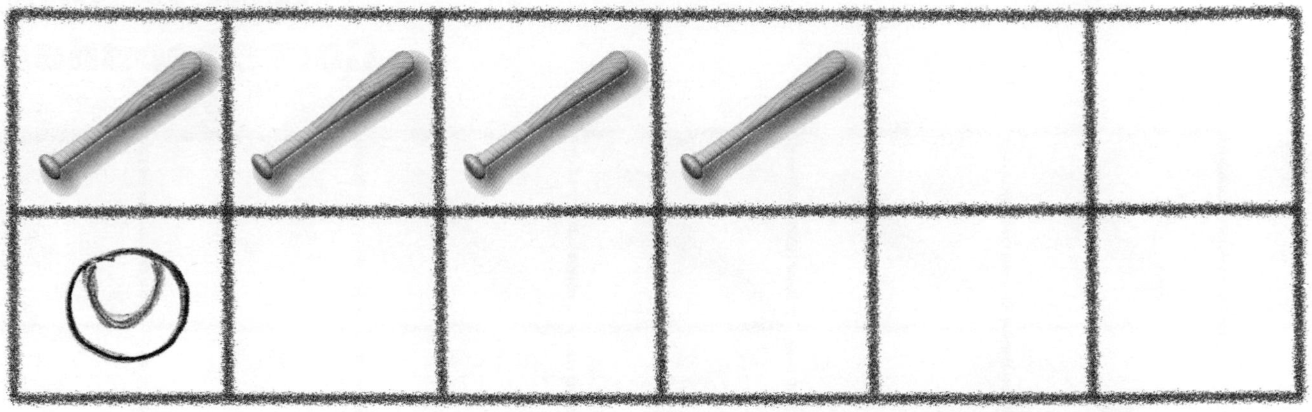

2

3

4
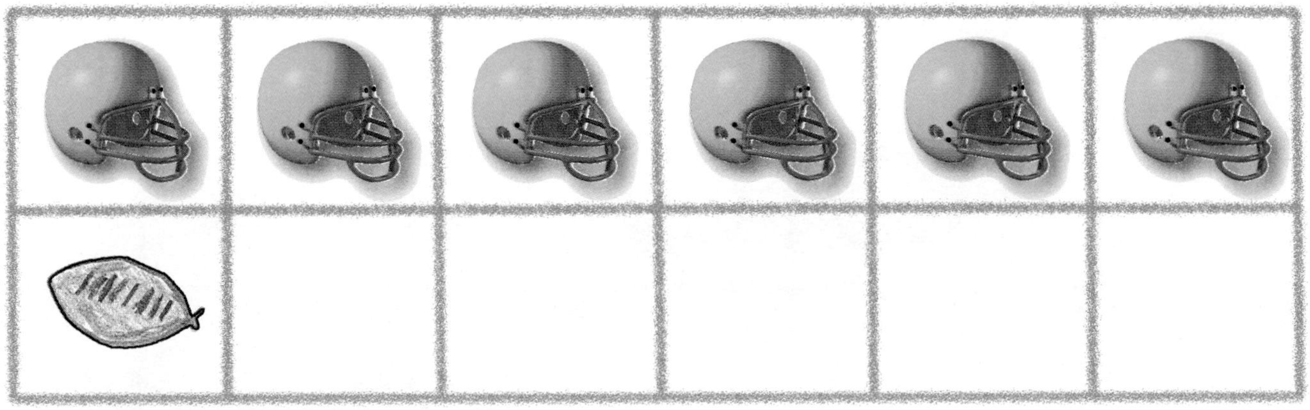

1. Draw one baseball for each bat.
2. Draw one basketball for each basket.
3. Draw one tennis ball for each racket.
4. Draw one football for each helmet.

Home Note **Your child used one-to-one correspondence to make equal groups.**
ACTIVITY **Set out a group of objects. Have your child set out a group that has one object for each object in your group.**

Name ____________________

More and Fewer

1

2

3
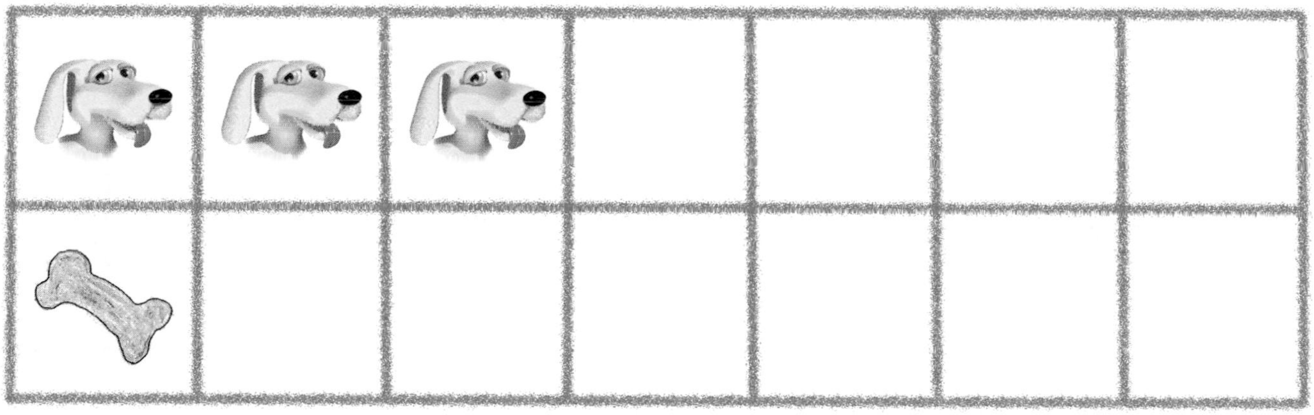

4

Draw bones to show more bones than dogs.

REVIEW

REVIEW

1

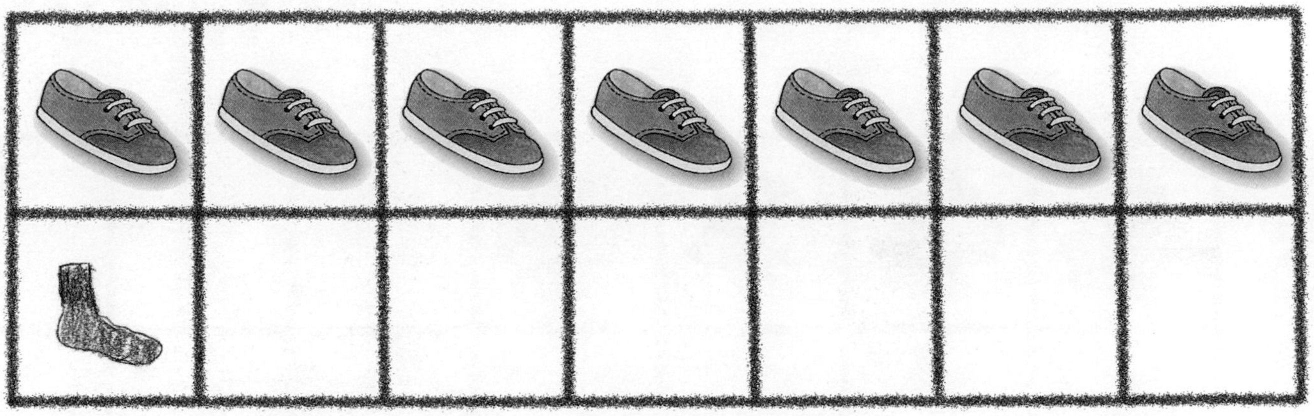

2

3

4

Draw socks to show fewer socks than shoes.

Home Note Your child used one-to-one correspondence to make groups with more or fewer. **ACTIVITY** Set out a group of small objects. Have your child set out a group that has more and then a group that has fewer objects than your group.

Name ______________________________

Patterns • Numbers Through 5

1

1
one

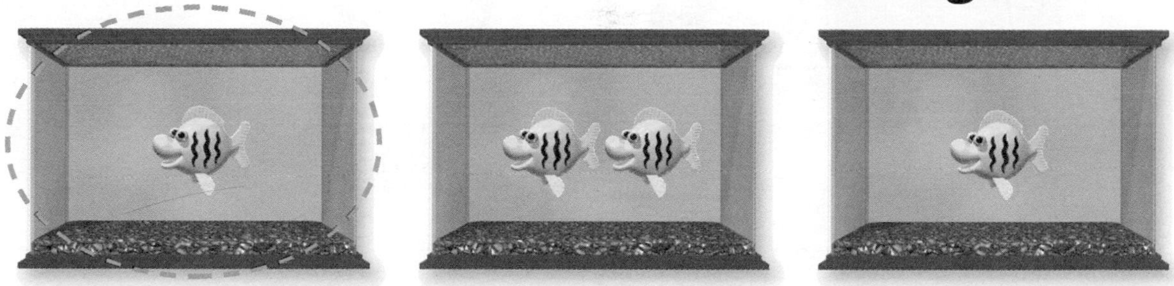

2

2
two

3

3
three

4

4
four

5

5
five

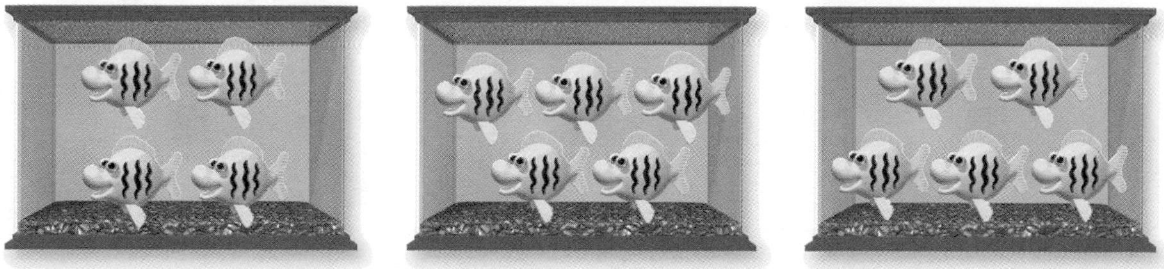

6

0
zero

Count. Circle the groups of fish that show the number.

REVIEW

REVIEW

1 0 1 2 3

4 5

2  2

3 

4 

5 

6 

7 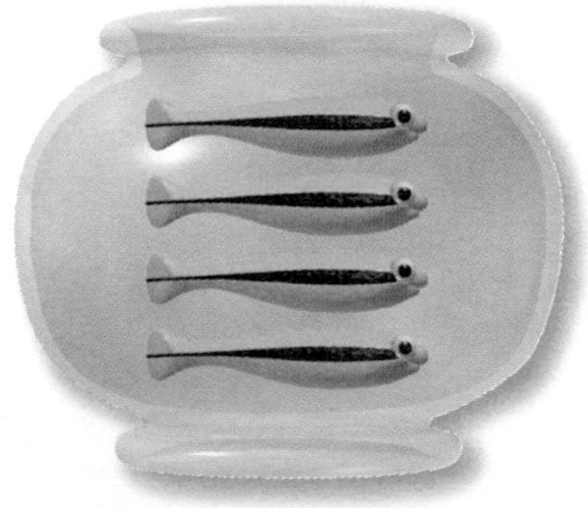

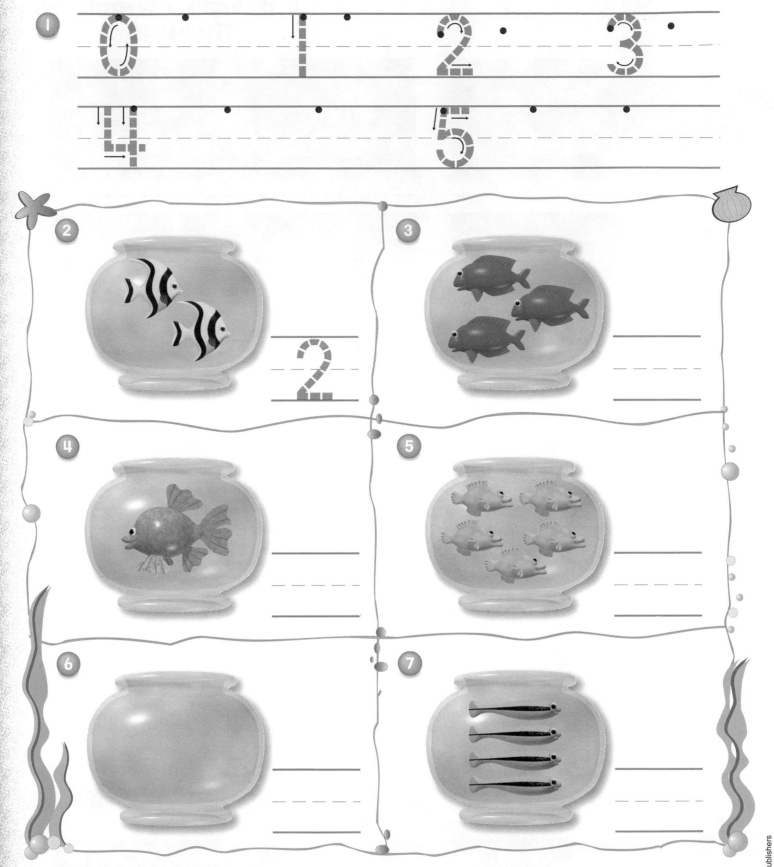

1. Write the numbers.
2.–7. Count. Write the number that tells how many fish.

Home Note Your child identified groups of 0 to 5 objects.
ACTIVITY Put out groups of 1 to 5 objects. Have your child count and tell how many. Put 2 objects in a container, and then take both of them out. Have your child tell how many are then in the container (zero).

Name ______________________________

Patterns • Numbers Through 9

1

6

six

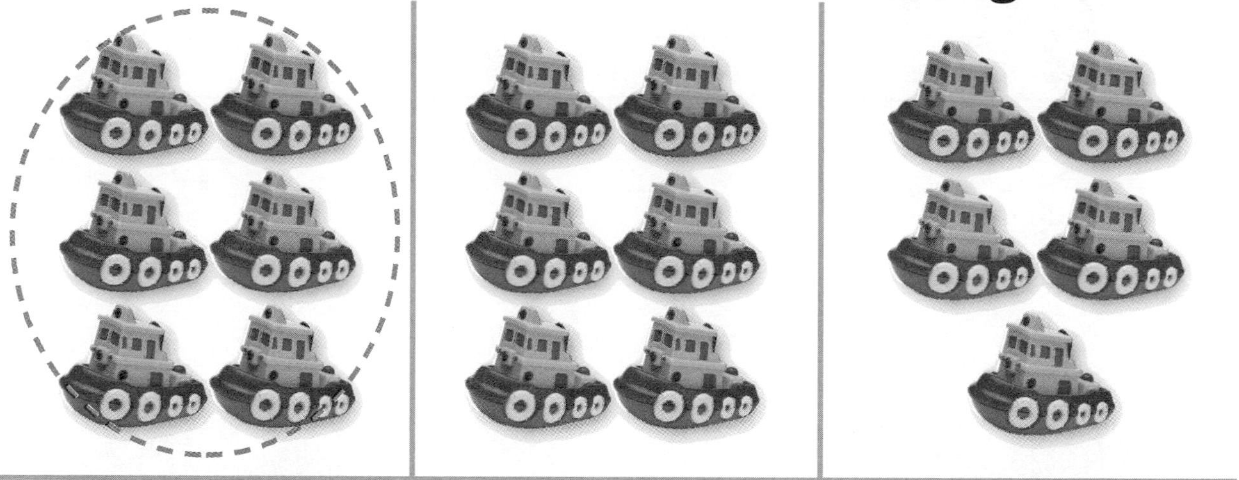

2

7

seven

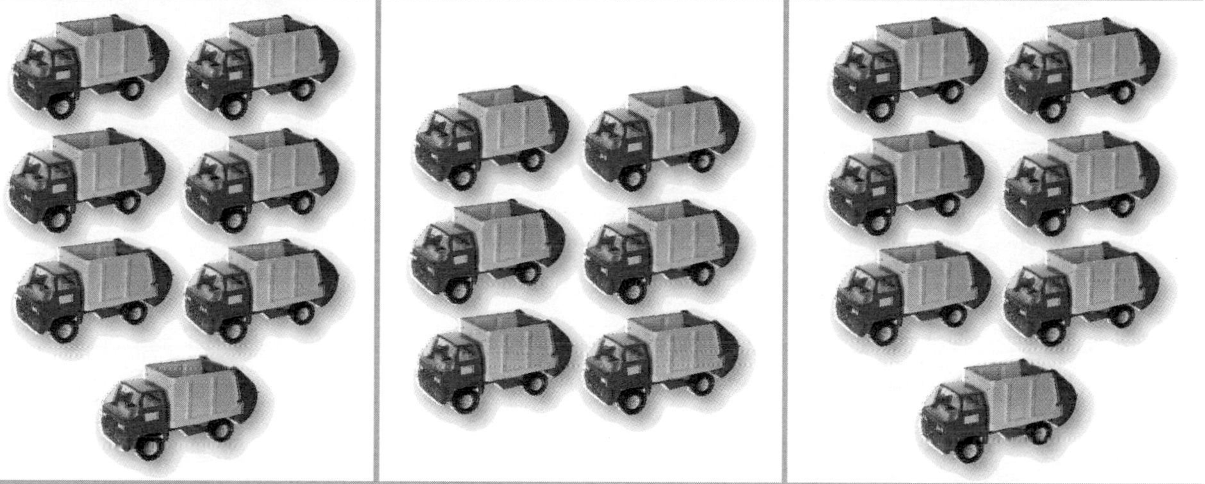

3

8

eight

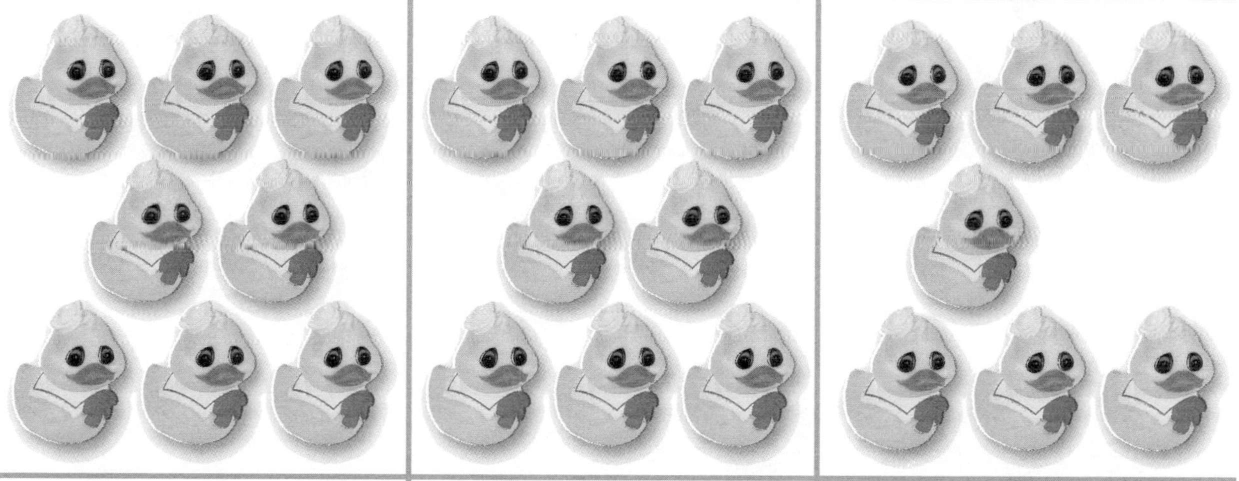

4

9

nine

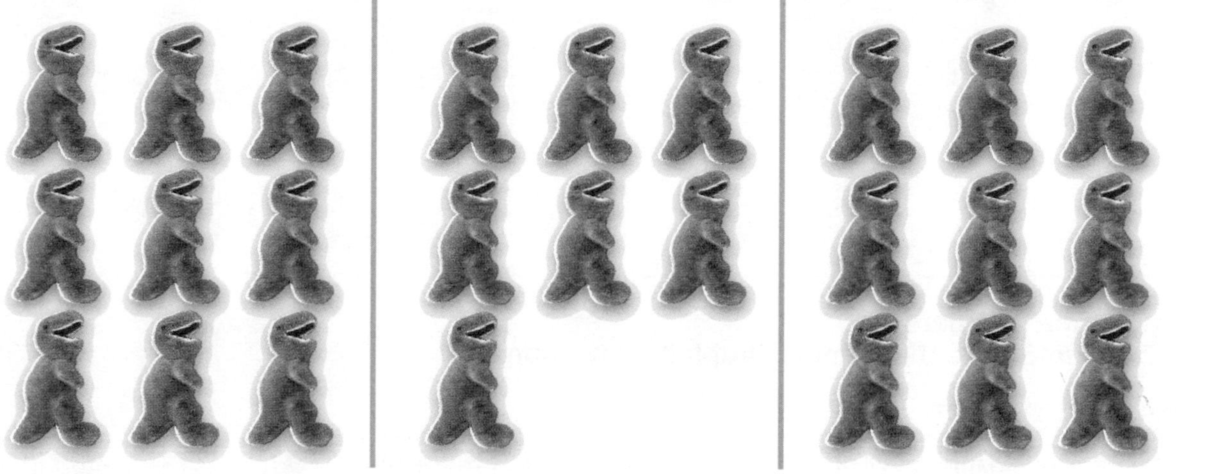

Count. Circle the groups that show the number.

REVIEW

REVIEW

1 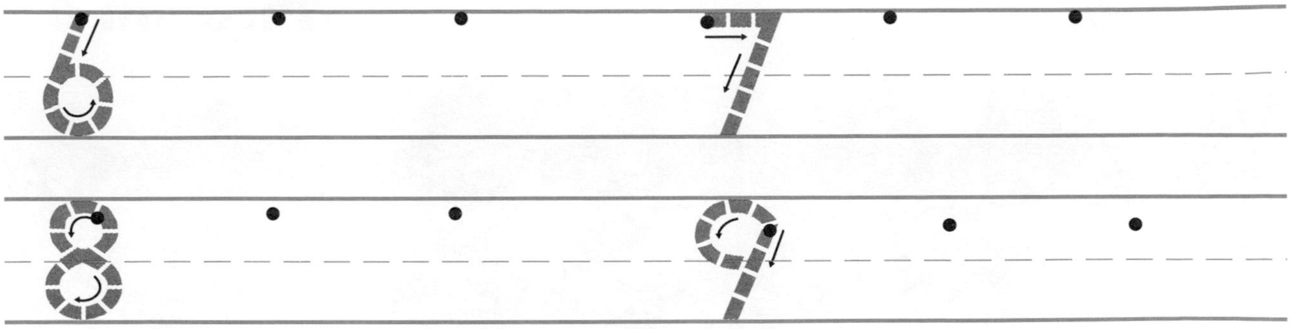

2 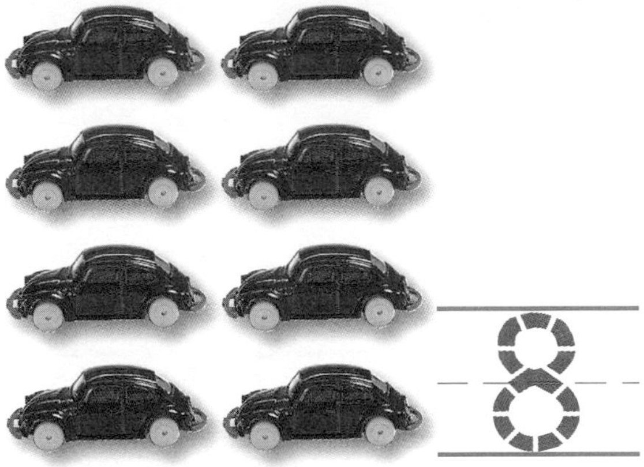

3 

4 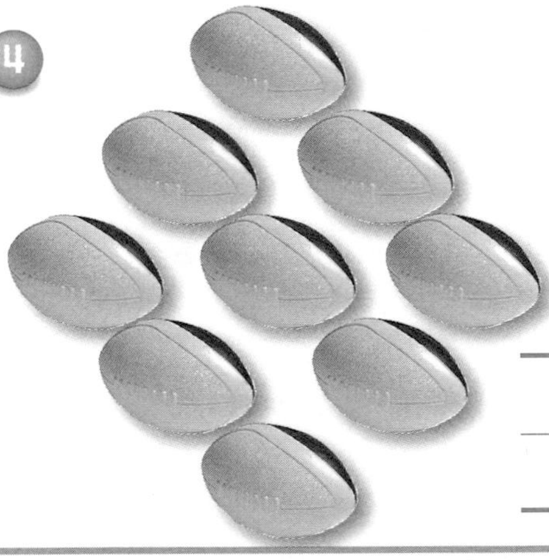

5 

6 

7 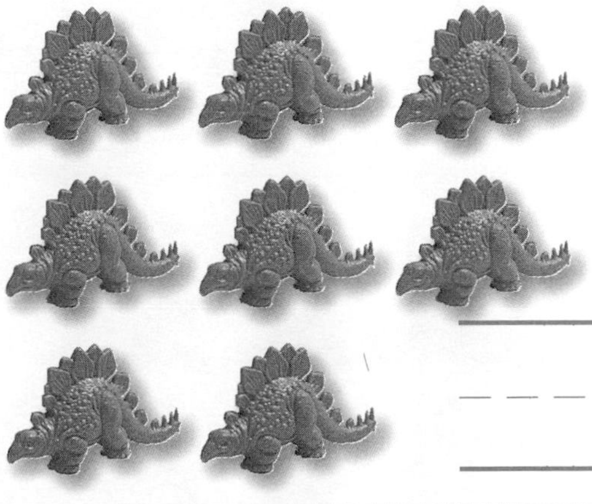

1. Write the numbers.
2.–7. Count. Write the number that tells how many.

Home Note Your child identified groups of 6 to 9 objects.
ACTIVITY Have your child make groups of objects to show the numbers 6 through 9.

Name ______________________________

Ten

1

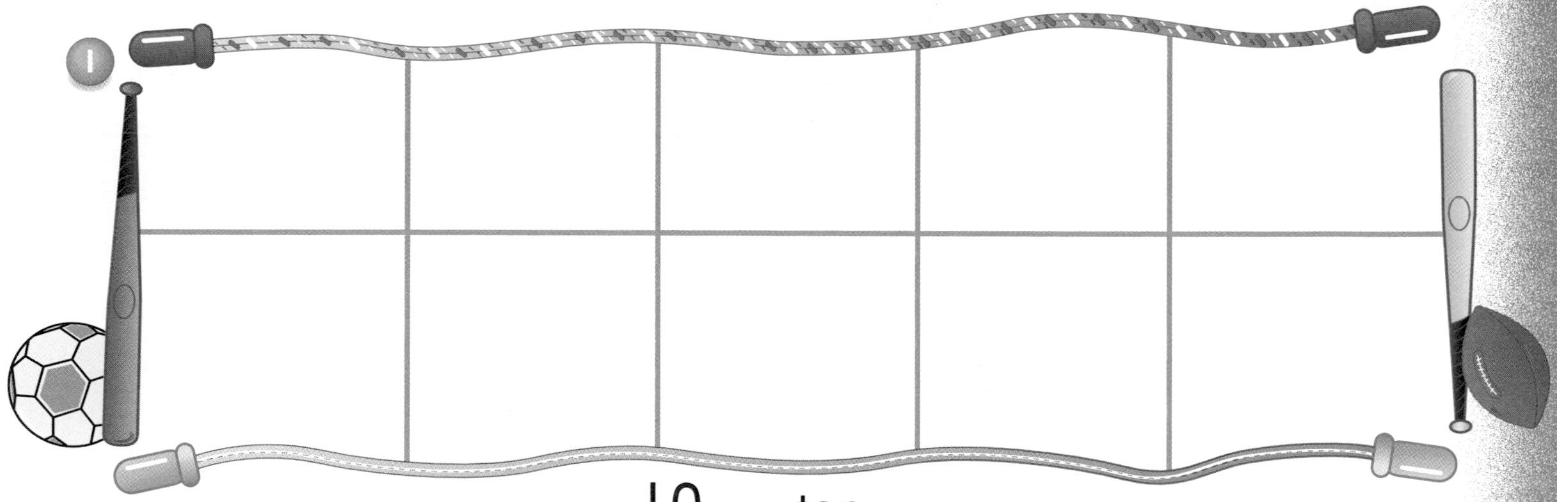

10 ten

2

8 9 10

3

8 9 10

4

8 9 10

5

8 9 10

6

8 9 10

7

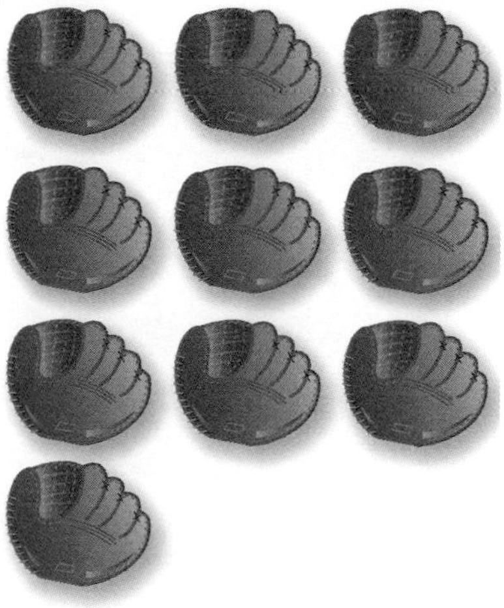

8 9 10

1. Put a counter in each square. Count.
2.–7. Count. Circle the number that tells how many.

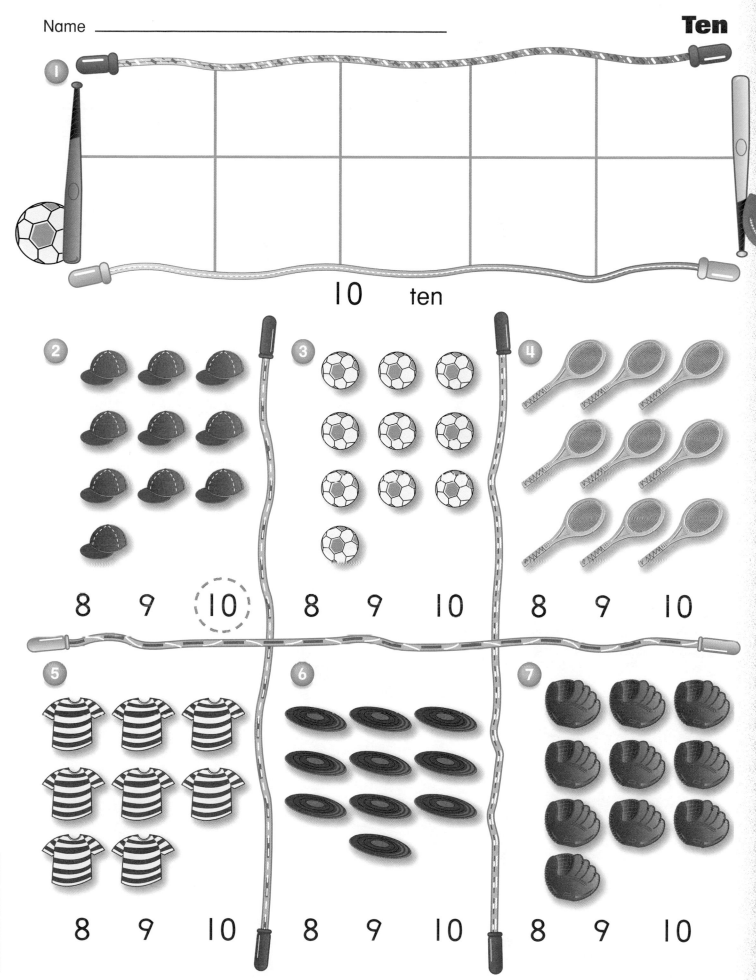

REVIEW

REVIEW

1 10

2
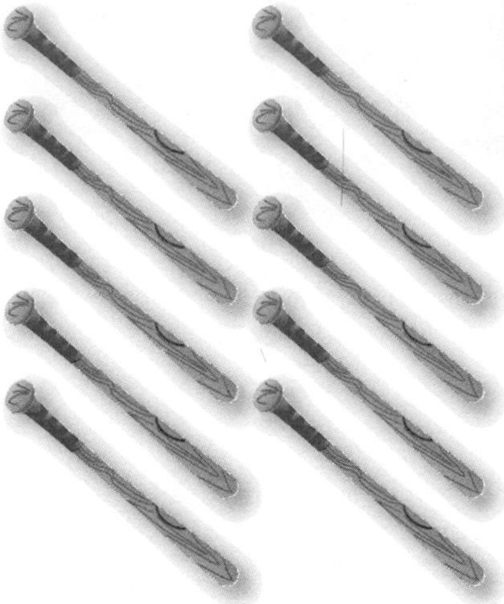
10

3

4
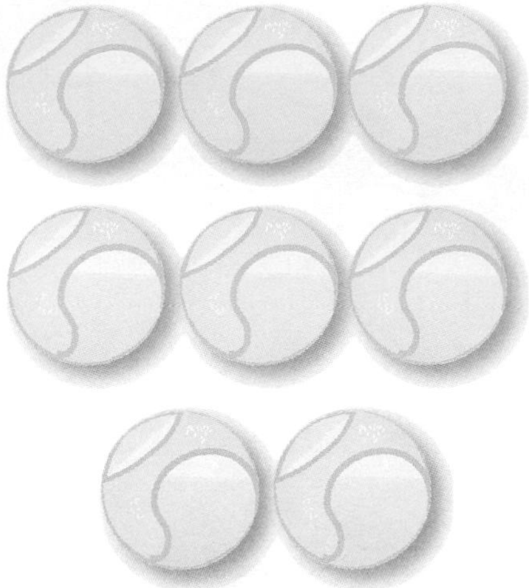

5

6

7
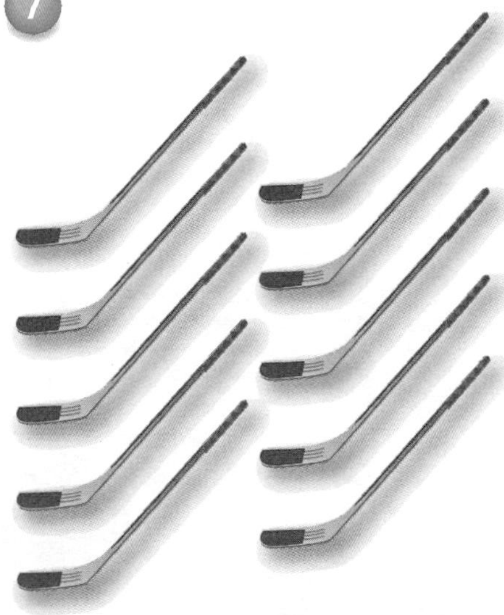

1. Write the number.
2.–7. Count. Write the number that tells how many.

Home Note Your child identified groups of 10.
ACTIVITY Have your child make groups of small objects to show 10.

Name ______________________________

Greater Than

1

4	◯ ◯ ◯ ◯
3	◯ ◯ ◯

2

6	
7	

3

7	
5	

4

4	
6	

Use counters to show both numbers. Draw the counters. Compare the groups. Circle the number that is greater.

REVIEW

REVIEW

1  5 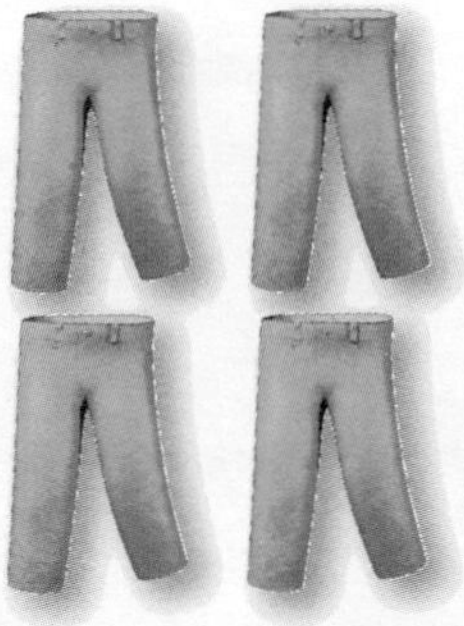 4

2 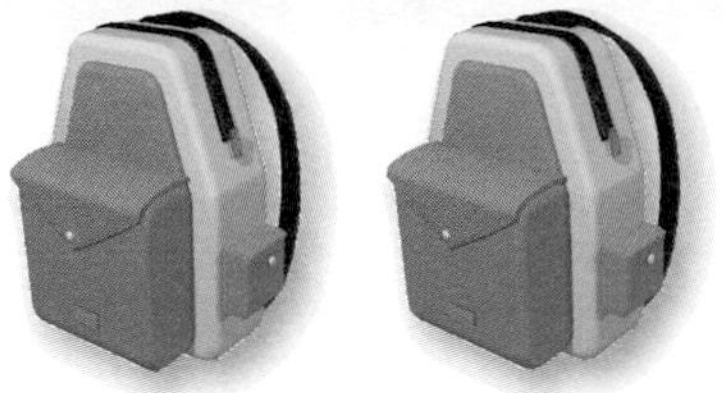 ___ 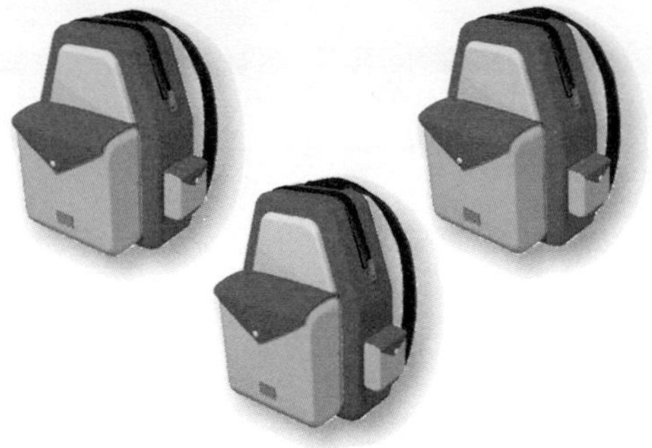 ___

3 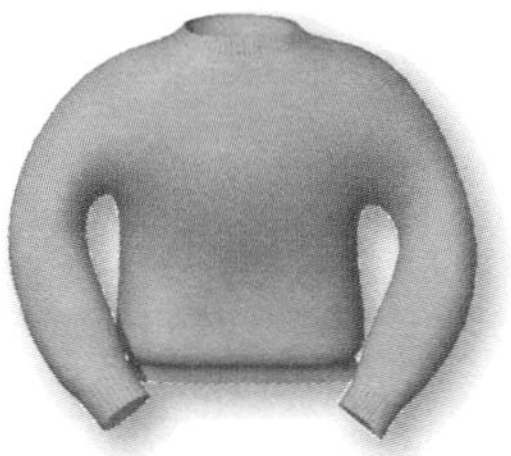 ___ 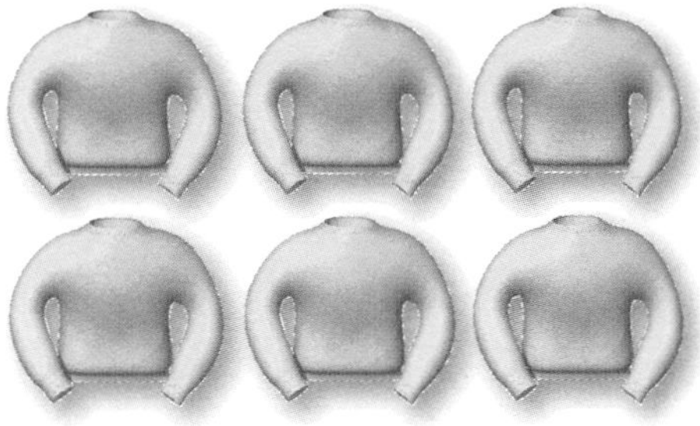 ___

4 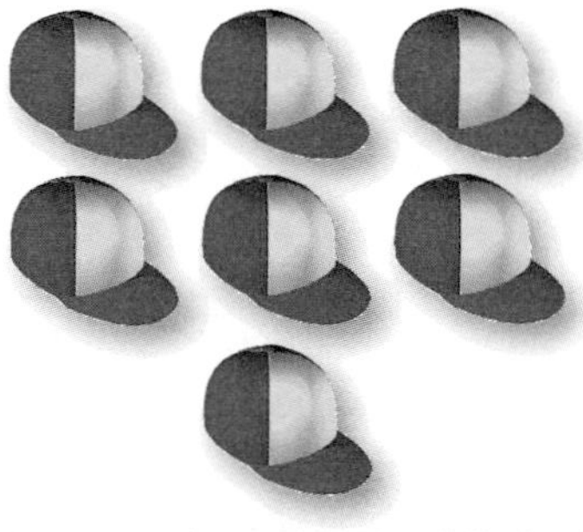 ___ 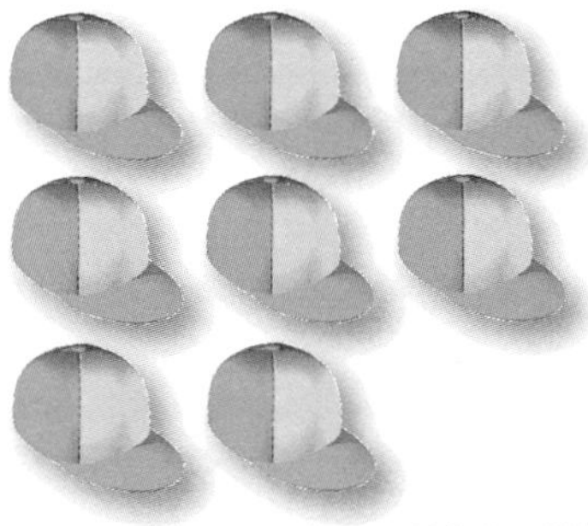 ___

5  ___  ___

Count. Write the numbers. Compare the groups.
Circle the number that is greater.

Home Note **Your child identified which of two numbers is greater.**
ACTIVITY **Set out a group of objects. Have your child set out objects to show a group that has more.**

Name ____________________

Less Than

1

5	
6	

2

4	
3	

3

7	
6	

4

5	
7	

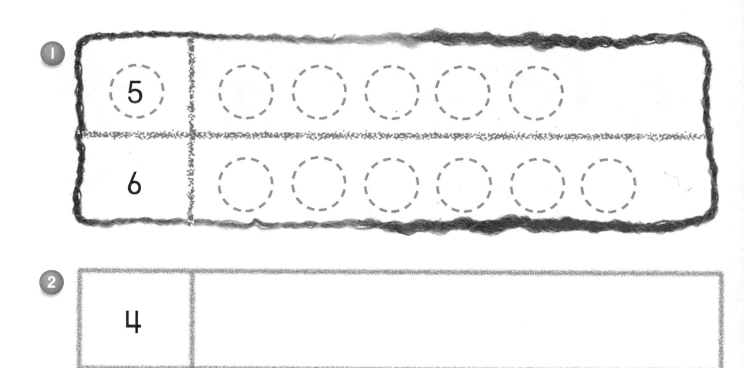

Use counters to show both numbers. Draw the counters. Compare the groups. Circle the number that is less.

REVIEW

REVIEW

1

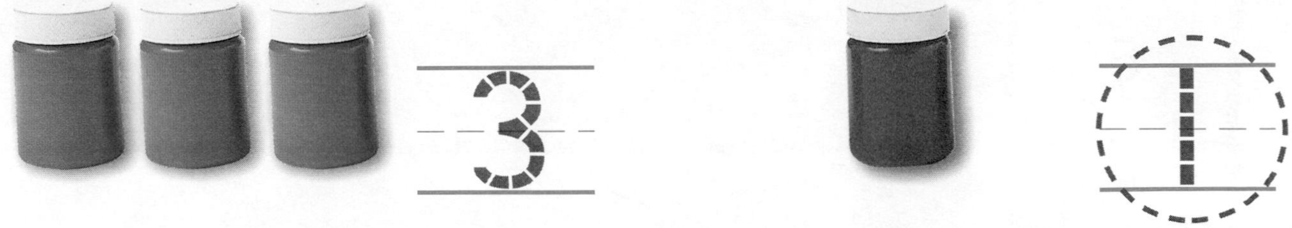

2

3

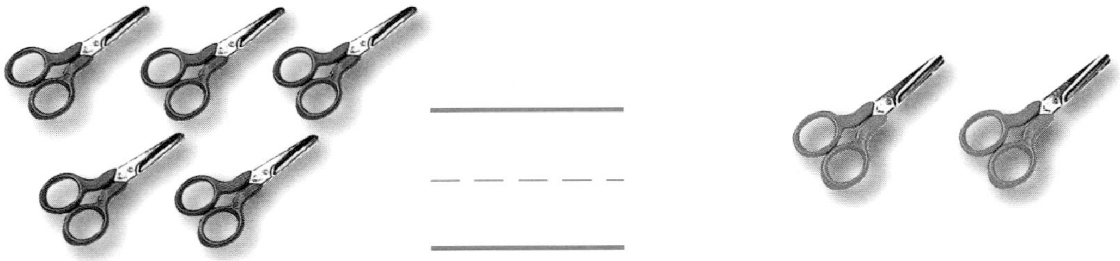

4

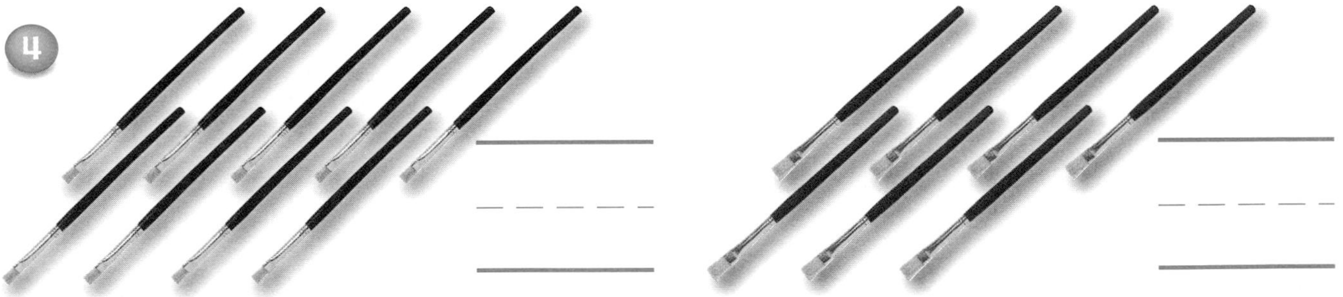

5

Count. Write the numbers. Compare the groups.
Circle the number that is less.

Home Note Your child identified which of two numbers is less.
ACTIVITY Say two numbers such as 7 and 9, and ask your child to tell which one is less. Have him or her show the numbers with objects to check the answer.

Name ____________________

Patterns • Order Through 10

1

1

2

3

4 1 2 __ __ 5 __ __ __ __ 10

5 10 9 8 __ __ __ __ 3 __ 1

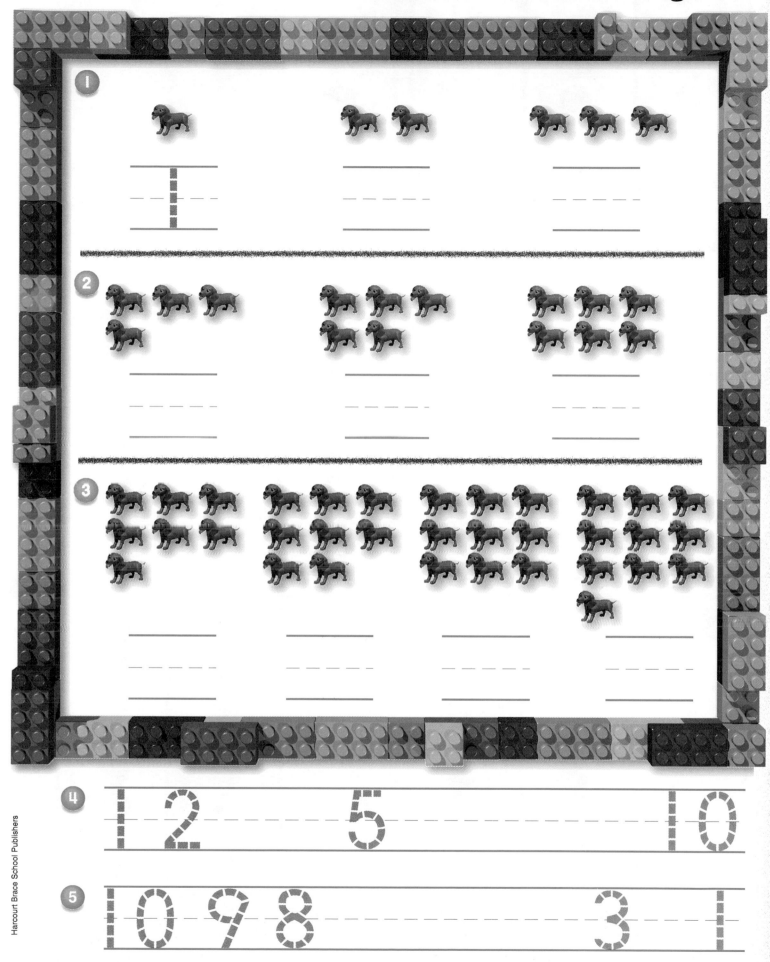

1.–3. Count. Write how many.
4.–5. Write the missing numbers.

REVIEW

REVIEW

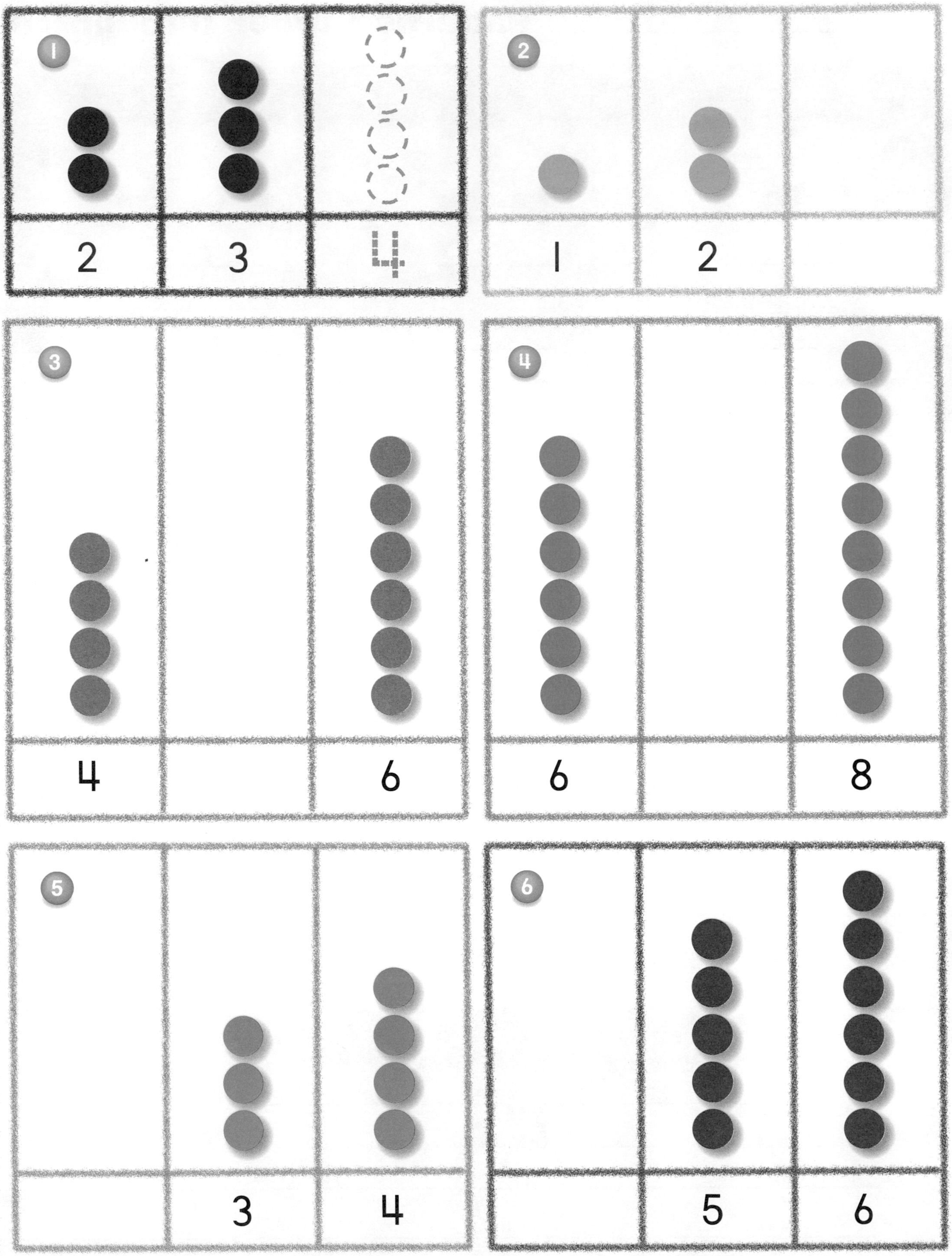

Draw circles to show the missing number.
Write the number.

Home Note Your child put three numbers in order.
ACTIVITY Have your child show three numbers in order—for example, 4, 5, and 6—with objects.

Name ______________________

Ordinal Numbers

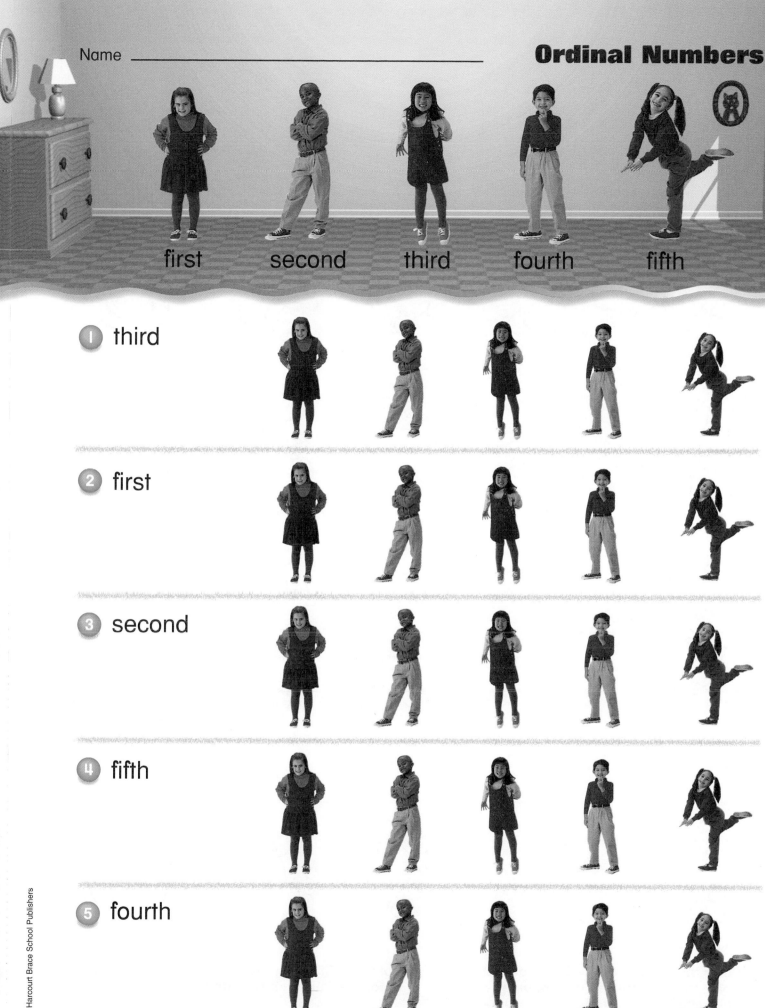

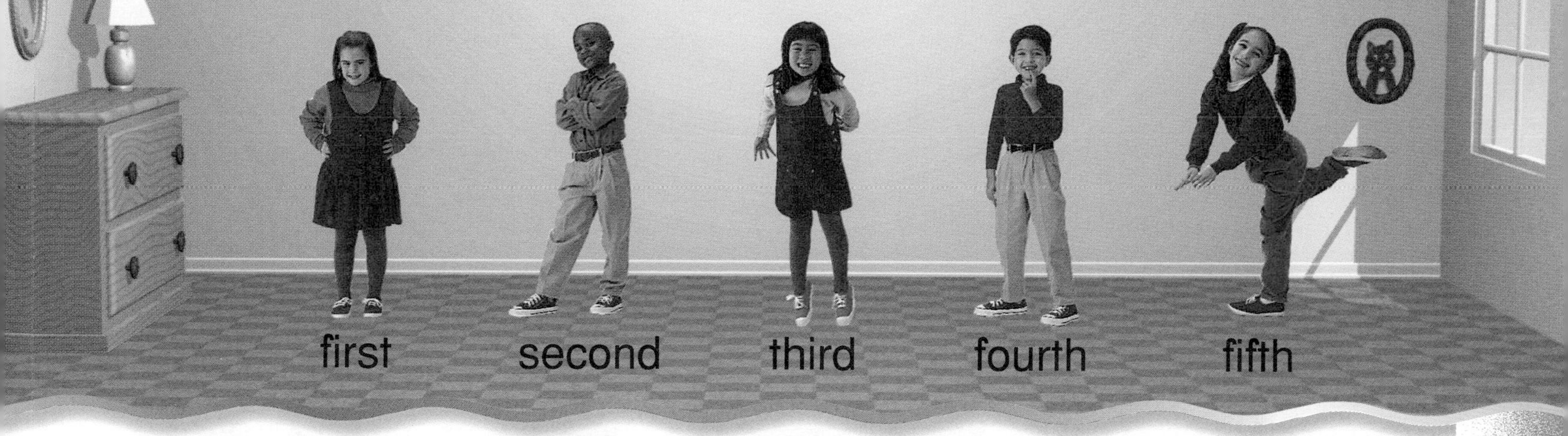

1 third

2 first

3 second

4 fifth

5 fourth

1. Circle the child who is third.
2. Circle the child who is first.
3. Circle the child who is second.
4. Circle the child who is fifth.
5. Circle the child who is fourth.

REVIEW

first second third fourth fifth

1 

2 

3 

4 

5 

REVIEW

1. Color the second stuffed animal brown.
2. Color the first stuffed animal blue.
3. Color the third stuffed animal orange.
4. Color the fifth stuffed animal yellow.
5. Color the fourth stuffed animal red.

Home Note Your child identified positions to *fifth*.
ACTIVITY Place five objects in a row. Have your child give the order of the objects, starting from the left.

Name ______________________

Chapter Review/Test

1
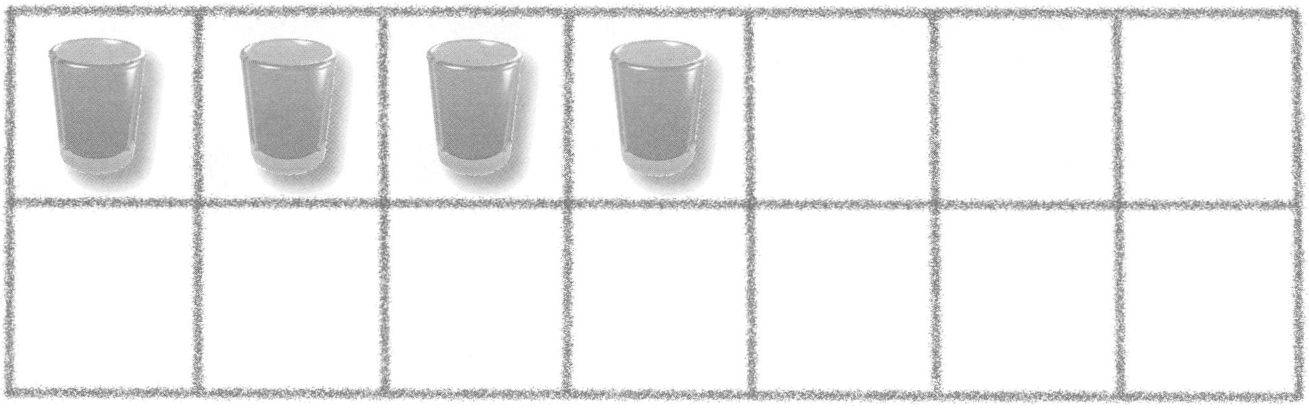

2

3
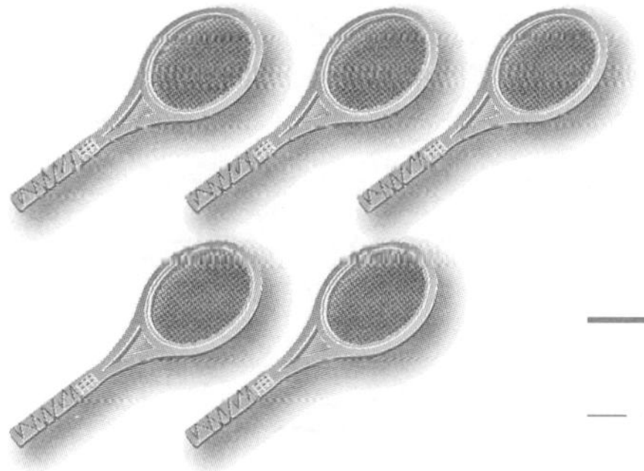

______

4
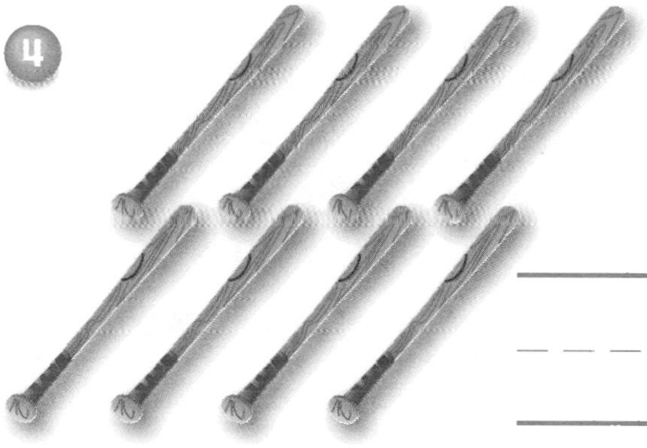

______

5
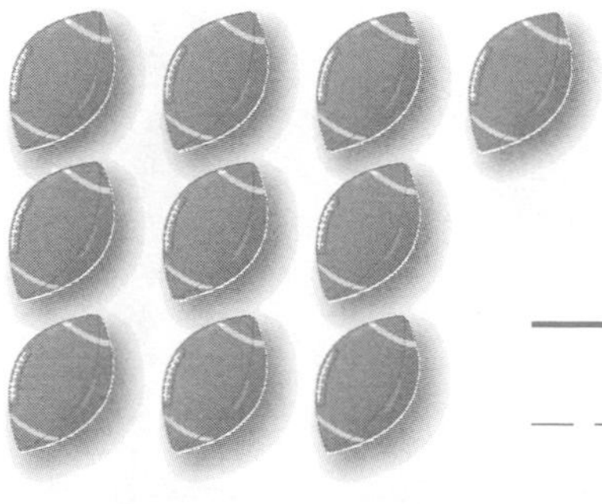

______

6

______

1. Draw one straw for each cup.
2. Draw footballs to show more footballs than helmets.
3.–6. Count. Write the number that tells how many.

REVIEW

REVIEW

7

8

9

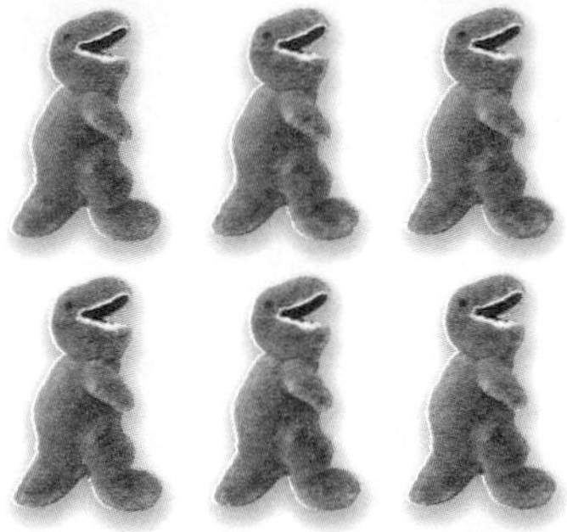

10

11

1 2 4 8 10

12

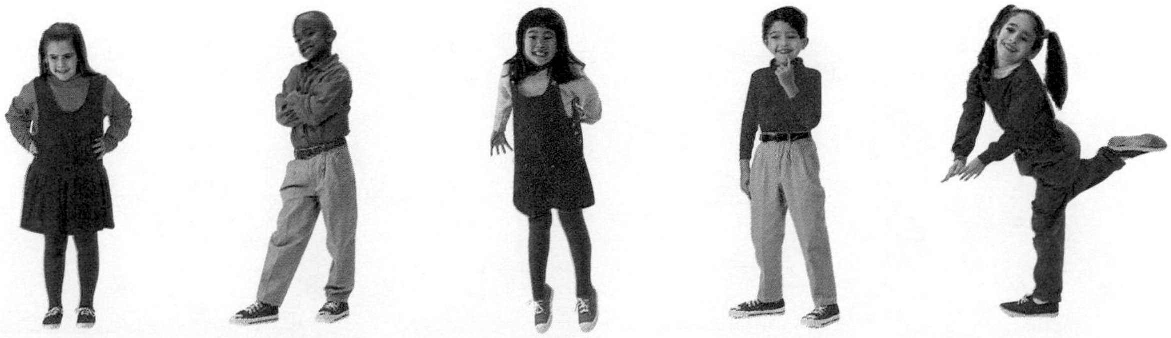

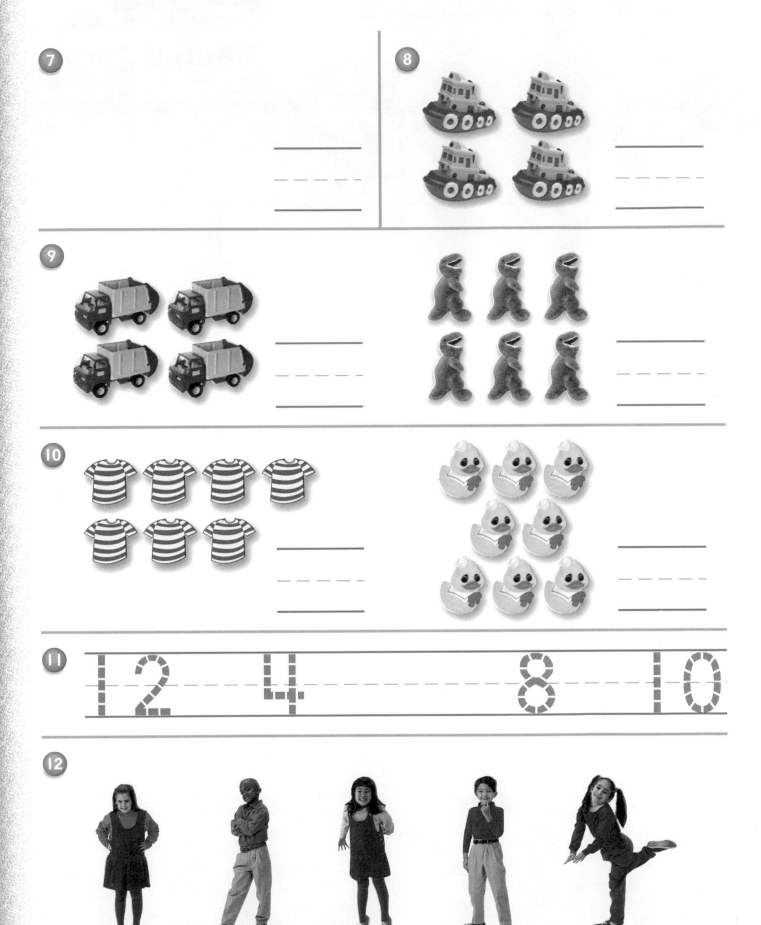

7.–8. Count. Write the number that tells how many.
9. Count. Write the numbers. Compare the groups. Circle the number that is greater.
10. Count. Write the numbers. Compare the groups. Circle the number that is less.
11. Write the missing numbers.
12. Circle the child who is second.

CHAPTER 1

Understanding Addition

What addition sentences can you tell about the animals?

Home Note In this chapter, your child will learn addition facts to 6.
ACTIVITY Have your child make up addition sentences using the pictures on this page.

SCHOOL-HOME CONNECTION

Dear Family,

Today we started Chapter 1. We will learn to act out story problems, use pictures to add, and write addition sentences. Here are the new vocabulary words and an activity for us to do together at home.

Love,

____

Vocabulary

$$4 + 1 = 5$$

plus (points to +) **equals** (points to =)

The **sum** is the answer.

ACTIVITY

Invite your child to act out addition sentences, using objects found at home. For example, your child can show you 2 shoes and then add 2 more shoes and tell you the addition sentence. Find other objects that can be used to practice addition.

Visit our Web site for additional activities and ideas.
http://www.hbschool.com

Name ______________________

Modeling Addition Story Problems

Listen to the story.
Use ● to model the story. Draw the ●.
Write how many there are in all.

There are 2 children in all.

1

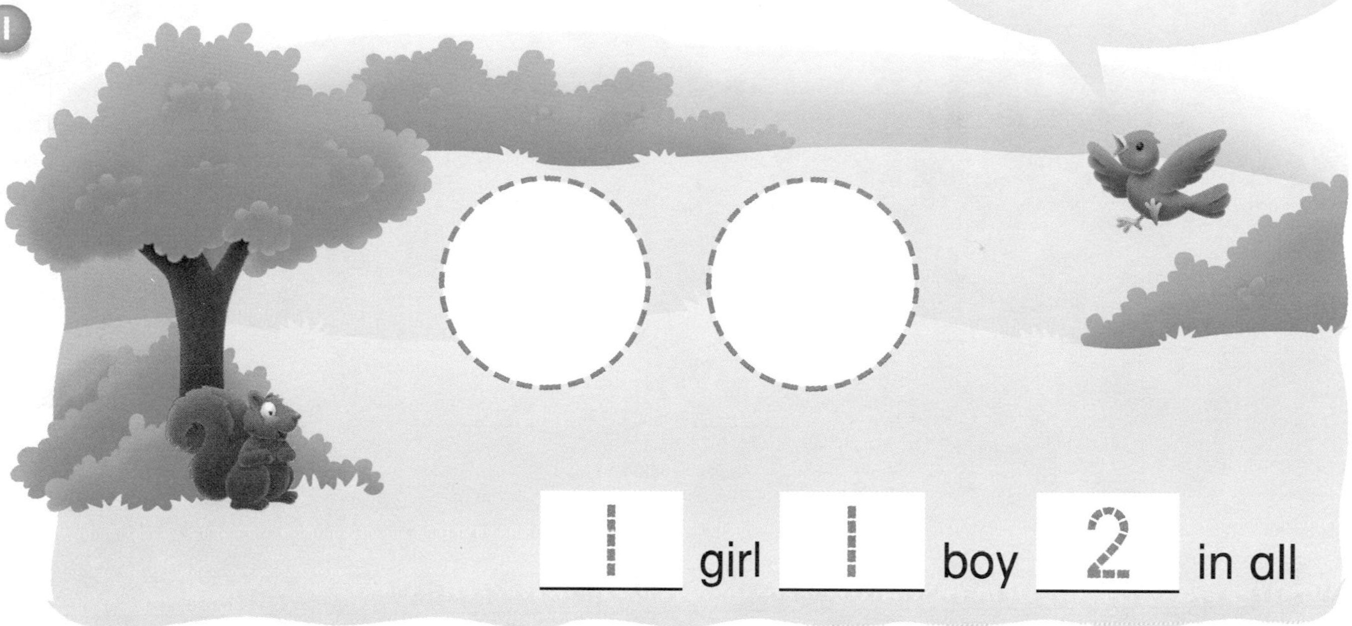

1 girl 1 boy 2 in all

2

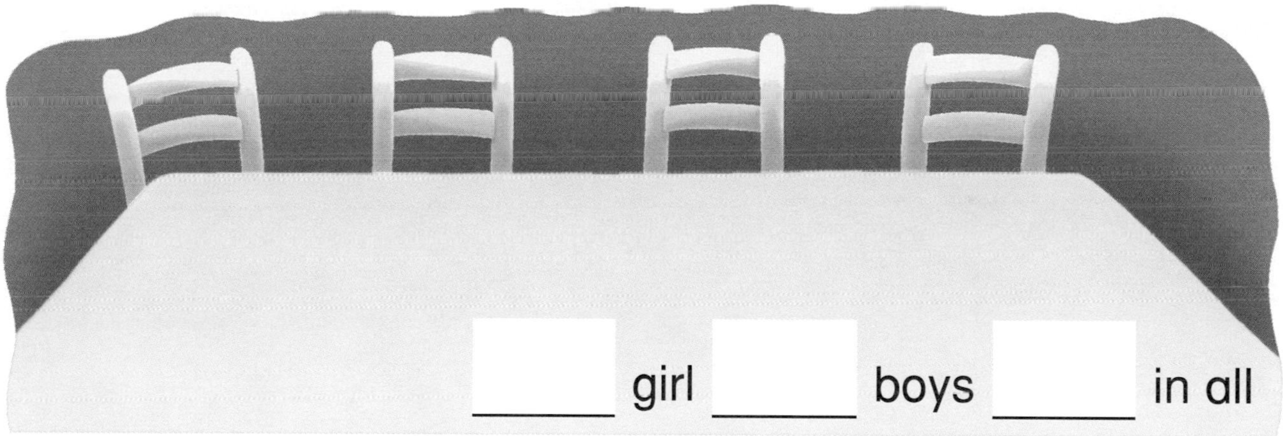

_____ girl _____ boys _____ in all

3

_____ girls _____ boys _____ in all

Practice

Listen to the story. Use ● to model the story.
Draw the ●. Write how many there are in all.

1

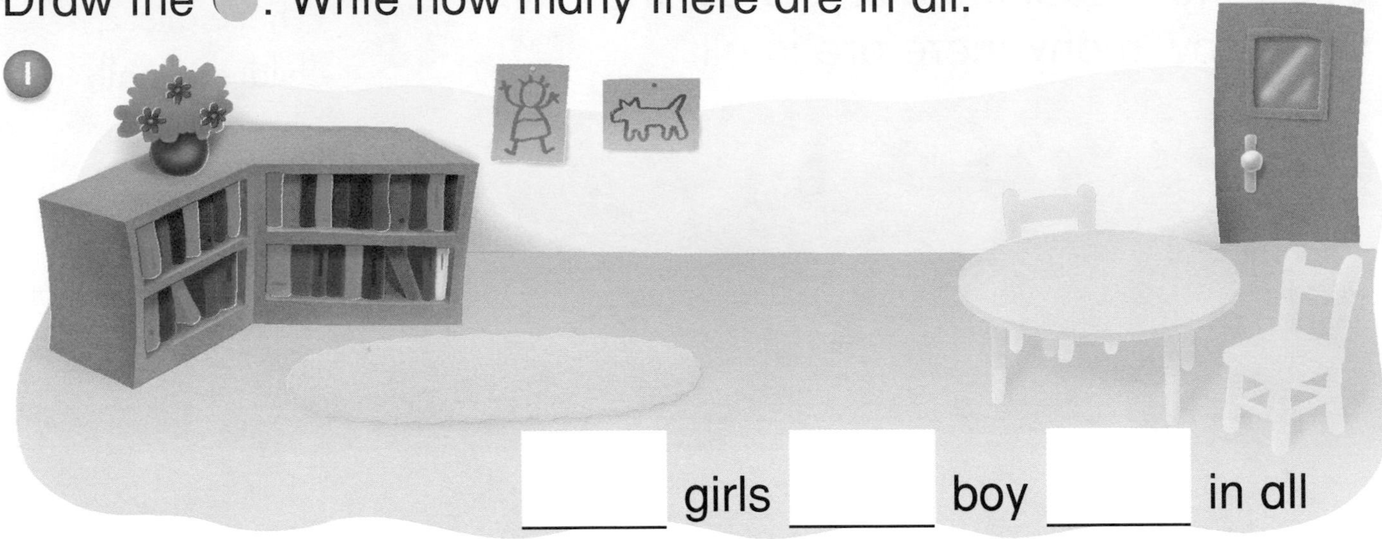

_____ girls _____ boy _____ in all

2

_____ girls _____ boys _____ in all

3

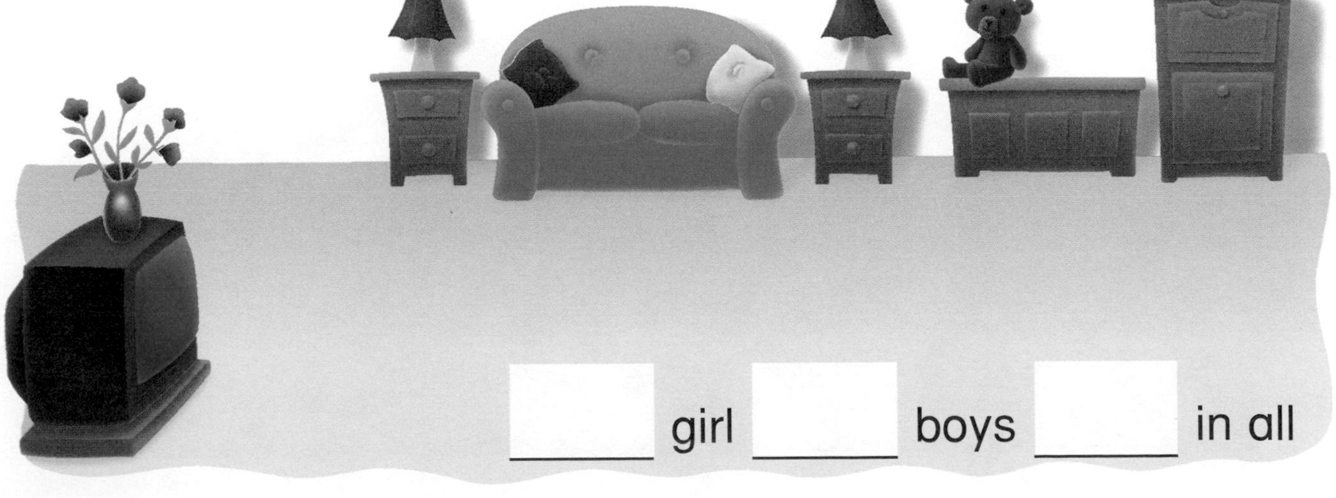

_____ girl _____ boys _____ in all

Home Note Your child modeled addition problems.
ACTIVITY Have your child tell an addition story for each picture.

Name ______________________

Adding 1

Two plus one equals three.

2 + 1 = 3
sum

Draw 1 more . Color it blue. Write the sum.

1

4 + 1 = ____

2

3 + 1 = ____

3

1 + 1 = ____

4

5 + 1 = ____

5

2 + 1 = ____

6

4 + 1 = ____

Talk About It • Critical Thinking

What happens when 1 more is added to a group?

Practice

Draw and color 1 more.
Write the sum.

1. 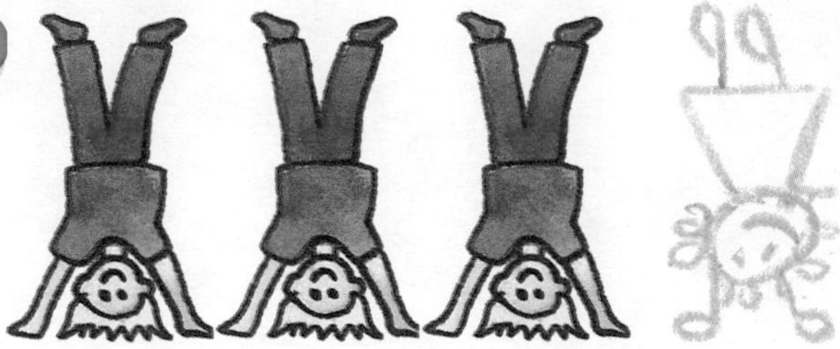

$3 + 1 = \underline{4}$
sum

2. 

$2 + 1 = $ _____

3. 

$5 + 1 = $ _____

4. 

$1 + 1 = $ _____

5. 

$4 + 1 = $ _____

6. 

$3 + 1 = $ _____

Home Note Your child used pictures to add 1 to a number.
ACTIVITY Have your child practice adding 1 to numbers 0–5.

Name ______________________

Adding 2

$3 + 2 =$ 5

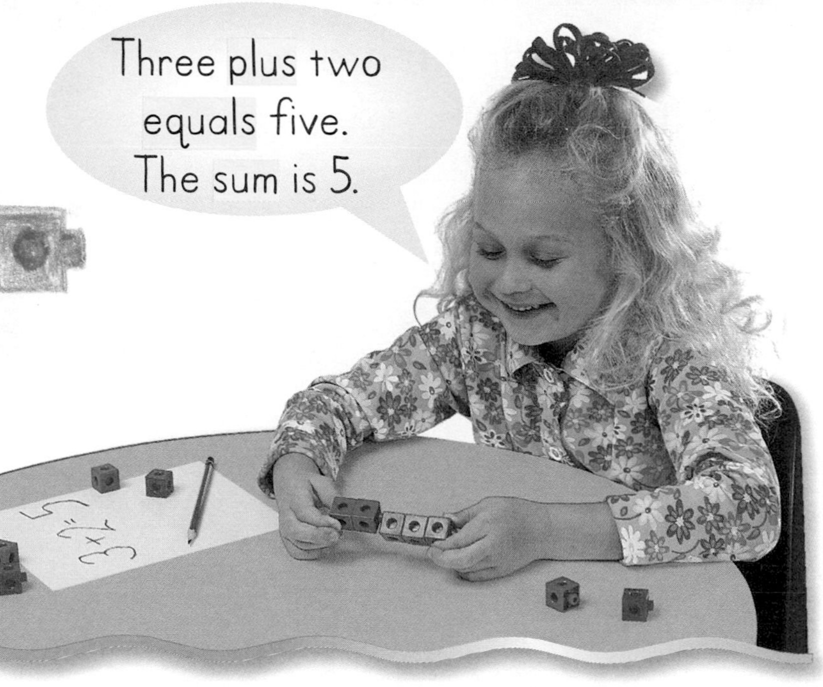

Three plus two equals five. The sum is 5.
3+2=5

Draw 2 more cubes.
Color them blue.
Write the sum.

1. 

$4 + 2 =$ ____

2. 

$2 + 2 =$ ____

3. 

$1 + 2 =$ ____

4. 

$3 + 2 =$ ____

5. 

$4 + 2 =$ ____

6. 

$1 + 2 =$ ____

Practice

Draw and color 2 more.
Write the sum.

1 

$1 + 2 = \underline{3}$
sum

2 

$3 + 2 =$ ____

3 

$4 + 2 =$ ____

4 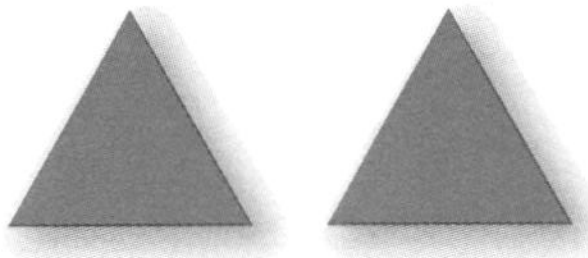

$2 + 2 =$ ____

Mixed Review

Write the number.

5 

6 

7 

Home Note Your child used pictures to add 2 to a number.
ACTIVITY Have your child add 2 to the numbers 0–4.

Name ______________________________

Using Pictures to Add

$3 + 2 = 5$ is an addition sentence.

$3 + 2 =$ 5

Write the sum.

1.

$4 + 2 =$ ____

2.
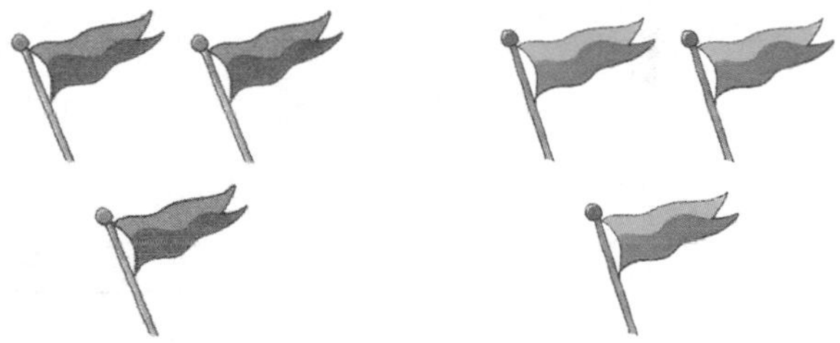

$3 + 3 =$ ____

3.

$2 + 2 =$ ____

4.
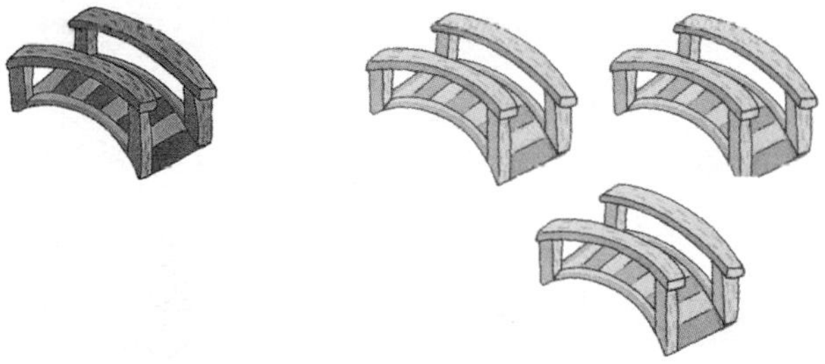

$1 + 3 =$ ____

5.

$2 + 1 =$ ____

6.

$5 + 1 =$ ____

Practice

Write the sum.

1 

$3 + 3 =$ 6

2 

$2 + 3 =$ ____

3 

$5 + 1 =$ ____

4 

$2 + 1 =$ ____

Write About It

5 Draw a picture of you
and a group of your friends.
Write an addition sentence
about your picture.

Home Note **Your child used pictures to find the sum.**
ACTIVITY Ask your child to tell how he or she found each sum.

Name ______________________________

Understand • Plan • Solve • Look Back

Problem Solving

Write an Addition Sentence

Write the addition sentence.

1. How many dogs in all?

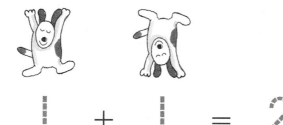

1 + 1 = 2 dogs

2. How many fish in all?

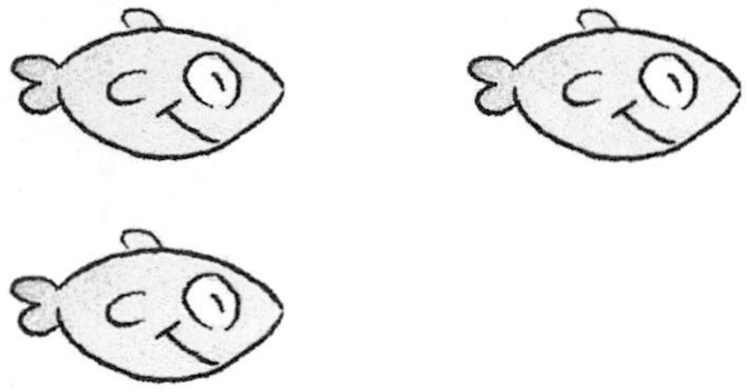

____ + ____ = ____ fish

3. How many sheep in all?

____ + ____ = ____ sheep

4. How many ducks in all?

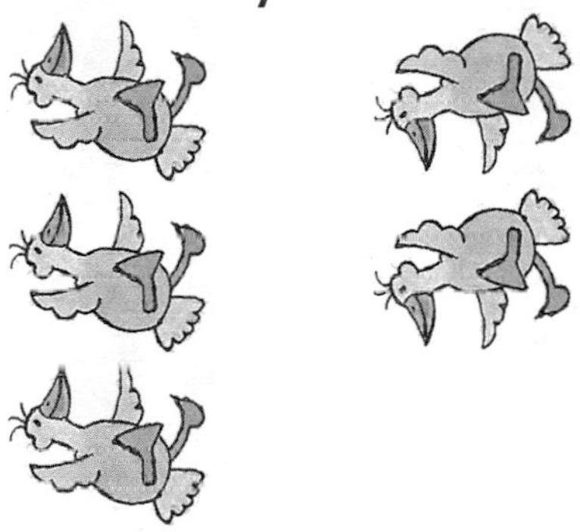

____ + ____ = ____ ducks

5. How many cats in all?

____ + ____ = ____ cats

6. How many bugs in all?

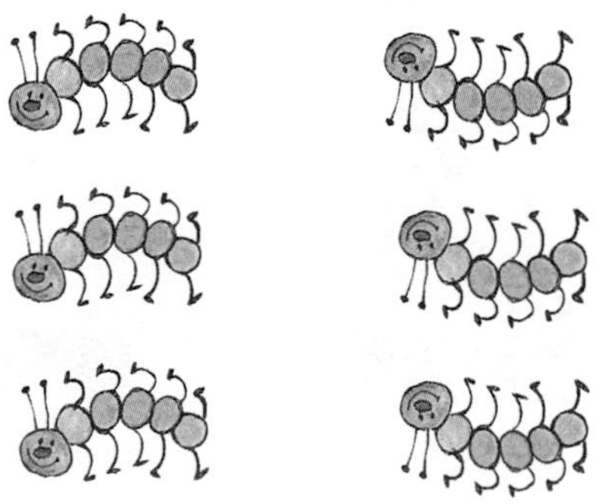

____ + ____ = ____ bugs

Talk About It • Critical Thinking

How did you decide which numbers to use?

Practice

Write the addition sentence.

1. How many goats in all?

2 + 3 = 5 goats

2. How many cows in all?

_____ + _____ = _____ cows

3. How many bears in all?

_____ + _____ = _____ bears

4. How many birds in all?

_____ + _____ = _____ birds

5. How many tigers in all?

_____ + _____ = _____ tigers

6. How many lions in all?

_____ + _____ = _____ lions

Home Note **Your child used pictures to write addition sentences.**
ACTIVITY Have your child use small objects to show addition stories. Then ask him or her to write the addition sentences.

Name ______________________________

Chapter 1

Review/Test

Concepts and Skills

Draw and color 1 more. Write the sum.

1 

2 + 1 = ____

2 

4 + 1 = ____

Draw and color 2 more. Write the sum.

3 

3 + 2 = ____

4 

2 + 2 = ____

Write the sum.

5 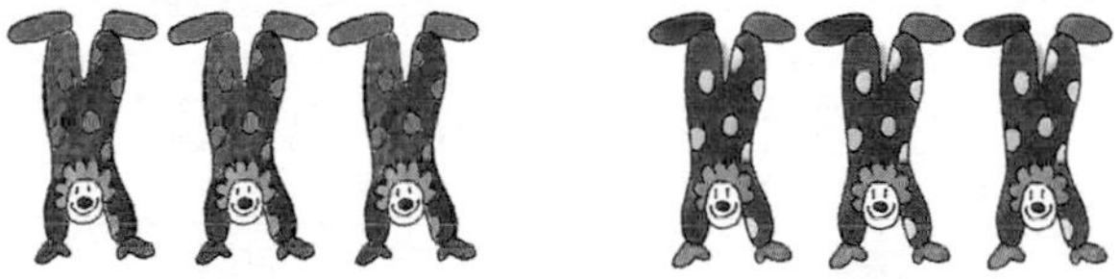

3 + 3 = ____

6 

4 + 2 = ____

Problem Solving

Write the addition sentence.

7 How many sheep in all?

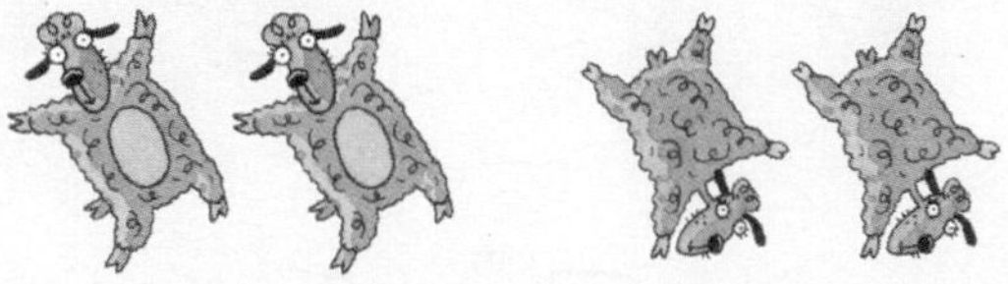

____ + ____ = ____ sheep

8 How many bugs in all?

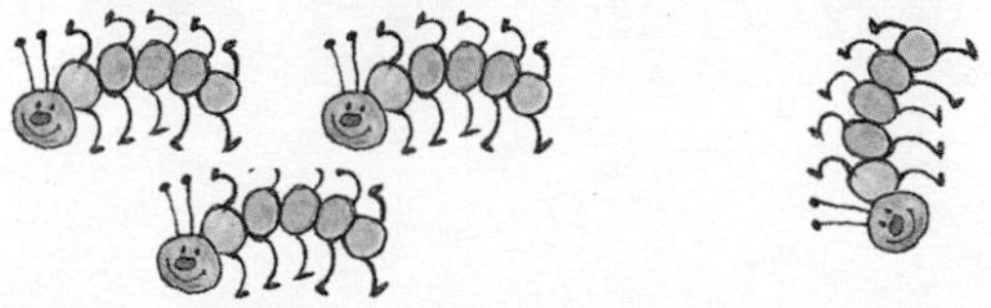

____ + ____ = ____ bugs

Name ____________________

TAAS Prep

Mark the best answer.

1. How many are there?

- ⬭ 3
- ⬭ 4
- ⬭ 5
- ⬭ 6

2. How many are there?

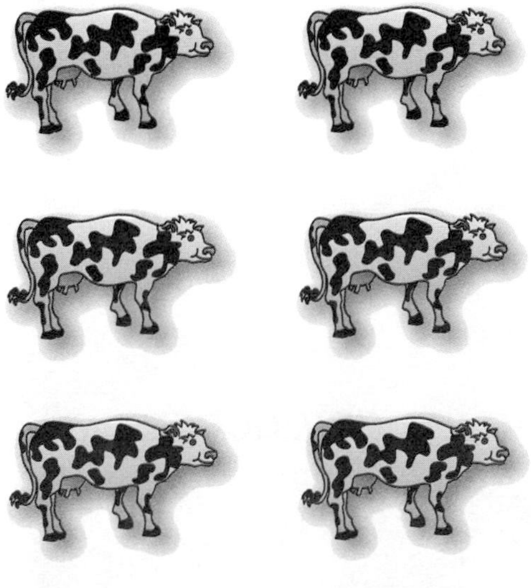

- ⬭ 6
- ⬭ 7
- ⬭ 8
- ⬭ 9

3. Look at the picture.
How many balloons in all?

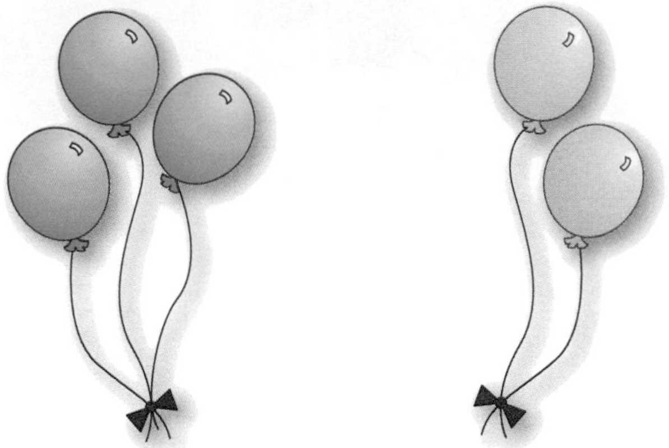

- ⬭ 3
- ⬭ 4
- ⬭ 5
- ⬭ 6

4. Which addition sentence tells how many in all?

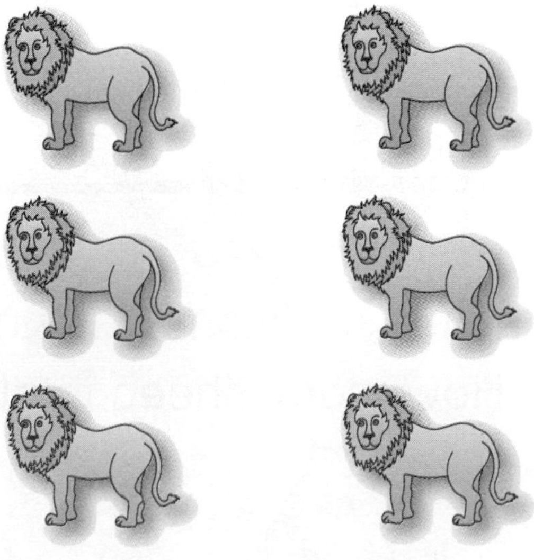

- ⬭ $1 + 3 = 4$
- ⬭ $4 + 1 = 5$
- ⬭ $3 + 2 = 5$
- ⬭ $3 + 3 = 6$

CHAPTER 2

Understanding Subtraction

Tell a subtraction story about the picture.

Home Note In this chapter, your child will learn subtraction facts from 6.
ACTIVITY Have your child make up a subtraction sentence using the picture on this page.

SCHOOL-HOME CONNECTION

Dear Family,

Today we started Chapter 2. We will model story problems, use pictures to subtract, and write subtraction sentences. Here are the new vocabulary words and an activity for us to do together at home.

Love,

____

Vocabulary

Use these pictures, symbols, and words when you talk with your child about subtraction.

This is a **subtraction sentence**.

subtract (points to 3 –) **minus** (points to –)

$$3 - 2 = 1$$

difference (points to 1)

3 minus 2 equals 1

3 subtract 2 equals 1

ACTIVITY

Have your child act out subtraction story problems while you shop together. For example, you might say, "Put 6 apples in a bag. Take 2 out. How many apples are left in the bag?"

Visit our Web site for additional ideas and activities.
http://www.hbschool.com

Name ______________________________

Modeling Subtraction Story Problems

Listen to the story.
Use ● to model the story.
Draw the ●. Cross out how many go away.
Write how many are left.

1

4 dogs 2 go away 2 are left

2

_____ ducks _____ go away _____ is left

3

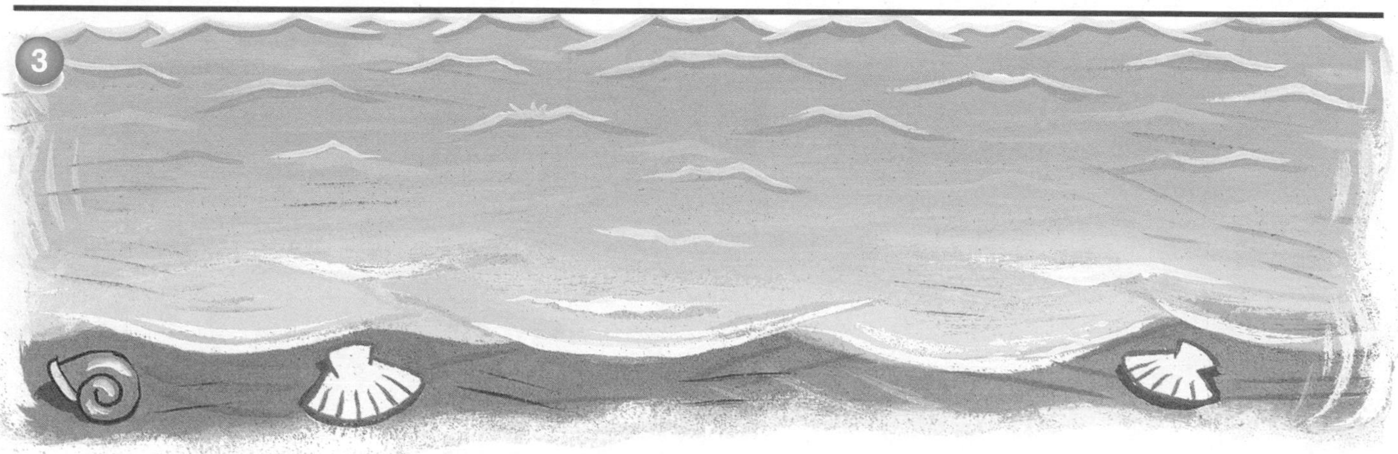

_____ fish _____ go away _____ are left

Practice

Listen to the story.
Use ● to model the story.
Draw the ●. Cross out how many go away.
Write how many are left.

1

_____ monkeys _____ go away _____ are left

2

_____ fish _____ go away _____ are left

3

_____ goats _____ goes away _____ is left

Home Note Your child modeled subtraction problems.
ACTIVITY Have your child model a subtraction story for each picture.

Name ____________________

Subtracting 1

4−1 = 3
minus
3 marbles are left.

$4 - 1 = 3$

Cross out 1 marble.
Write how many are left.

1. 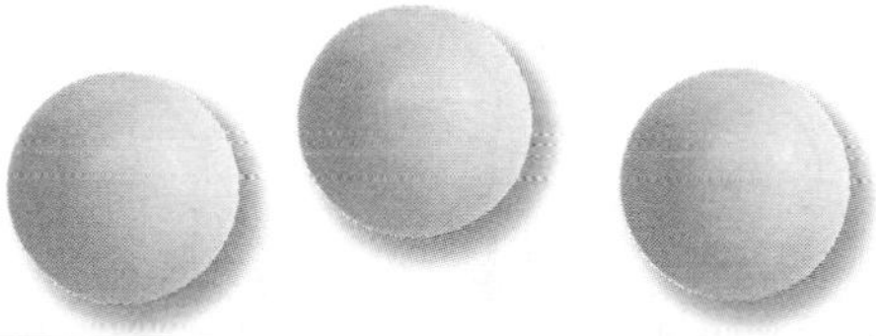

$3 - 1 =$ ____

2. 

$6 - 1 =$ ____

3. 

$5 - 1 =$ ____

4. 

$3 - 1 =$ ____

5. 

$1 - 1 =$ ____

6. 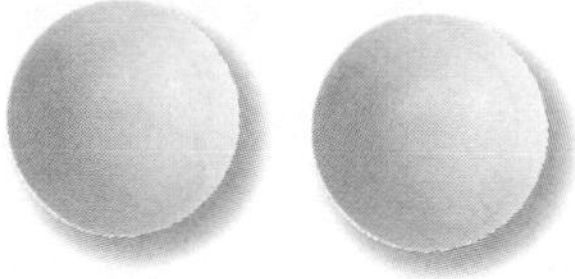

$2 - 1 =$ ____

Practice

Cross out 1 marble.
Write how many are left.

1

$6 - 1 = $ 5

2

$5 - 1 = $ ____

3

$4 - 1 = $ ____

4

$3 - 1 = $ ____

5

$2 - 1 = $ ____

6

$1 - 1 = $ ____

Home Note **Your child used pictures to subtract 1 from a number.**
ACTIVITY **Have your child use objects to model each problem.**

Name ________________________

Subtracting 2

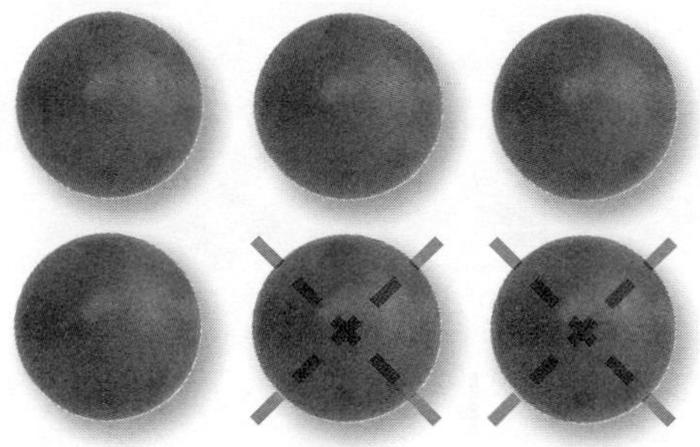

4 is the difference.
It tells how many are left.

$6 - 2 = \underline{4}$

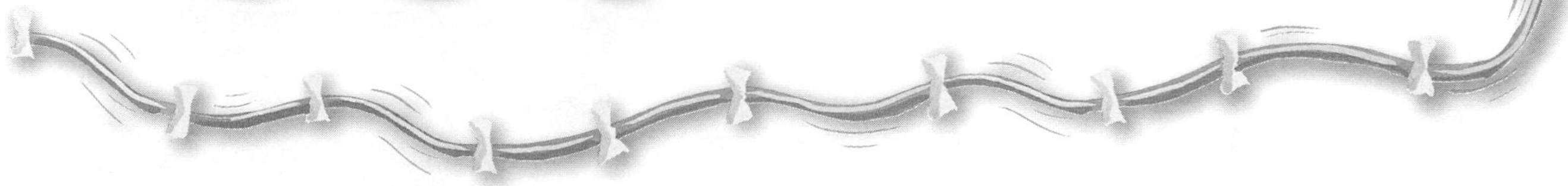

Cross out 2 pictures.
Write the difference.

1 

$4 - 2 = \_\_\_\_$

2 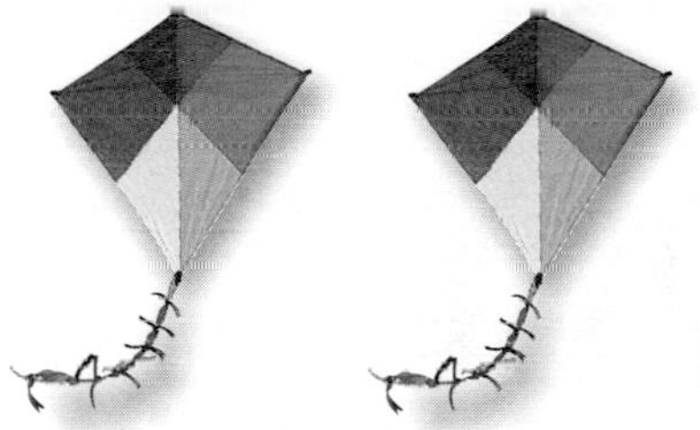

$2 - 2 = \_\_\_\_$

3 

$3 - 2 = \_\_\_\_$

4 

$5 - 2 = \_\_\_\_$

Talk About It • Critical Thinking

Explain to a classmate how crossing out helps you find the difference.

Practice

Cross out pictures to find how many are left.
Write the difference.

1 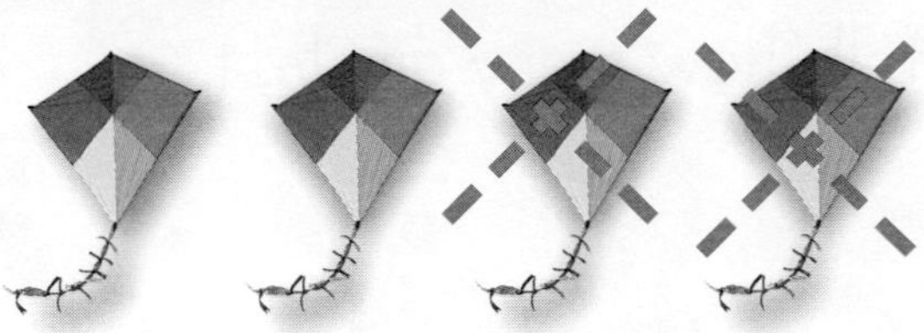

4 − 2 = 2

2 

6 − 1 = ____

3 

3 − 2 = ____

4 

5 − 2 = ____

5 

5 − 1 = ____

6 

4 − 2 = ____

7 

6 − 2 = ____

8 

2 − 1 = ____

Mixed Review

Write the addition sentence.

9 

____ + ____ = ____

10 

____ + ____ = ____

Home Note Your child used pictures to subtract 2 from a number.
ACTIVITY Have your child use objects to model the subtraction sentences.

Name ______________________

Writing Subtraction Sentences

6 − 2 = 4

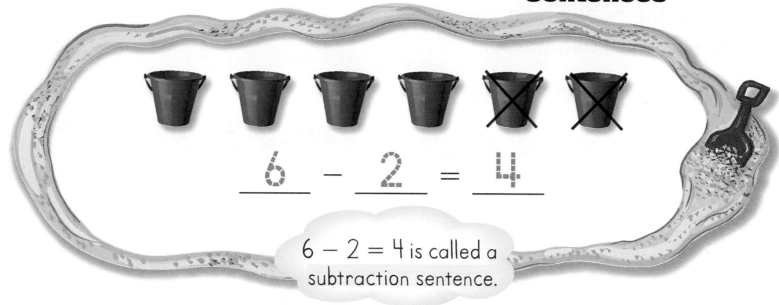

6 − 2 = 4 is called a subtraction sentence.

Write the subtraction sentence to show the difference.

1 

_____ − _____ = _____

2 

_____ − _____ = _____

3 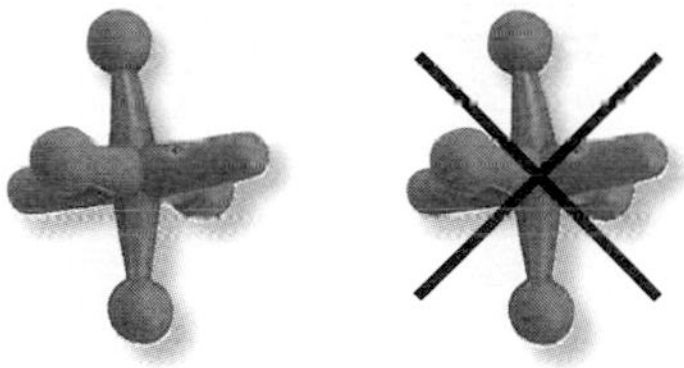

_____ − _____ = _____

4 

_____ − _____ = _____

5 

_____ − _____ = _____

6 

_____ − _____ = _____

Talk About It • Critical Thinking

How could you check that the subtraction sentence is correct?

Practice

Write the subtraction sentence to show the difference.

1

6 – 1 = 5

2

____ – ____ = ____

3

____ – ____ = ____

4
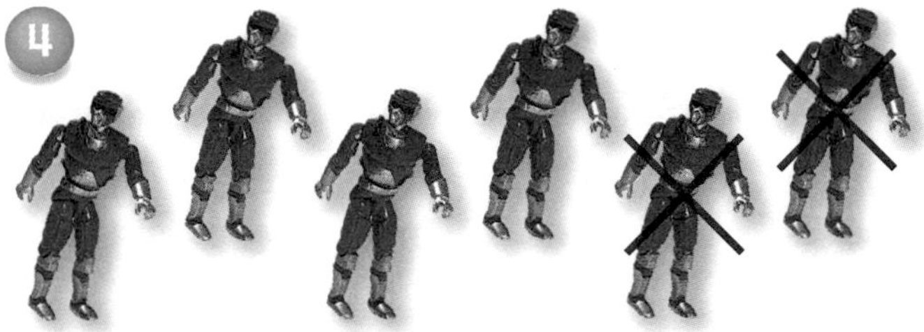

____ – ____ = ____

5
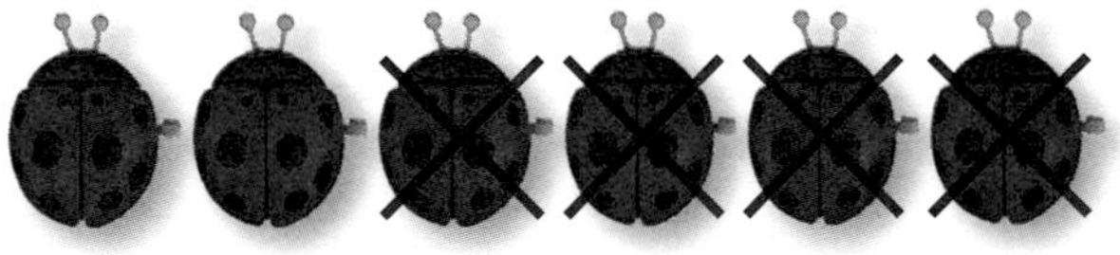

____ – ____ = ____

6

____ – ____ = ____

Write About It

7 Tell a story about the picture. Then write the subtraction sentence.

Home Note Your child used pictures to write subtraction sentences.
ACTIVITY Have your child use objects to show subtraction stories. Then ask him or her to write the subtraction sentences.

Name ____________________

Understand • Plan • Solve • Look Back

Problem Solving
Make a Model

Add or subtract. Use ●.
Draw the ●.

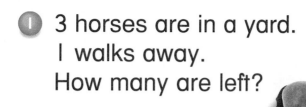

1. 3 horses are in a yard.
I walks away.
How many are left?

__2__ horses

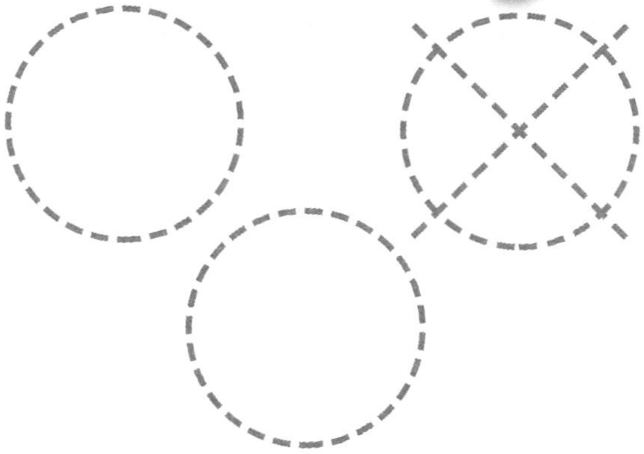

2. 3 bugs are on a rock.
2 more come.
How many in all?

______ bugs

3. 4 frogs are in a pond.
I jumps away.
How many are left?

______ frogs

4. 2 birds are on a branch.
2 more come.
How many in all?

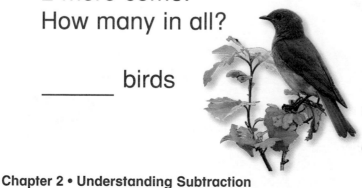

______ birds

Practice

Add or subtract. Use ●.
Draw the ●.

1. Tim saw 4 ants.
   1 walked away.
   How many are left?

   3 ants

2. Kathy has 2 dogs.
   She gets 2 more.
   How many in all?

   ______ dogs

3. Steve has 3 cats.
   1 runs away.
   How many are left?

   ______ cats

4. Greg has 5 fish.
   2 swim away.
   How many are left?

   ______ fish

Home Note **Your child solved addition and subtraction story problems.**
ACTIVITY Make up story problems like the ones in this lesson. Have your child use objects or pictures to solve the problems.

Name ____________________

Concepts and Skills

Cross out 1. Write how many are left.

1 

$4 - 1 =$ ____

2 

$5 - 1 =$ ____

Cross out 2. Write the difference.

3 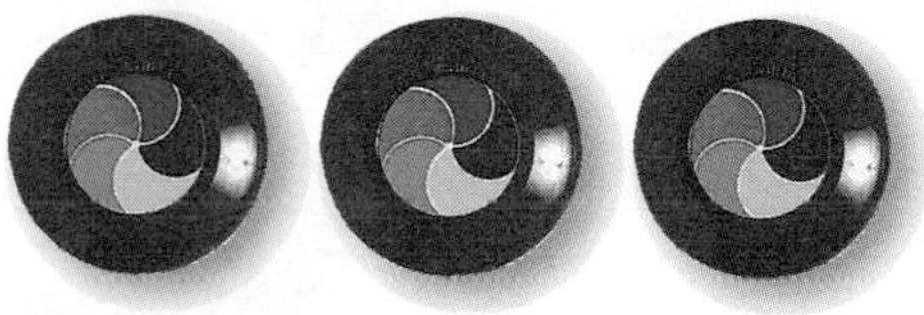

$3 - 2 =$ ____

4 

$6 - 2 =$ ____

Write the subtraction sentence.

5 

____ − ____ = ____

6 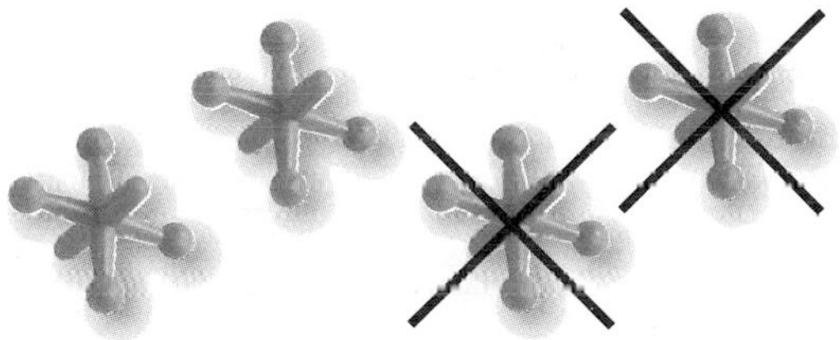

____ − ____ = ____

Problem Solving

Add or subtract. Use ●.
Draw the ●.

7 3 rabbits are in the grass.
2 rabbits hop away.
How many are left?

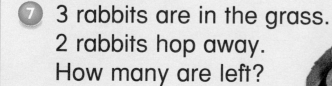

____ rabbit

Name ______________________________

Cumulative Chapters 1–2

TAAS Prep

Mark the best answer.

1. Cross out 2. Which number is the difference?

- ⬭ 0
- ⬭ 1
- ⬭ 2
- ⬭ 3

2. Which subtraction sentence tells how many are left?

- ⬭ 4 − 2 = 2
- ⬭ 5 − 3 = 2
- ⬭ 6 − 2 = 4
- ⬭ 6 − 4 = 2

3. Find the sum or the difference.

6 dogs are in the yard.
2 dogs run away.
How many are left?

- ⬭ 2
- ⬭ 3
- ⬭ 4
- ⬭ 5

4. Find the sum or the difference.

There are 2 black cats.
There are 2 brown cats.
How many cats in all?

- ⬭ 3
- ⬭ 4
- ⬭ 5
- ⬭ 6

Name ______________________________

MATH FUN

You will need:
,

Find the Sums and Differences

1. Play with a partner.
2. Spin the , and move to the right space. Read the problem. Say its sum or difference.
3. Have your partner check the sum or difference. Keep trying until you have the correct answer.
4. Take turns with your partner.
5. The first one to reach END is the winner.

START / **END**	$3 + 1$	$2 - 0$	$4 + 2$	$1 + 5$
$4 + 2$				$6 - 1$
$2 - 1$				$5 + 1$
$5 - 3$				$4 + 2$
$1 + 5$				$3 - 2$
$4 - 2$				$5 - 2$
$1 + 3$	$0 + 6$	$3 - 0$	$6 - 2$	$4 + 1$

Home Note Your child has been learning addition and subtraction facts to six.
ACTIVITY Play this game at home to help your child practice these facts.

Name ____________________

Calculator | Computer

Using a Calculator

1. Color the [ON/C] key (crayon).
2. Color the [+] key (crayon).
3. Color the [−] key (crayon).
4. Color the [=] key (crayon).

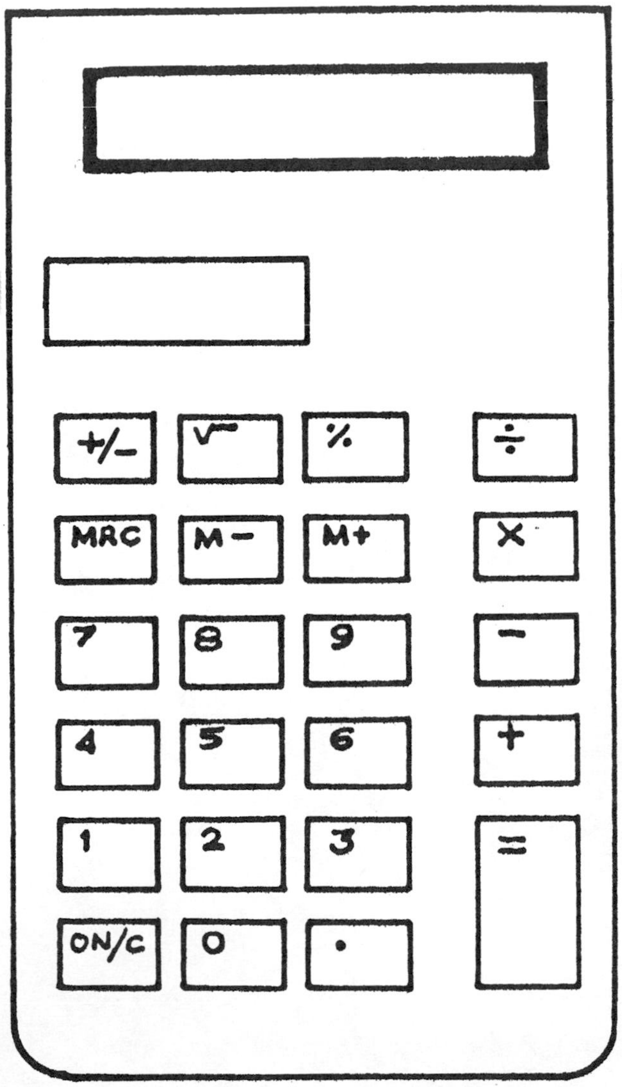

Use a calculator.
Press the keys.
Write what you see.

5. [ON/C] [0]
6. [ON/C] [3] [+] [1] [=] [ ]
7. [ON/C] [5] [−] [2] [=] [ ]
8. [ON/C] [3] [+] [3] [=] [ ]

How Many Frogs?

written by Lucy Floyd
illustrated by Nathan Jarvis

This book will help me review stories about addition and subtraction concepts.

This book belongs to ______________________.

_____ little frog sits by the tree.

Some frogs come along.

Now there are ____.

_____ little frogs sit by the gate.

Some frogs come along.

Now there are ____.

_____ little frogs sit in the sun.

Some frogs go away.

Now there is ____.

How many frogs hide in the sticks?
Look for the frogs!

Can you find _____?

Name ____________________

Chapters 1–2 Review/Test

Concepts and Skills

Draw 1 more. Color it blue. Write the sum.

1 

2 + 1 = ____

2 

5 + 1 = ____

Cross out 1. Write how many are left.

3 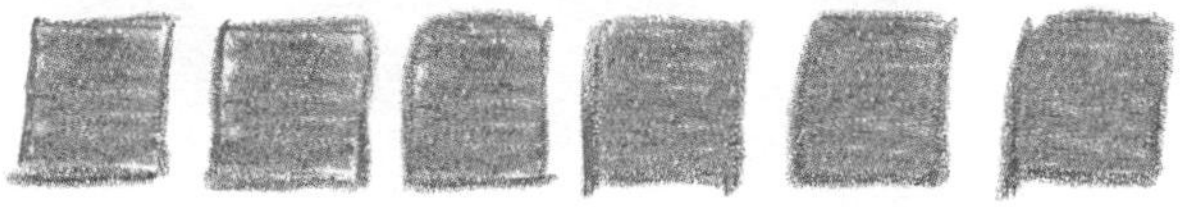

6 − 1 = ____

4 

3 − 1 = ____

Draw 2 more. Color them blue. Write the sum.

5 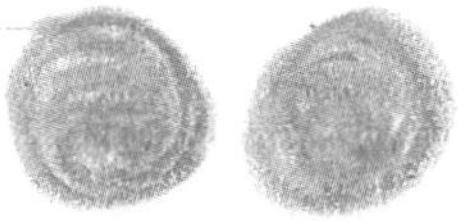

2 + 2 = ____

6 

3 + 2 = ____

Cross out 2. Write the difference.

7 

3 − 2 = ____

8 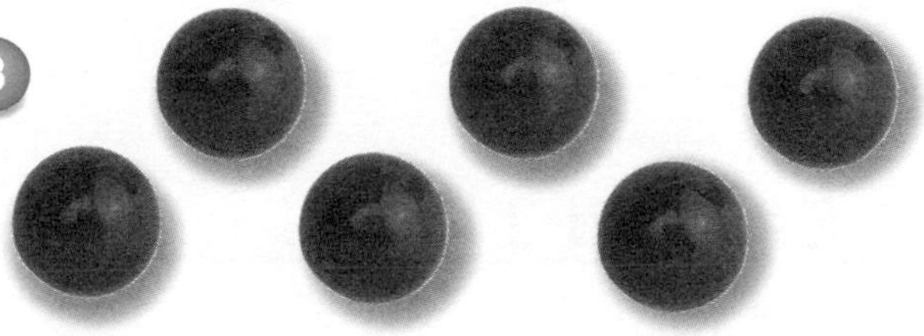

6 − 2 = ____

Write the addition sentence.

9 

_____ + _____ = _____

10 

_____ + _____ = _____

Write the subtraction sentence.

11 

_____ − _____ = _____

12 

_____ − _____ = _____

Problem Solving

Add or subtract. Use ◯.
Draw the ◯.

13 4 ducks are in a pond.
2 more come.
How many in all?

_____ ducks

14 Ben had 3 boats.
He lost 1.
How many are left?

_____ boats

Name ______________________________

Performance Assessment

Use ■ and ■.
Make a train of 6.
Color the cubes to match your train.
Write the addition sentence.

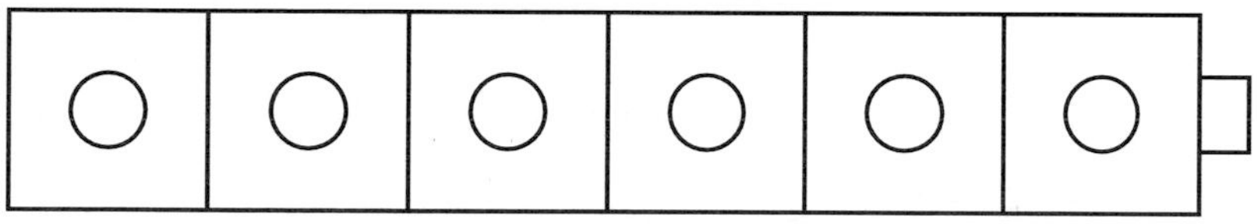

1 ____ + ____ = ____

Make a train of 6.
Color the cubes to match your train.
Cross out some of the ■.
Write the subtraction sentence.

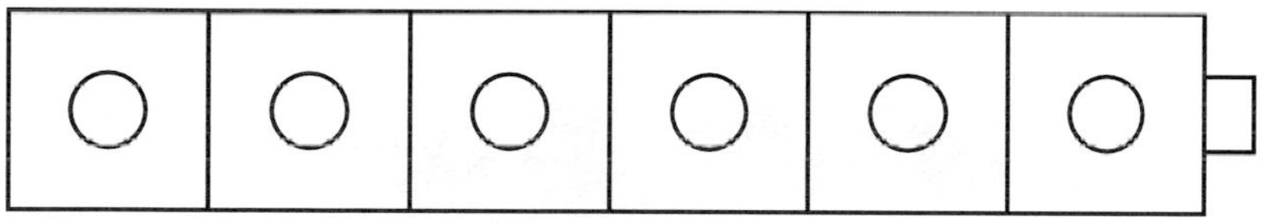

2 ____ − ____ = ____

Write About It

3 Draw a picture to solve the problem.

5 − 2 = ____

Name ______________________

Chapters 1–2 Cumulative Review

Fill in the ⬭ for the correct answer.

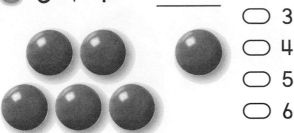

1. 5 + 1 = ____

- ⬭ 3
- ⬭ 4
- ⬭ 5
- ⬭ 6

2. 2 + 2 = ____

- ⬭ 2
- ⬭ 4
- ⬭ 6
- ⬭ Not Here

3. 2 − 1 = ____

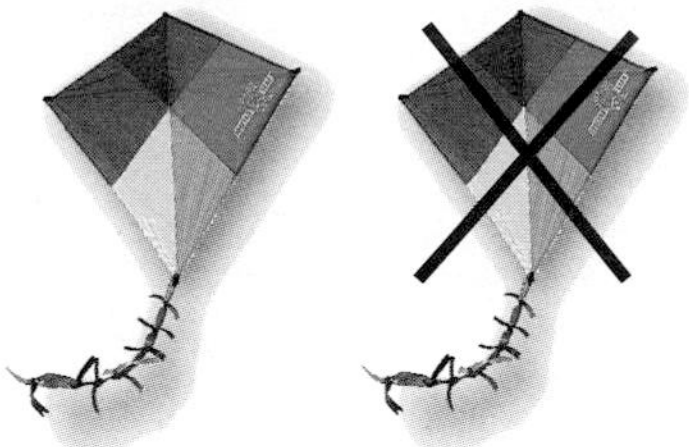

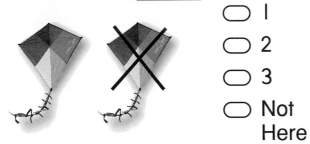

- ⬭ 1
- ⬭ 2
- ⬭ 3
- ⬭ Not Here

4. 6 − 1 = ____

- ⬭ 3
- ⬭ 4
- ⬭ 5
- ⬭ Not Here

5. 5 − 2 = ____

- ⬭ 1
- ⬭ 2
- ⬭ 3
- ⬭ 7

6. 4 − 2 = ____

- ⬭ 2
- ⬭ 3
- ⬭ 4
- ⬭ 5

7. Add or subtract. Use ●.

3 cats play.
2 more come.
How many in all?

- ⬭ 2
- ⬭ 3
- ⬭ 4
- ⬭ 5

8. Add or subtract. Use ●.

4 dogs sit.
1 runs away.
How many are left?

- ⬭ 2
- ⬭ 3
- ⬭ 5
- ⬭ 6

CHAPTER 3 Addition Combinations

What addition stories can you tell about the number 7?

Home Note In this chapter, your child will learn to add numbers to 10.
ACTIVITY Have your child make up addition stories about the pictures on this page.

SCHOOL-HOME CONNECTION

Dear Family,

Today we started Chapter 3. We will practice adding numbers to make 10. Here are the new vocabulary words and an activity for us to do together at home.

Love,

____

Vocabulary

Order Property—The order in which you add numbers doesn't change the sum.

$2 + 3 = 5$

$3 + 2 = 5$

ACTIVITY

Draw a line to divide a sheet of paper in half. Then give your child 6 small objects, such as buttons. Have him or her place some of the objects on each side of the paper and say the addition sentence. Then have him or her turn the paper around and say the new addition sentence.

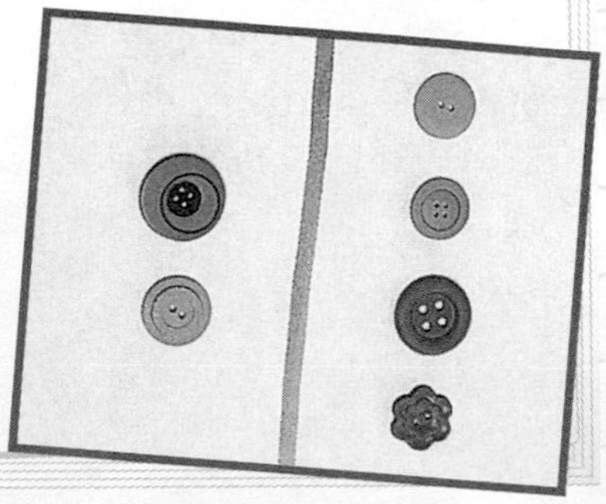

Visit our Web site for additional ideas and activities.
http://www.hbschool.com

Name ______________________

Order Property

You can add in any order and get the same sum.

$2 + 1 = \underline{3}$ $1 + 2 = \underline{3}$

Use Workmat 2 and ● to find each sum.
Color these counters to match.

1. ○○○ $3 + 0 =$ ____ ○○○ $0 + 3 =$ ____

2. ○○○○○ $4 + 1 =$ ____ ○○○○○ $1 + 4 =$ ____

3. ○○○○○ $3 + 2 =$ ____ ○○○○○ $2 + 3 =$ ____

Talk About It • Critical Thinking

What happens to the sum when you change the order of the numbers?

Practice

Use Workmat 2 and ● to find each sum.
Circle the two addition sentences that have the same sum.

1. (3 + 1 = 4)   2 + 4 = 6   (1 + 3 = 4)

2. 4 + 1 = ____   1 + 4 = ____   1 + 2 = ____

3. 3 + 2 = ____   4 + 0 = ____   0 + 4 = ____

4. 4 + 3 = ____   5 + 1 = ____   1 + 5 = ____

5. 2 + 3 = ____   2 + 2 = ____   3 + 2 = ____

Problem Solving ● **Reasoning**

Use ■ and ■.

6. Bob has 2 ■ and 3 ■.
Pat has 3 ■ and 2 ■.
Do they have the same number of cubes?

Yes   No

Draw a picture to show your answer.

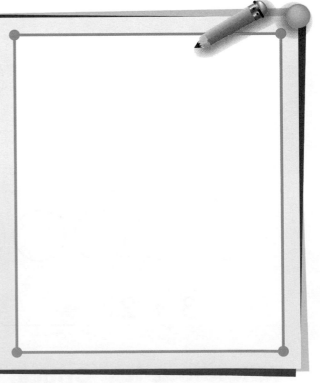

Home Note Your child identified the order property of addition.
ACTIVITY Have your child use small objects to show 3 + 1 and 1 + 3 and then tell you why the two sums are the same.

Name ____________________

Addition Combinations

Use ●.

Write all the ways to make 7 and 8.

1. 3 + 4 = 7
2. ____ + ____ = 7
3. ____ + ____ = 7
4. ____ + ____ = 7
5. ____ + ____ = 7
6. ____ + ____ = 7
7. ____ + ____ = 7
8. ____ + ____ = 7

---

9. ____ + ____ = 8
10. ____ + ____ = 8
11. ____ + ____ = 8
12. ____ + ____ = 8
13. ____ + ____ = 8
14. ____ + ____ = 8
15. ____ + ____ = 8
16. ____ + ____ = 8
17. ____ + ____ = 8

Practice

Use Workmat 2 and ● to find each sum.

1. 1 + 5 = 6	2. 6 + 0 = ____	3. 3 + 3 = ____
4. 2 + 3 = ____	5. 2 + 4 = ____	6. 6 + 1 = ____
7. 4 + 3 = ____	8. 3 + 0 = ____	9. 3 + 1 = ____
10. 0 + 6 = ____	11. 5 + 2 = ____	12. 4 + 4 = ____
13. 4 + 2 = ____	14. 5 + 3 = ____	15. 8 + 0 = ____
16. 1 + 5 = ____	17. 2 + 5 = ____	18. 3 + 5 = ____

Mixed Review

Write the addition or subtraction sentence.

19. 

____ − ____ = ____

20. 

____ + ____ = ____

Home Note Your child identified sums for combinations to 7 and 8.
ACTIVITY Have your child use small objects to show different ways to make 7 and 8.

Name ______________________________

More Addition Combinations

Use [cube] and [cube].

Write all the ways to make 9.

1. 8 + 1 = 9

2. ____ + ____ = ____

3. ____ + ____ = ____

4. ____ + ____ = ____

5. ____ + ____ = ____

6. ____ + ____ = ____

7. ____ + ____ = ____

8. ____ + ____ = ____

9. ____ + ____ = ____

10. ____ + ____ = ____

Practice

Use ■ and ■.
Write all the ways to make 10.

1. 9 + 1 = 10
2. ___ + ___ = ___
3. ___ + ___ = ___
4. ___ + ___ = ___
5. ___ + ___ = ___
6. ___ + ___ = ___
7. ___ + ___ = ___
8. ___ + ___ = ___
9. ___ + ___ = ___
10. ___ + ___ = ___
11. ___ + ___ = ___

Home Note Your child identified sums for combinations to 9 and 10.
ACTIVITY Have your child use small objects to show different ways to make 9 and 10.

Name ______________________

Horizontal and Vertical Addition

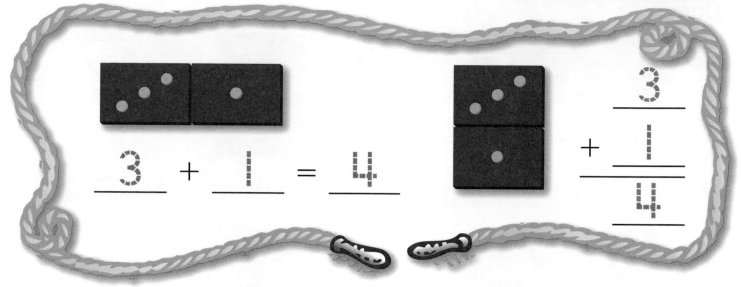

3 + 1 = 4

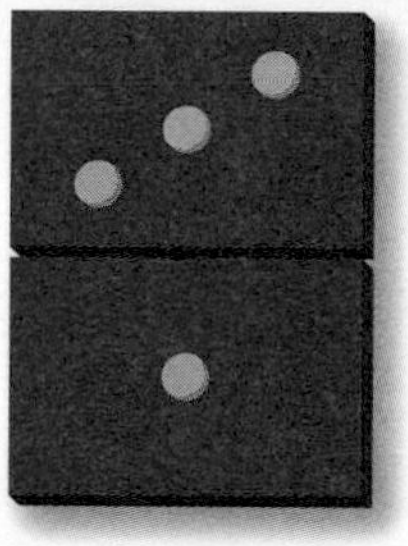

  3
+ 1

  4

Complete.

1.

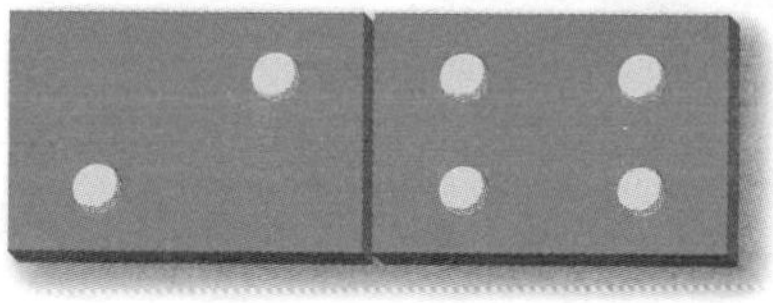

___ + ___ =

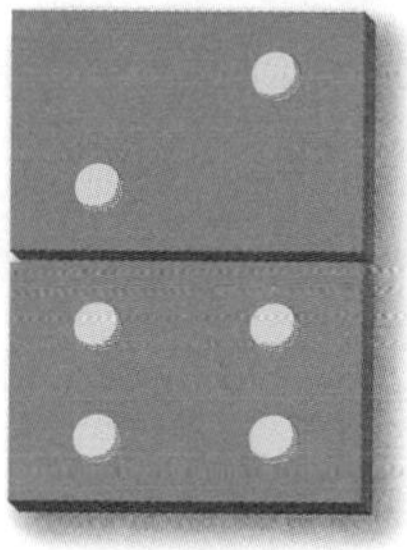

  ___
+ ___

  ___

2.

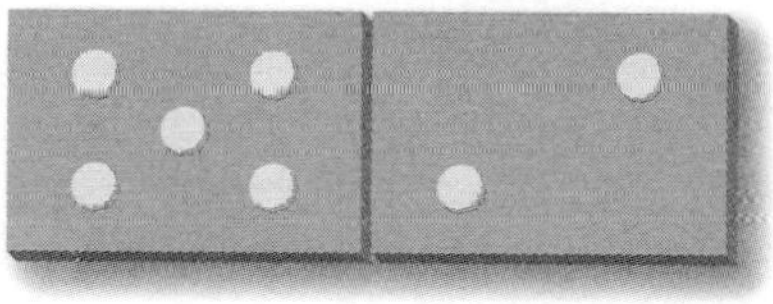

___ + ___ = ___

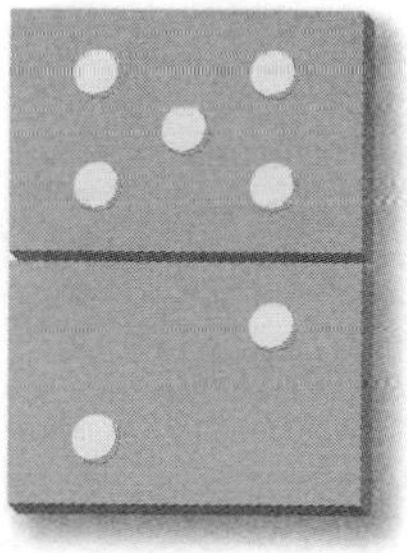

  ___
+ ___

  ___

3.

___ + ___ = ___

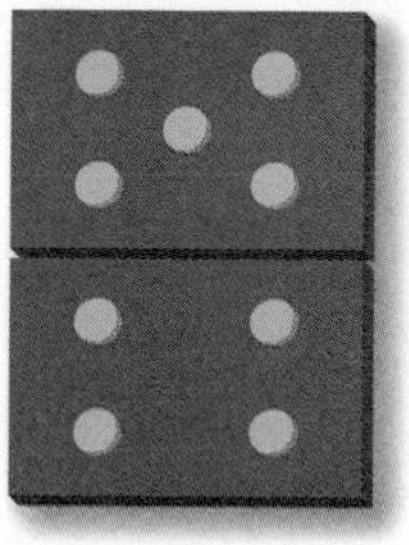

  ___
+ ___

  ___

Talk About It • **Critical Thinking**

Why is the sum the same both ways?

Practice

Complete.

1. 

4 + 3 = 7

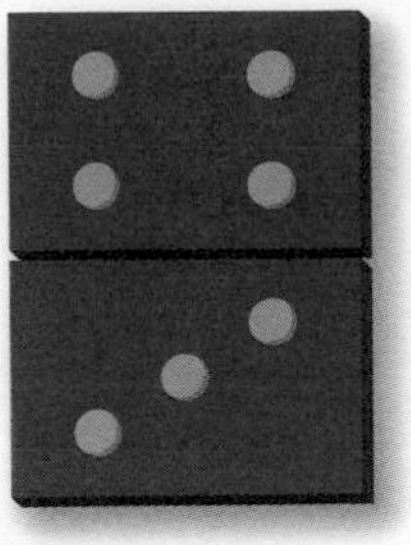

$$\begin{array}{r} 4 \\ +\ 3 \\ \hline 7 \end{array}$$

2. 

____ + ____ = ____

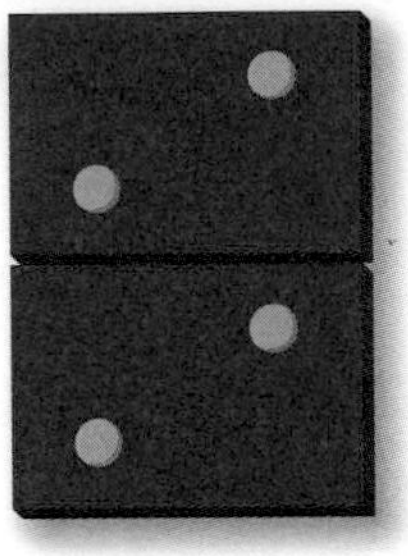

$$\begin{array}{r} \_\_ \\ +\ \_\_ \\ \hline \_\_ \end{array}$$

3. 

____ + ____ = ____

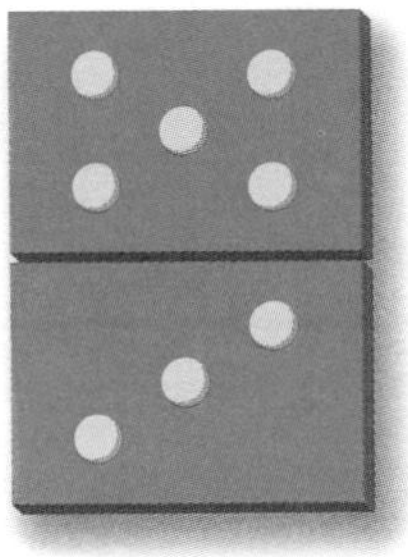

$$\begin{array}{r} \_\_ \\ +\ \_\_ \\ \hline \_\_ \end{array}$$

Write About It

4. Tell a story to a classmate about the picture. Then write the addition sentence.

Home Note Your child solved horizontal and vertical addition problems.
ACTIVITY Write horizontal and vertical addition problems for your child to solve.

Name ______________________

Understand • Plan • Solve • Look Back

Problem Solving
Make a Model

Use (penny) to show each price.
Draw them. Write how many in all.

1

6 ¢

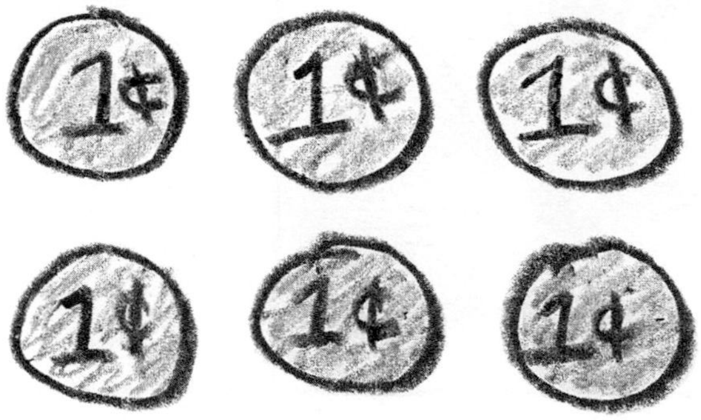

2

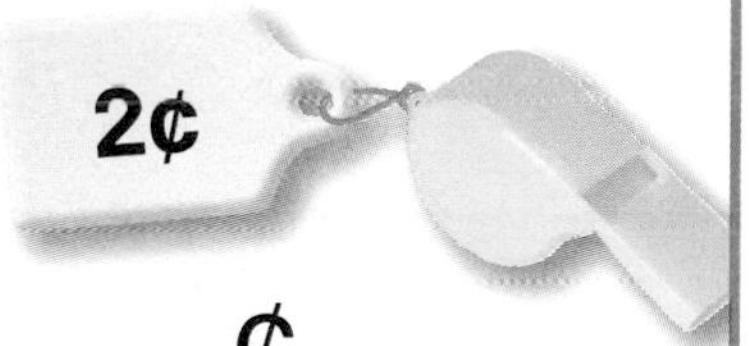

_____ ¢

3

_____ ¢

Practice

Use  to show each price.
Draw them. Write how many in all.

1
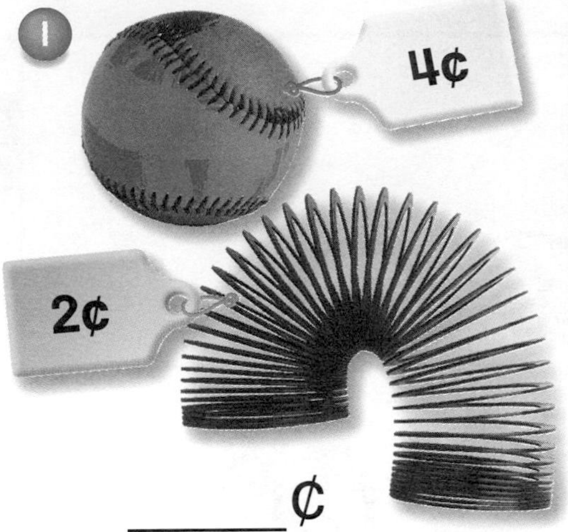

______¢

2

______¢

3

______¢

Home Note **Your child solved problems using money.**
ACTIVITY Choose two objects from pages 67 and 68. Have your child use pennies to show the total cost.

Name ______________________________

Chapter 3
Review/Test

Concepts and Skills

Find each sum. Circle the two addition sentences that have the same sum.

1. $4 + 2 =$ _____ $2 + 4 =$ _____ $1 + 3 =$ _____

2. $2 + 1 =$ _____ $2 + 3 =$ _____ $1 + 2 =$ _____

Write the ways to make 8.

3. _____ + _____ = 8
4. _____ + _____ = 8
5. _____ + _____ = 8
6. _____ + _____ = 8
7. _____ + _____ = 8
8. _____ + _____ = 8
9. _____ + _____ = 8
10. _____ + _____ = 8

Complete.

11. 

_____ + _____ = _____

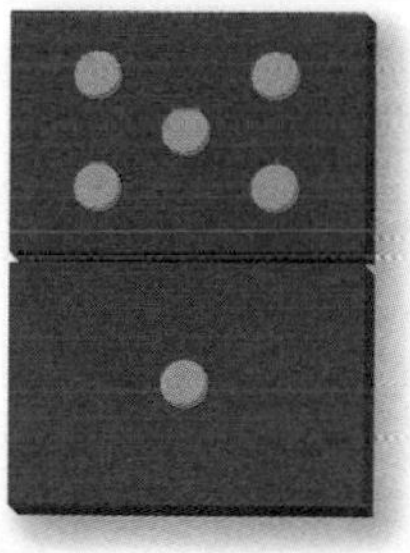

_____
\+ _____
_____

Problem Solving

12. Use  to show each price.
Draw them.
Write the total amount.

_____ ¢

Harcourt Brace School Publishers

Name ______________________________

Cumulative Chapters 1–3

TAAS Prep

Mark the best answer.

1. Which addition sentence tells how many in all?

- ⬭ $2 + 3 = 5$
- ⬭ $2 + 5 = 7$
- ⬭ $4 + 3 = 7$
- ⬭ $5 + 3 = 8$

2. Look at the picture. How many in all?

- ⬭ 5
- ⬭ 6
- ⬭ 7
- ⬭ 8

3. Which subtraction sentence tells how many are left?

- ⬭ $4 - 2 = 2$
- ⬭ $5 - 1 = 4$
- ⬭ $6 - 3 = 3$
- ⬭ $6 - 2 = 4$

4. Look at the picture. How many are left?

- ⬭ 2
- ⬭ 3
- ⬭ 4
- ⬭ 5

CHAPTER 4

Addition Facts to 10

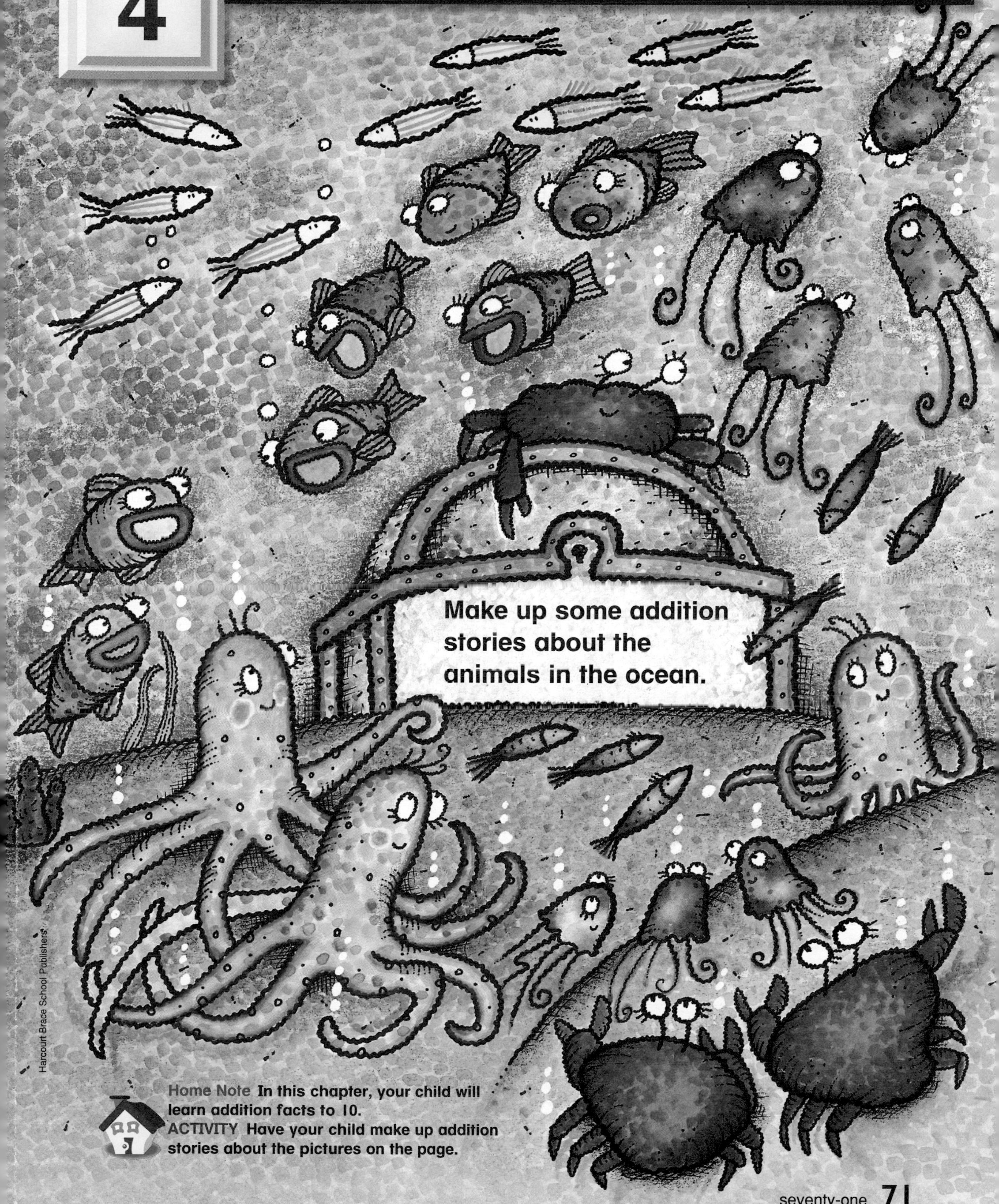

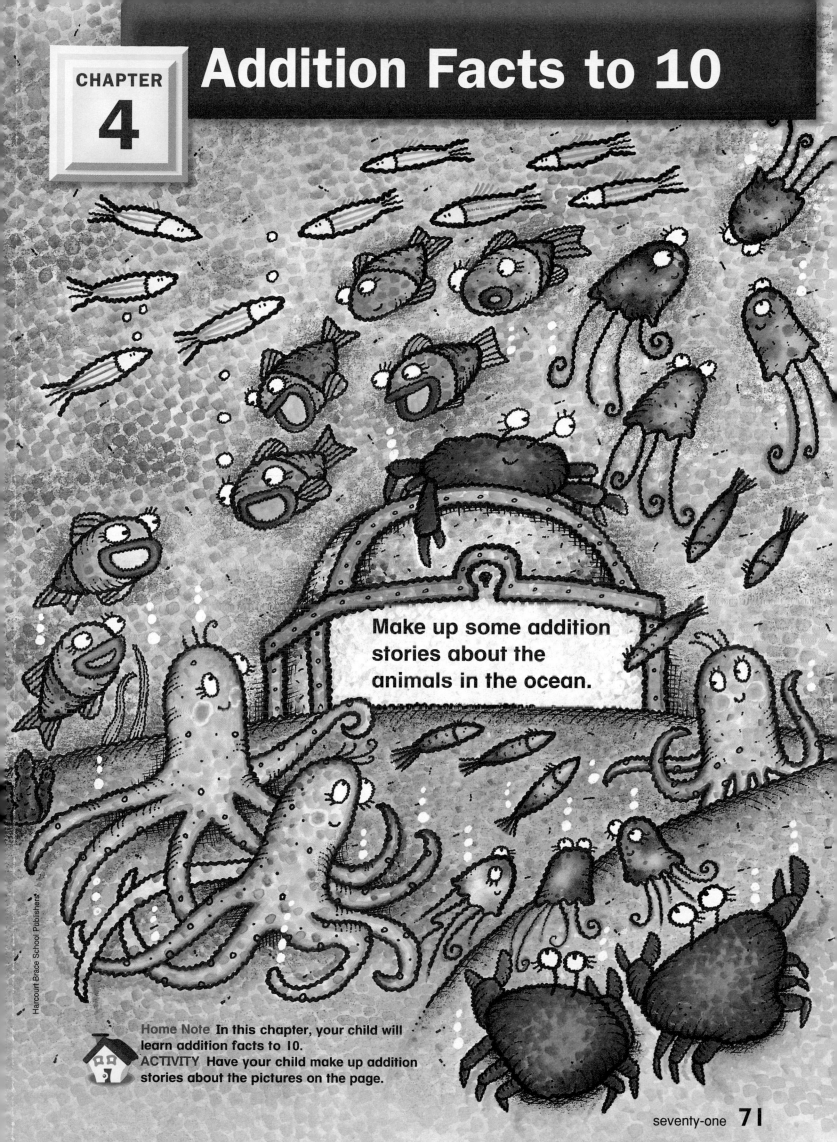

Make up some addition stories about the animals in the ocean.

Home Note In this chapter, your child will learn addition facts to 10.
ACTIVITY Have your child make up addition stories about the pictures on the page.

SCHOOL-HOME CONNECTION

Dear Family,

Today we started Chapter 4. We will learn addition facts to 10. Here are the new vocabulary words and an activity for us to do together at home.

Love,

____

Vocabulary

doubles Two numbers to add that are the same.

$4 + 4 = 8, 5 + 5 = 10$

counting on A way to add by counting on from the larger number.

Start at 3.
Count on 2.
4,5

$3 + 2 = 5$

ACTIVITY

Have your child find objects that show doubles facts. For example, he or she may find a pair of shoes to show 1 + 1, a carton of eggs to show 6 + 6, or a six-pack of juice cans to show 3 + 3. Ask your child to say those doubles facts.

Visit our Web site for additional activities and ideas.
http://www.hbschool.com

Name ______________________

Counting On 1 and 2

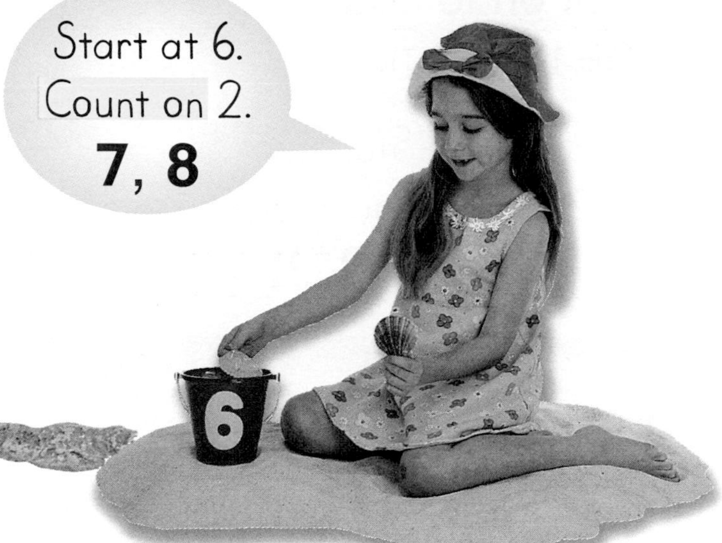

6 + 2 = 8

How many shells are there?
Count on to add. Write the sum.

1 

5 + 1 = ____

2 

4 + 1 = ____

3 

3 + 2 = ____

4 

8 + 1 = ____

5 

7 + 2 = ____

6 

2 + 2 = ____

Talk About It • Critical Thinking

How do you use counting on to find the sum?

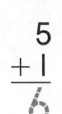

Practice

Count on to add.
Write the sum.

1. $\begin{array}{r} 2 \\ +2 \\ \hline 4 \end{array}$ Start at 2. Count on 2. **3, 4** $\begin{array}{r} 5 \\ +1 \\ \hline 6 \end{array}$ Start at 5. Count on 1. **6** $\begin{array}{r} 3 \\ +2 \\ \hline 5 \end{array}$ Start at 3. Count on 2. **4, 5**

2. $\begin{array}{r} 8 \\ +2 \\ \hline \end{array}$ $\begin{array}{r} 2 \\ +1 \\ \hline \end{array}$ $\begin{array}{r} 6 \\ +1 \\ \hline \end{array}$ $\begin{array}{r} 1 \\ +2 \\ \hline \end{array}$ $\begin{array}{r} 7 \\ +2 \\ \hline \end{array}$ $\begin{array}{r} 9 \\ +1 \\ \hline \end{array}$

3. $\begin{array}{r} 1 \\ +1 \\ \hline \end{array}$ $\begin{array}{r} 5 \\ +2 \\ \hline \end{array}$ $\begin{array}{r} 4 \\ +1 \\ \hline \end{array}$ $\begin{array}{r} 3 \\ +2 \\ \hline \end{array}$ $\begin{array}{r} 3 \\ +1 \\ \hline \end{array}$ $\begin{array}{r} 7 \\ +1 \\ \hline \end{array}$

4. $\begin{array}{r} 2 \\ +1 \\ \hline \end{array}$ $\begin{array}{r} 5 \\ +2 \\ \hline \end{array}$ $\begin{array}{r} 6 \\ +1 \\ \hline \end{array}$ $\begin{array}{r} 8 \\ +2 \\ \hline \end{array}$ $\begin{array}{r} 5 \\ +1 \\ \hline \end{array}$ $\begin{array}{r} 3 \\ +2 \\ \hline \end{array}$

Problem Solving • **Mental Math**

Write the sum.

5. I have 3 starfish in my bucket.
I find 2 more.
How many do I have in all?

_____ starfish

Home Note Your child used counting on 1 and 2 to find sums to 10.
ACTIVITY Choose numbers from 0 to 7, and have your child count on 1 or 2.

Name ______________________

Counting On 3

How many fish are there?
Count on to add.

Start at 4.
Count on 3.
5, 6, 7

1.

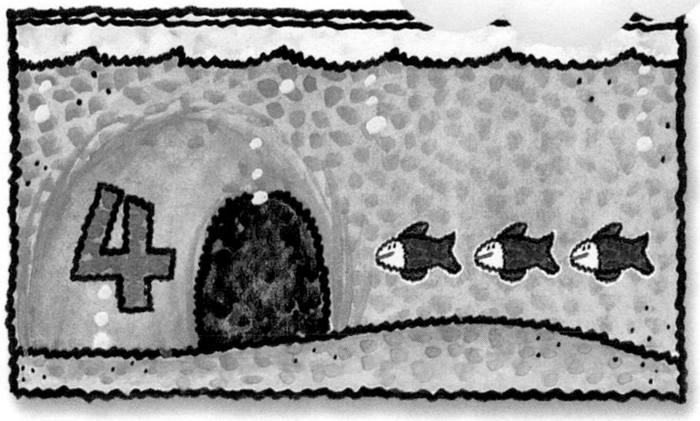

$4 + 3 = \underline{7}$

2.

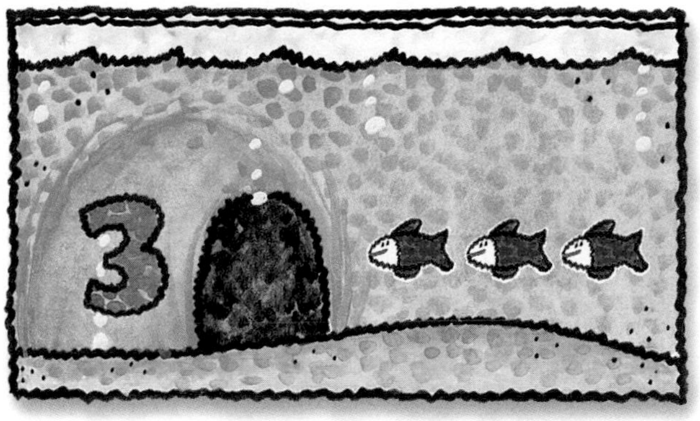

$3 + 3 =$ ____

3.

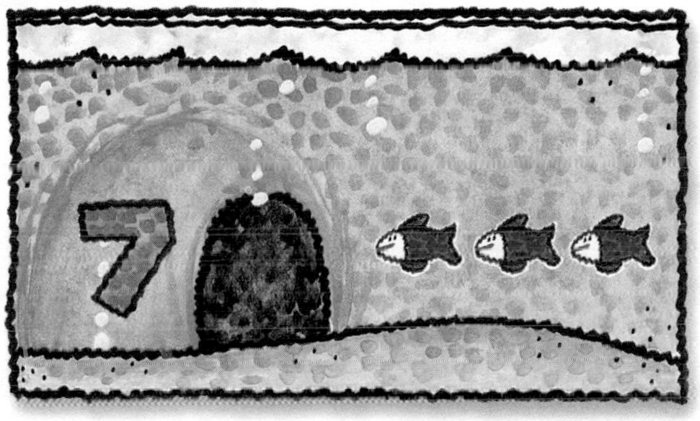

$7 + 3 =$ ____

4.

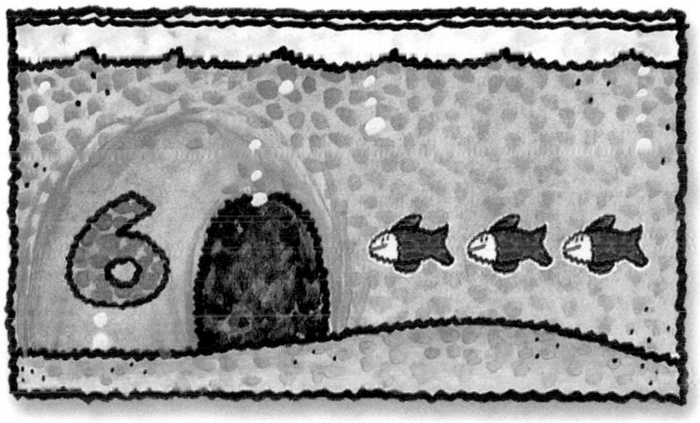

$6 + 3 =$ ____

5.

$1 + 3 =$ ____

6.

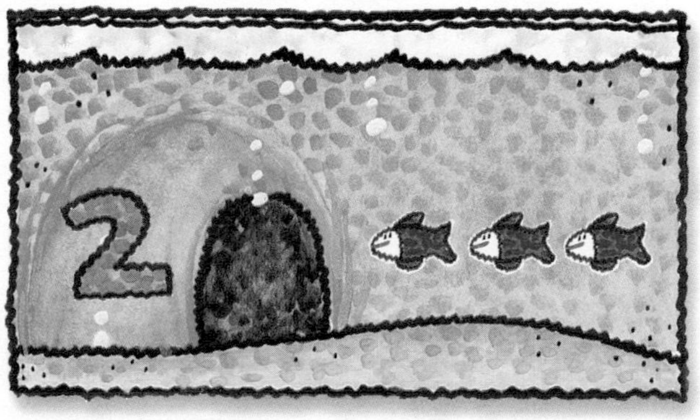

$2 + 3 =$ ____

Practice

Count on to add.

1. $5 + 3 = 8$ (Start at 5. Count on 3. **6, 7, 8**) | $6 + 3$ | $5 + 3$ | $2 + 3$ | $1 + 3$

2. $7 + 3$ | $8 + 2$ | $1 + 2$ | $6 + 2$ | $3 + 2$ | $5 + 2$

3. $6 + 2$ | $4 + 2$ | $7 + 2$ | $1 + 2$ | $6 + 1$ | $8 + 1$

4. $5 + 1$ | $4 + 1$ | $2 + 1$ | $7 + 1$ | $3 + 1$ | $8 + 1$

Mixed Review

Find ways to make 7.

5. ___ + ___ = 7
6. ___ + ___ = 7

Find ways to make 8.

7. ___ + ___ = 8
8. ___ + ___ = 8

Find ways to make 9.

9. ___ + ___ = 9
10. ___ + ___ = 9

Home Note Your child used counting on 3 to find sums to 10.
ACTIVITY Have your child count a group of objects, tell you the number, and then count on to add 3.

Name ______________________________

Doubles

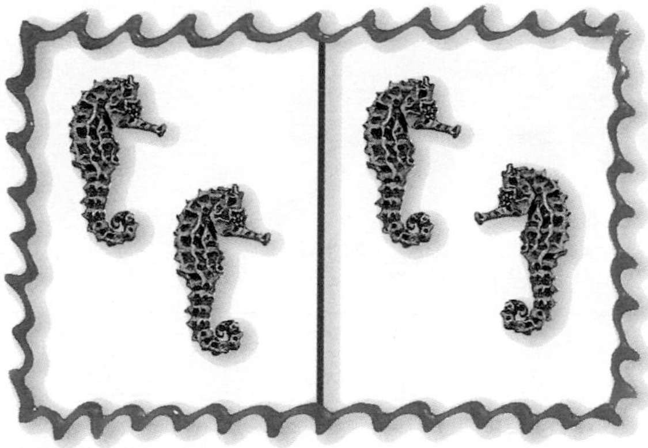

Two numbers to add that are the same are called doubles.

2 + 2 = 4

Make each picture show a double.
Write the doubles fact.

1

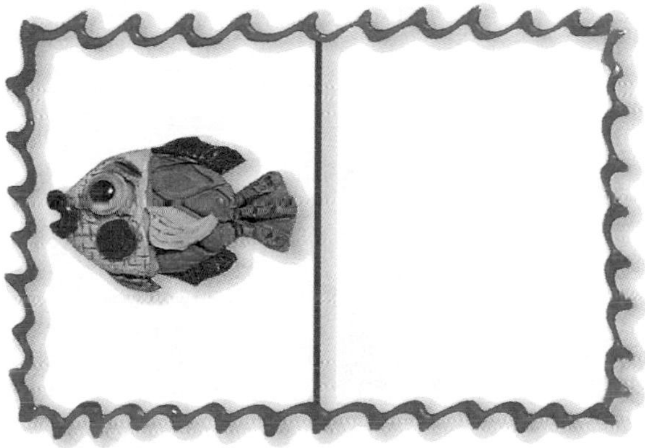

____ + ____ = ____

2

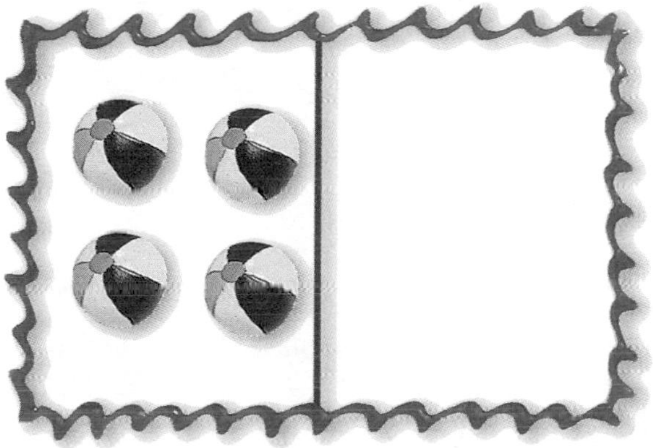

____ + ____ = ____

3

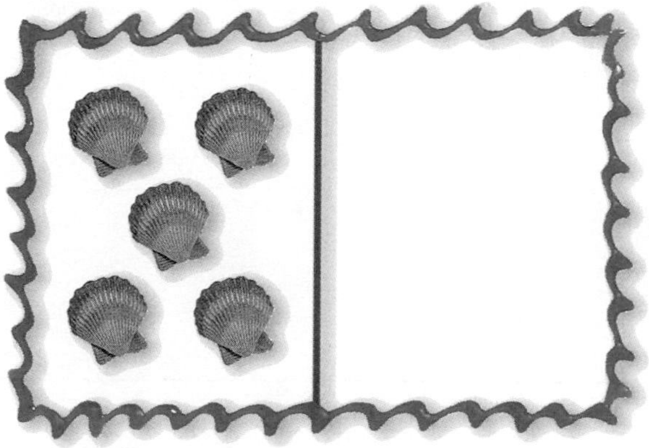

____ + ____ = ____

4

____ + ____ = ____

Talk About It • Critical Thinking

What are doubles?

Practice

Make each picture show a double.
Write the doubles fact.

1 

2 + 2 = 4

2 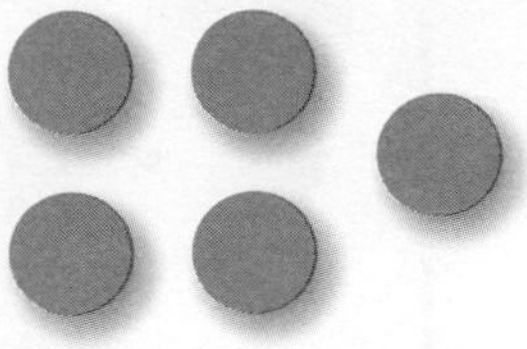

___ + ___ = ___

3 

___ + ___ = ___

4 

___ + ___ = ___

5 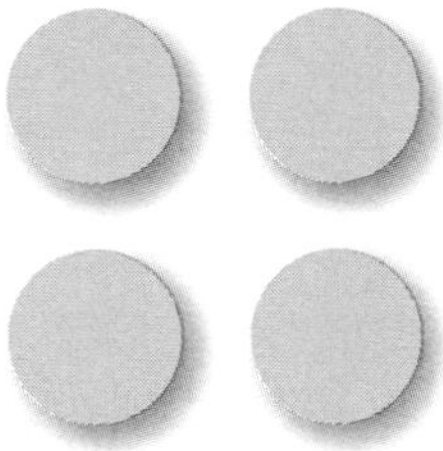

___ + ___ = ___

6 

___ + ___ = ___

Problem Solving

Solve.
Draw a picture to show your answer.

7 Beth has 4 fish.
Rico has 4 fish.
How many fish do they have in all?

___ fish

Home Note Your child used doubles to find sums to 10.
ACTIVITY Have your child use small objects to show doubles.

Name ________________________________

Addition Facts Practice

Write the missing sums.
Circle the sums of doubles.

+	0	1	2	3	4	5	6	7	8
0	0	1	2	3	4	5			8
1	1	2				6	7		9
2					6	7			
3	3		5	6				10	
4	4		6			9			
5		6	7			10			
6	6			9					
7		8	9						
8	8		10						

Talk About It • Critical Thinking

What patterns do you see?

Practice

Write the sum. Circle the doubles facts.

1

3 + 2 = ____   (2 + 2 = 4)   7 + 1 = ____

2

0 + 2 = ____   6 + 2 = ____   0 + 0 = ____

3

4 + 1 = ____   3 + 3 = ____   2 + 4 = ____

4

6 + 3 = ____   5 + 1 = ____   3 + 4 = ____

5

4	4	7	1	8	4
+2	+4	+2	+1	+1	+3
____	____	____	____	____	____

Write About It

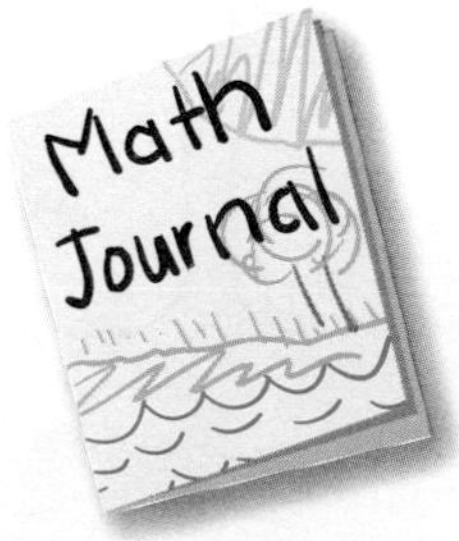

6 Tell a story about the picture.
Then write the addition sentence.

Home Note Your child practiced addition facts to 10.
ACTIVITY Cut apart these problems to make 18 small cards. Pick a card, and read the problem for your child to answer.

Name ______________________________

Understand • Plan • Solve • Look Back

Problem Solving
Draw a Picture

Draw pictures to solve.

1. 3 dogs run.
   3 dogs walk.
   How many dogs in all? 6

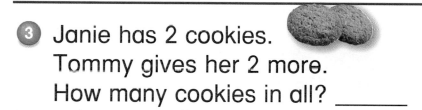

2. I have 3 balloons.
   I lose 1.
   How many are left? ______

3. Janie has 2 cookies.
   Tommy gives her 2 more.
   How many cookies in all? ______

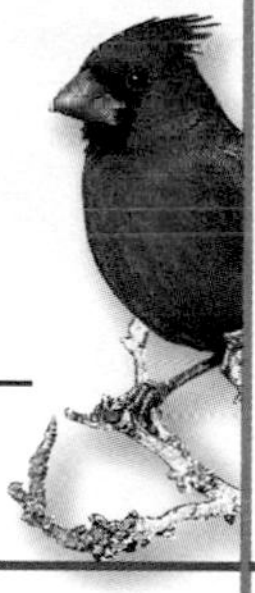

4. There are 5 blue birds.
   There are 2 red birds.
   How many birds in all? ______

5. I have 4 red apples.
   I have 3 green apples.
   How many apples in all? ______

6. Juan has 6 peanuts.
   He eats 2.
   How many are left? ______

Practice

Draw pictures to solve.

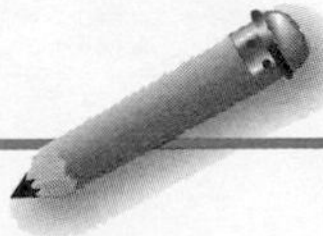

1. 3 pink shells are here.
   2 black shells are here.
   How many shells in all? 5

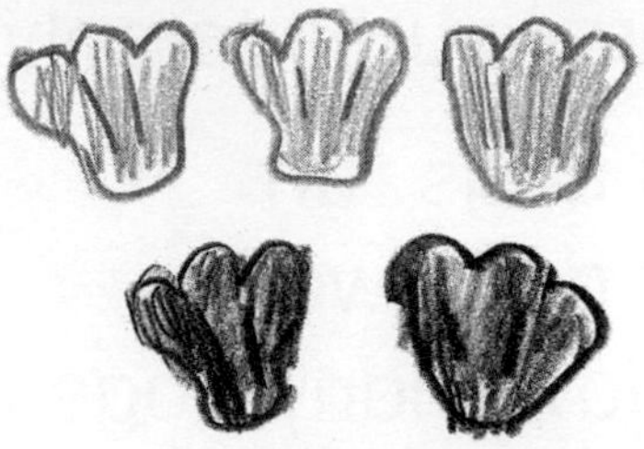

2. 4 fish are yellow.
   2 fish are green.
   How many fish in all? ____

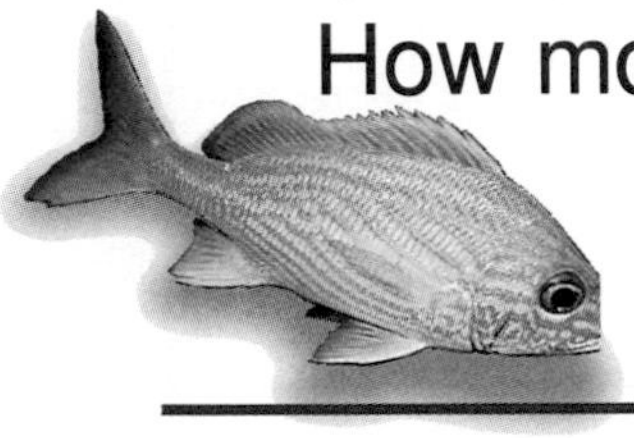

3. Mary has 6 toy boats.
   She gives away 1.
   How many are left? ____

4. Mario sees 6 cars
   in the lot.
   Soon 3 are gone.
   How many cars are left? ____

5. Here are 6 smooth rocks.
   Here is 1 bumpy rock.
   How many rocks in all? ____

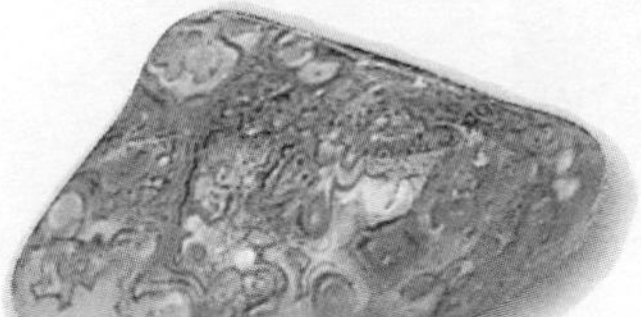

Home Note Your child used the strategy Draw a Picture to solve problems.
ACTIVITY Make up more problems, and have your child draw pictures to solve them.

Name ______________________

Concepts and Skills

How many fish are there? Count on to add.

1 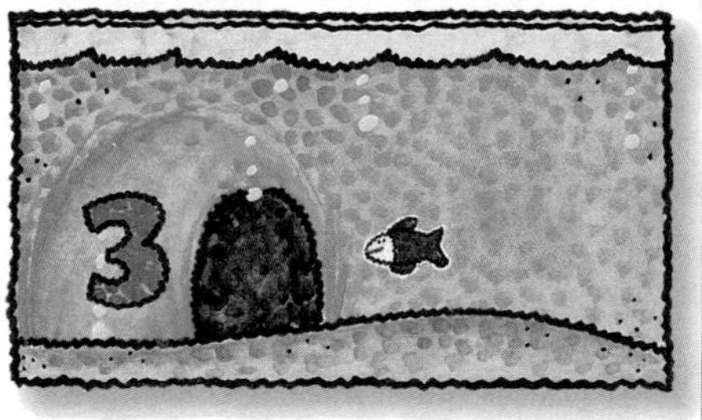

3 + 1 = ____

2 

4 + 2 = ____

3 

2 + 3 = ____

Write the doubles fact.

4 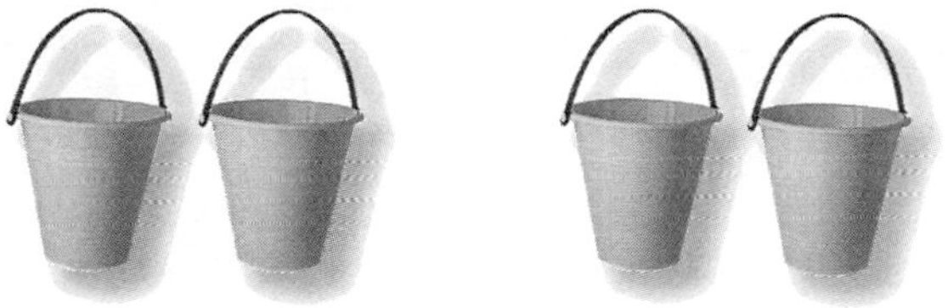

____ + ____ = ____

5 

____ + ____ = ____

Write the sum. Circle the doubles facts.

6

3	4	8	4	5	5
+3	+2	+1	+4	+2	+5

Problem Solving

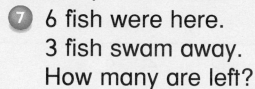

Draw a picture to solve.

7 6 fish were here.
3 fish swam away.
How many are left?

____ fish

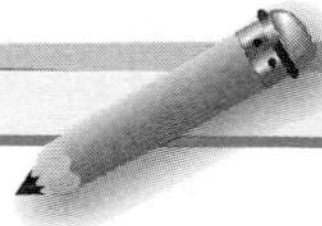

Name ____________________

Cumulative Chapters 1–4

TAAS Prep

Mark the best answer.

1. Which tells how many?

- ⬭ 5
- ⬭ 6
- ⬭ 7
- ⬭ 8

2. Which number is greater?

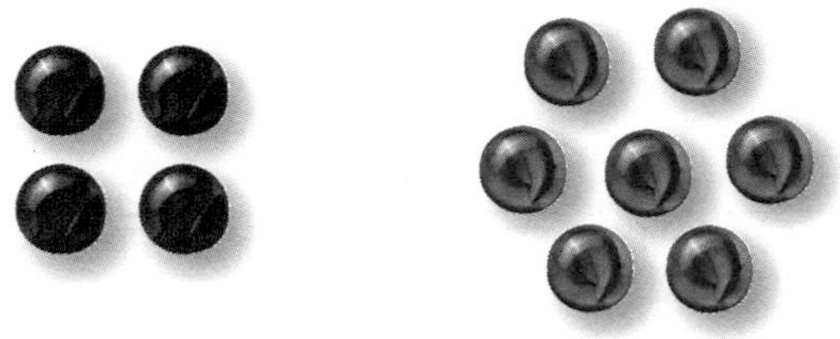

4 7

- ⬭ 4
- ⬭ 7

3. Which tells how many in all?

- ⬭ 1
- ⬭ 2
- ⬭ 3
- ⬭ 4

4. Which is a way to make 9?

- ⬭ 5 + 4
- ⬭ 6 + 2
- ⬭ 3 + 7
- ⬭ 4 + 3

5. Which number is less?

3 8

- ⬭ 3
- ⬭ 8

6. Which subtraction sentence tells how many are left?

- ⬭ 5 − 3 = 2
- ⬭ 8 − 2 = 6
- ⬭ 7 − 4 = 3
- ⬭ 6 − 3 = 3

CHAPTER 5 Subtraction Combinations

Tell some subtraction stories about the animals.

Home Note In this chapter, your child is learning subtraction combinations for numbers through 8.
ACTIVITY Have your child make up subtraction stories about the pictures on this page.

SCHOOL-HOME CONNECTION

Dear Family,

Today we started Chapter 5. We will learn subtraction combinations and fact families. Here are the new vocabulary words and an activity for us to do together at home.

Love,

____

Vocabulary

subtraction combinations

All the possible numbers that can be subtracted from a number.

3 − 0 = 3	3 − 2 = 1
3 − 1 = 2	3 − 3 = 0

fact families

Addition and subtraction problems that use the same numbers.

4 + 3 = 7	7 − 4 = 3
3 + 4 = 7	7 − 3 = 4

ACTIVITY

Have your child write subtraction sentences on separate pieces of scrap paper. Then have him or her cut each paper in a different way after the equals sign so that a match is correct when the pieces fit. Your child should mix the pieces and match them to show the correct sentences.

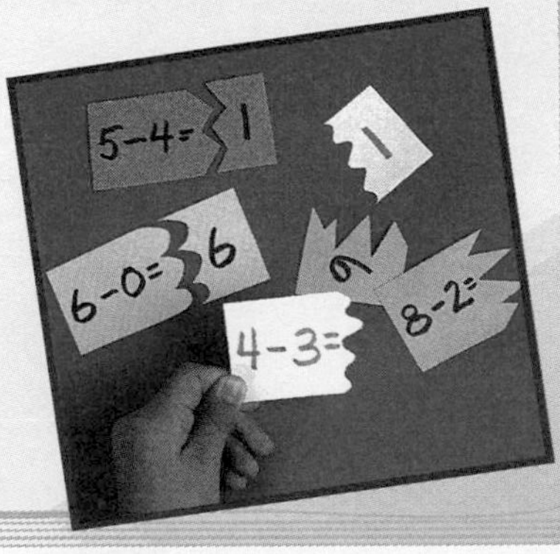

Visit our Web site for additional ideas and activities.
http://www.hbschool.com

Name ______________________________

Subtraction Combinations

Use ●.
Write ways to subtract from 7.

1 7 − 0 = 7	2 7 − ____ = ____
3 7 − ____ = ____	4 7 − ____ = ____
5 7 − ____ = ____	6 7 − ____ = ____
7 7 − ____ = ____	8 7 − ____ = ____

Practice

Use Workmat I and ●.
Write ways to subtract from 8.

1. 8 − 0 = 8
2. 8 − ____ = ____
3. 8 − ____ = ____
4. 8 − ____ = ____
5. 8 − ____ = ____
6. 8 − ____ = ____
7. 8 − ____ = ____
8. 8 − ____ = ____

9. 8 − ____ = ____

Problem Solving ● Reasoning

10. Solve.
There are 8 acorns.
Squirrels eat 6 of them.
How many are left?

____ acorns

Are there more or fewer acorns?
Circle the answer.

more

fewer

Home Note **Your child modeled subtraction combinations for numbers to 8.**
ACTIVITY Have your child use small objects to show all the ways to subtract from numbers up to 8.

Name ______________________

More Subtraction Combinations

Use ●.
Write ways to subtract from 9.

1. 9 − 0 = 9	2. 9 − ___ = ___
3. 9 − ___ = ___	4. 9 − ___ = ___
5. 9 − ___ = ___	6. 9 − ___ = ___
7. 9 − ___ = ___	8. 9 − ___ = ___
9. 9 − ___ = ___	10. 9 − ___ = ___

Practice

Use ●.

Write ways to subtract from 10.

1. 10 − 0 = 10
2. 10 − ____ = ____
3. 10 − ____ = ____
4. 10 − ____ = ____
5. 10 − ____ = ____
6. 10 − ____ = ____
7. 10 − ____ = ____
8. 10 − ____ = ____
9. 10 − ____ = ____
10. 10 − ____ = ____
11. 10 − ____ = ____

Home Note Your child modeled subtraction combinations for 9 and 10.
ACTIVITY Have your child use small objects to show these subtraction problems.

Name ______________________

Vertical Subtraction

Complete.

1. 

6 − 2 = 4

```
  6
− 2
---
  4
```

2. 

___ − ___ = ___

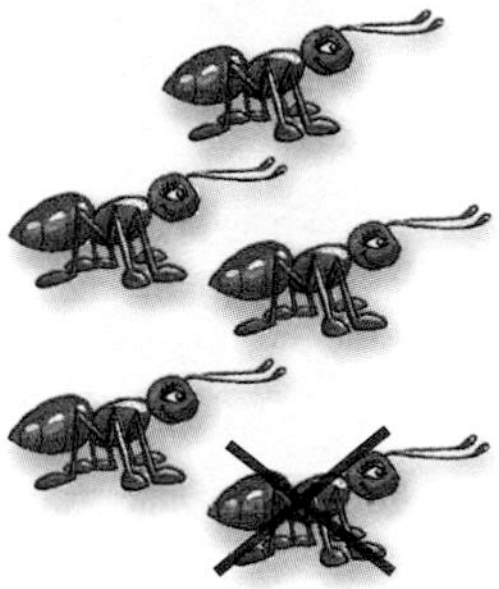

```
  ___
− ___
-----
```

3. 

___ − ___ = ___

```
  ___
− ___
-----
```

4. 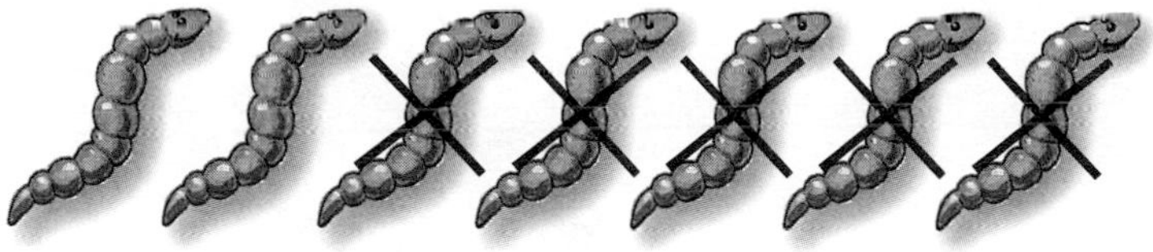

___ − ___ = ___

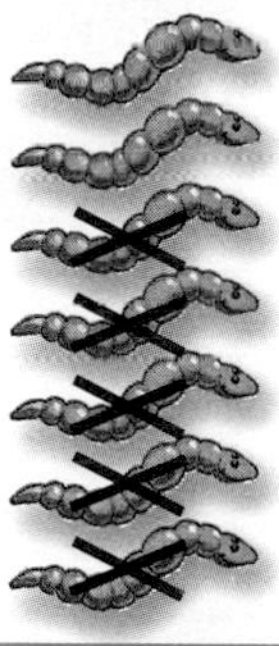

```
  ___
− ___
-----
```

5. 

___ − ___ = ___

```
  ___
− ___
-----
```


Talk About It • Critical Thinking

Why is the difference the same both ways?

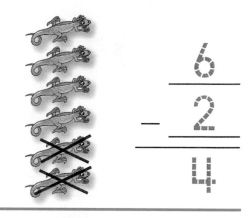

Practice

Complete.

1. 

5 − 5 = 0

$$\begin{array}{r} 5 \\ -\ 5 \\ \hline 0 \end{array}$$

2. 

____ − ____ = ____

$$\begin{array}{r} \_\_\_ \\ -\ \_\_\_ \\ \hline \end{array}$$

3. 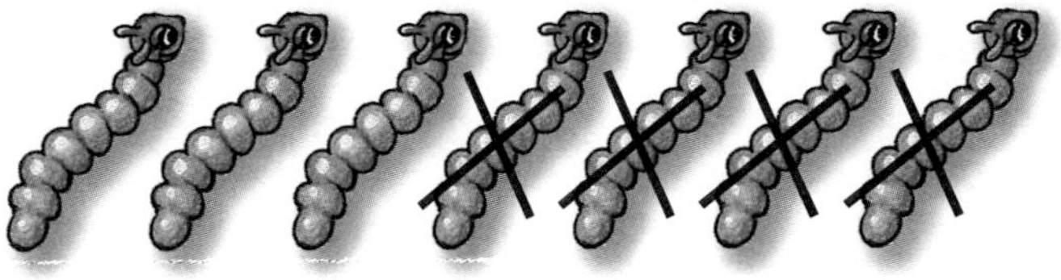

____ − ____ = ____

$$\begin{array}{r} \_\_\_ \\ -\ \_\_\_ \\ \hline \end{array}$$

4. 

____ − ____ = ____

$$\begin{array}{r} \_\_\_ \\ -\ \_\_\_ \\ \hline \end{array}$$

Mixed Review

5. Draw and color 1 more.
Write the addition sentence.

____ + ____ = ____

Home Note Your child solved vertical and horizontal subtraction problems.
ACTIVITY Write horizontal and vertical problems for your child to solve.

Name ______________________

Fact Families

Make a cube train to show addition facts.

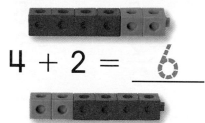

4 + 2 = 6

2 + 4 = 6

Break the train to show subtraction facts.

6 − 2 = 4

6 − 4 = 2

2 | 4 | 6 are the numbers in this fact family.

Use Workmat I and cube. Add or subtract.
Write the numbers in each fact family.

1.
3 + 2 = ____
2 + 3 = ____
5 − 2 = ____
5 − 3 = ____
☐ ☐ ☐

2.
4 + 1 = ____
1 + 4 = ____
5 − 1 = ____
5 − 4 = ____
☐ ☐ ☐

3.
7 + 2 = ____
2 + 7 = ____
9 − 2 = ____
9 − 7 = ____
☐ ☐ ☐

4.
8 + 1 = ____
1 + 8 = ____
9 − 1 = ____
9 − 8 = ____
☐ ☐ ☐

Practice

Add or subtract.
Write the numbers in each fact family.

1

$$\begin{array}{r}3\\+2\\\hline 5\end{array}\quad\begin{array}{r}2\\+3\\\hline 5\end{array}\quad\begin{array}{r}5\\-2\\\hline 3\end{array}\quad\begin{array}{r}5\\-3\\\hline 2\end{array}$$

3	2	5

2

$$\begin{array}{r}4\\+1\\\hline \end{array}\quad\begin{array}{r}1\\+4\\\hline \end{array}\quad\begin{array}{r}5\\-1\\\hline \end{array}\quad\begin{array}{r}5\\-4\\\hline \end{array}$$

3

$$\begin{array}{r}4\\+2\\\hline \end{array}\quad\begin{array}{r}2\\+4\\\hline \end{array}\quad\begin{array}{r}6\\-2\\\hline \end{array}\quad\begin{array}{r}6\\-4\\\hline \end{array}$$

4

$$\begin{array}{r}5\\+1\\\hline \end{array}\quad\begin{array}{r}1\\+5\\\hline \end{array}\quad\begin{array}{r}6\\-1\\\hline \end{array}\quad\begin{array}{r}6\\-5\\\hline \end{array}$$

Home Note Your child modeled fact families and wrote the sums and differences.
ACTIVITY Have your child write some fact families.

Name ____________________

Subtracting to Compare

Draw lines to match.
Subtract to compare.
Write how many more.

1

$4 - 3 = 1$

1 more 

2

$5 - 3 =$ ____

____ more 

3

$7 - 5 =$ ____

____ more 

4

$4 - 2 =$ ____

____ more 

5

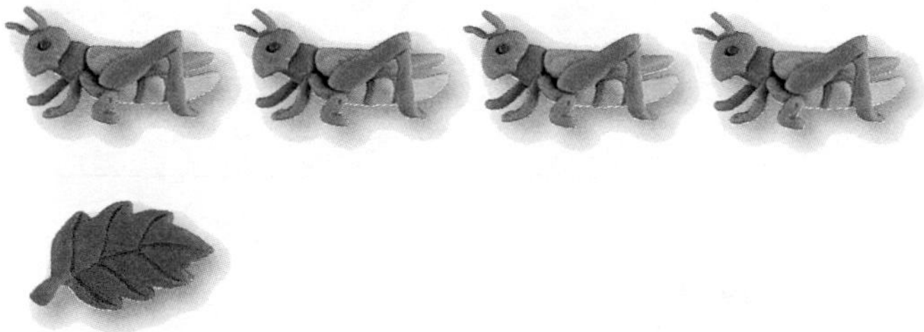

$4 - 1 =$ ____

____ more 

6

$8 - 4 =$ ____

____ more 

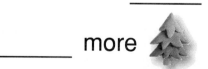

Practice

Draw lines to match.
Subtract to compare.
Write how many more.

1 

$5 - 4 = \underline{1}$

$\underline{1}$ more

2 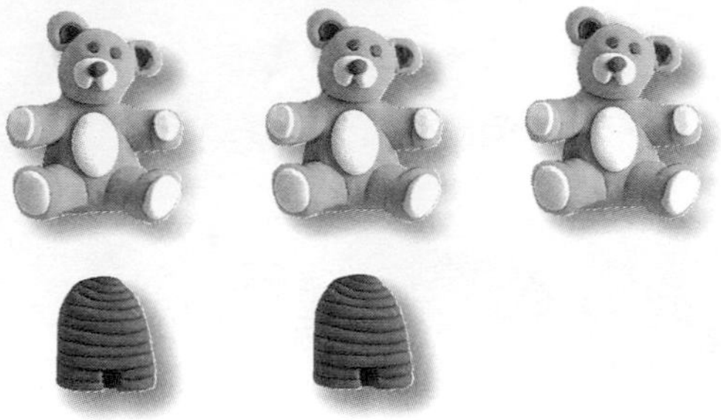

$3 - 2 = \underline{\quad}$

____ more

3 

$6 - 1 = \underline{\quad}$

____ more

4 

$7 - 5 = \underline{\quad}$

____ more

Write About It

5 Draw a picture about these three numbers

5	3	2

.

Write two addition sentences and two subtraction sentences to show the fact family.

Home Note Your child modeled comparative subtraction and completed subtraction sentences.
ACTIVITY Have your child tell you a subtraction story about each problem.

Name ______________________________

Chapter 5

Review/Test

Concepts and Skills

Write ways to subtract from 5.

1. 5 − ____ = ____
2. 5 − ____ = ____
3. 5 − ____ = ____
4. 5 − ____ = ____
5. 5 − ____ = ____
6. 5 − ____ = ____

Complete.

7. 

____ − ____ = ____

____
− ____
____

Add or subtract. Write the numbers in the fact family.

8. 

5 + 2 = ____ 7 − 5 = ____

2 + 5 = ____ 7 − 2 = ____

☐ ☐ ☐

Draw lines to match.
Subtract to compare.
Write how many more.

9. 

6 − 3 = ____

____ more 

10. 

4 − 2 = ____

____ more 

Name ________________________________

Cumulative Chapters 1–5

TAAS Prep

Mark the best answer.

1. Which subtraction sentence tells how many are left?

- ⬭ 5 − 2 = 3
- ⬭ 6 − 2 = 4
- ⬭ 7 − 2 = 5
- ⬭ 7 − 3 = 4

2. Which number tells the sum or difference?

9 cats are in the house.
4 cats go out.
How many are left?

- ⬭ 4
- ⬭ 5
- ⬭ 6
- ⬭ 7

3. Which numbers are in this fact family?

2 + 3 = 5   5 − 3 = 2
3 + 2 = 5   5 − 2 = 3

- ⬭ 2, 3, 5
- ⬭ 5, 2, 7
- ⬭ 5, 3, 8
- ⬭ 9, 4, 5

4. Look at the picture. How many in all?

- ⬭ 3
- ⬭ 4
- ⬭ 7
- ⬭ 8

5. What picture shows a doubles fact?

- ⬭ 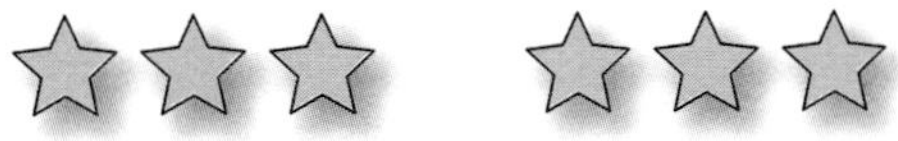
- ⬭ 
- ⬭ 
- ⬭ 

6. Look at the pictures. Which number shows how many?

- ⬭ 4
- ⬭ 6
- ⬭ 8
- ⬭ 10

CHAPTER 6

Subtraction Facts to 10

What subtraction stories can you see out the window?

Home Note In this chapter, your child is learning subtraction facts to 10.
ACTIVITY Have your child make up subtraction stories about the pictures on the page.

SCHOOL-HOME CONNECTION

Dear Family,

Today we started Chapter 6. We will learn more ways to solve subtraction problems. Here are the vocabulary words and an activity for us to do together at home.

Love,

____

Vocabulary

number line

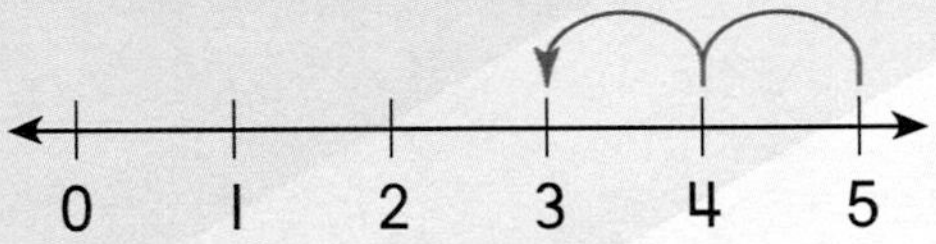

$5 - 2 = 3$

counting back
A way to subtract by counting back from the larger number.

Count back 2.
4, 3

ACTIVITY

Give your child a set of 10 small snack items, such as raisins or peanuts. Let him or her eat the treats one at a time, telling you the new subtraction sentence each time.

Visit our Web site for additional ideas and activities.
http://www.hbschool.com

Name ______________________________

Counting Back 1 and 2

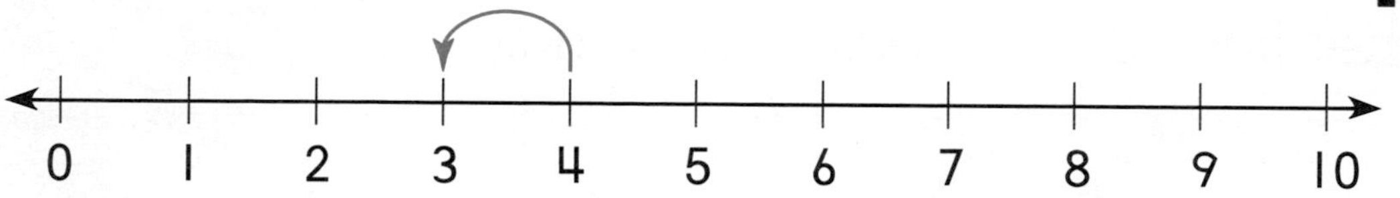

4 − 1 = 3

Start at 4 on the number line.
Count back 1. Where are you?

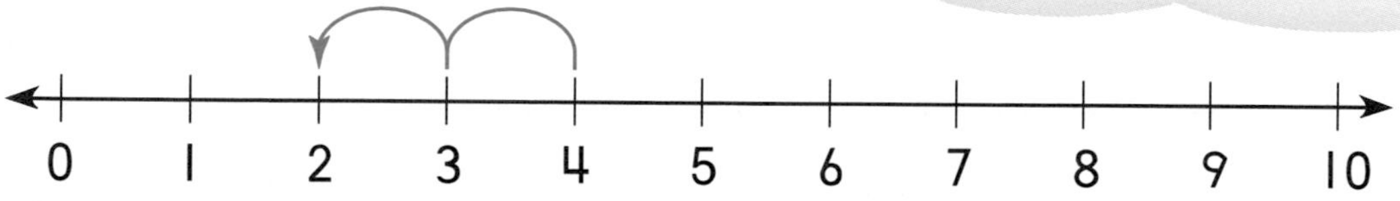

4 − 2 = 2

Start at 4 on the number line.
Count back 2. Where are you?

Use the number line.
Count back to subtract.

1.
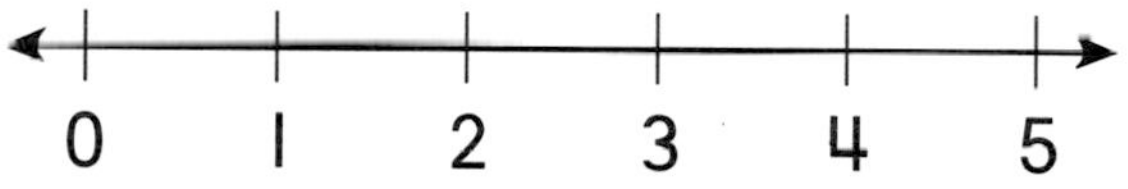

2 − 1 = ____

2.
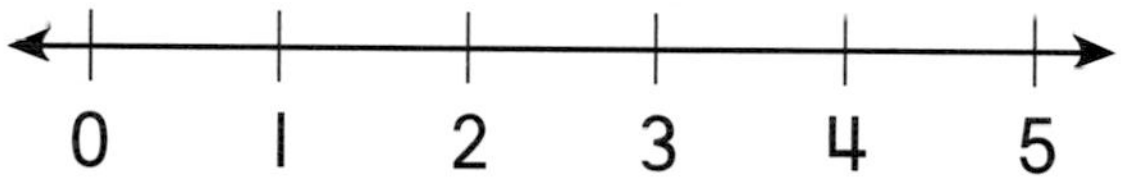

3 − 1 = ____

3.
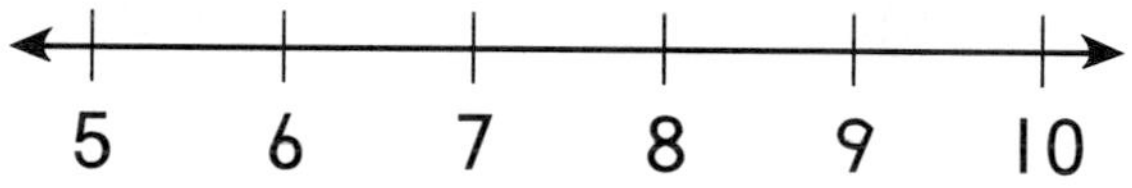

10 − 1 = ____

4.
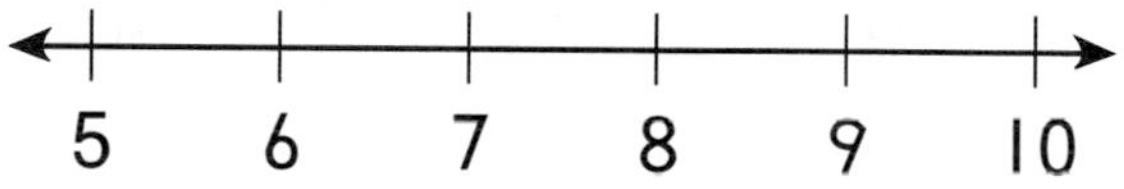

8 − 2 = ____

5.
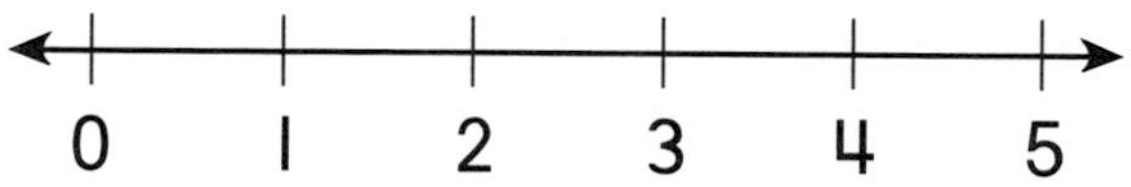

3 − 2 = ____

6.
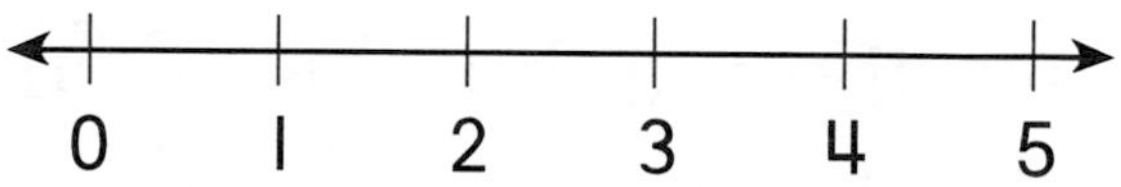

5 − 2 = ____

Talk About It • Critical Thinking

How do you use the number line to help you subtract?

Practice

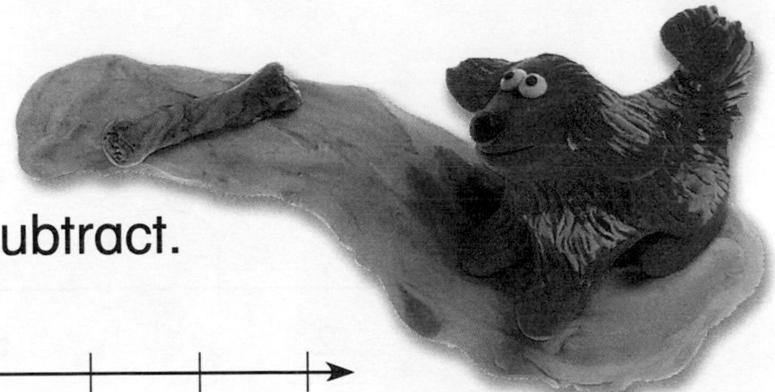

Use the number line. Count back to subtract.

1

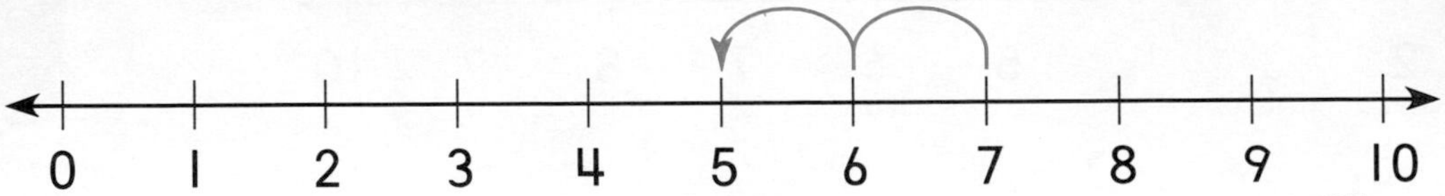

7 − 2 = 5

Start at 7. Count back 2.
Where are you?

2

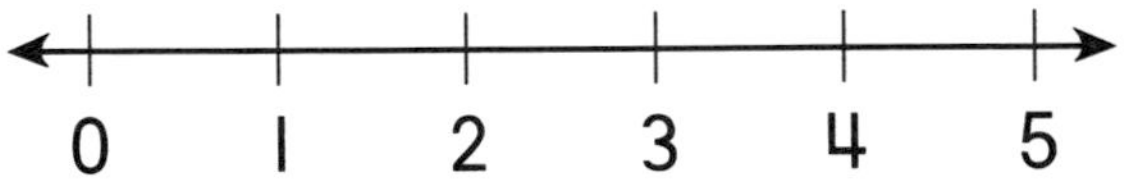

3 − 1 = ____

3

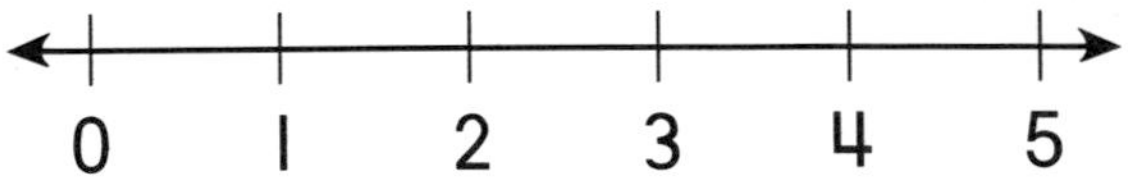

5 − 2 = ____

4

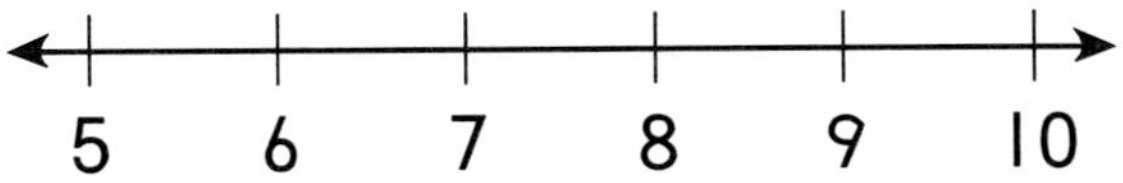

10 − 2 = ____

5

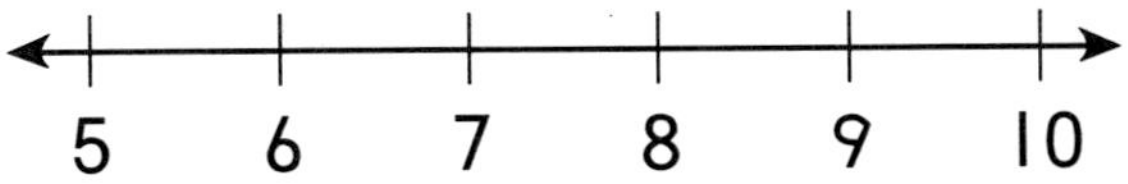

9 − 2 = ____

6

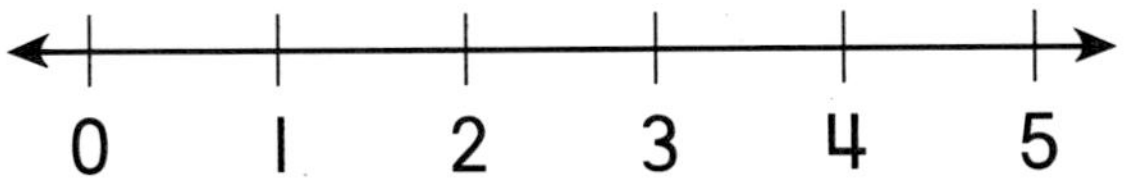

4 − 1 = ____

7

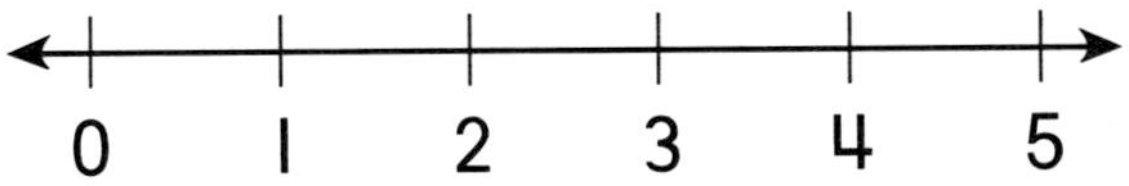

5 − 2 = ____

Problem Solving

Use the number line. Count back.

8 6 birds are in the tree.
2 birds fly away.
How many now? ____

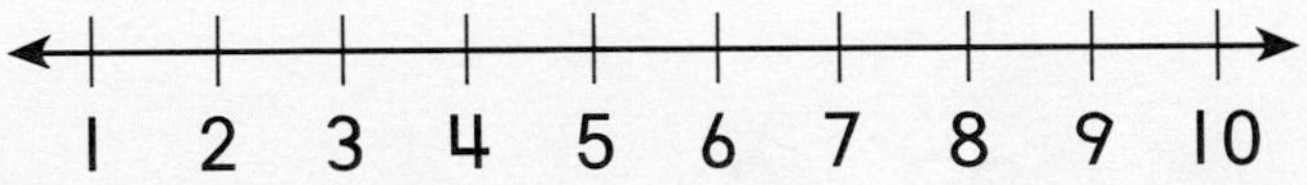

Home Note Your child used the number line to count back 1 and 2 to subtract.
ACTIVITY Have your child use his or her finger to show counting back jumps on the number line.

Name ______________________

Counting Back 3

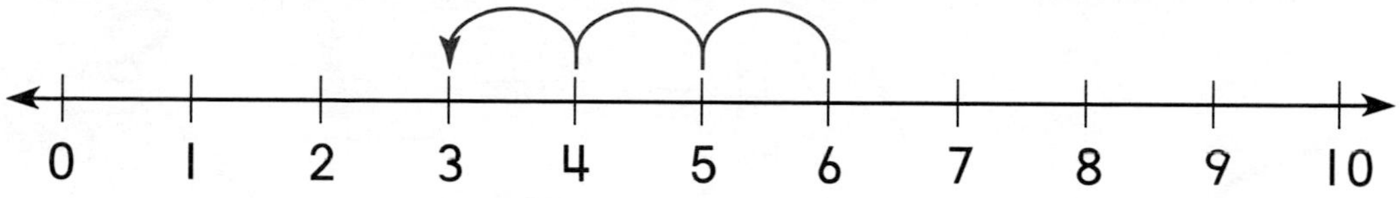

$6 - 3 = \underline{3}$

Start at 6. Count back 3.
Where are you?

Use the number line. Count back to subtract.

1.

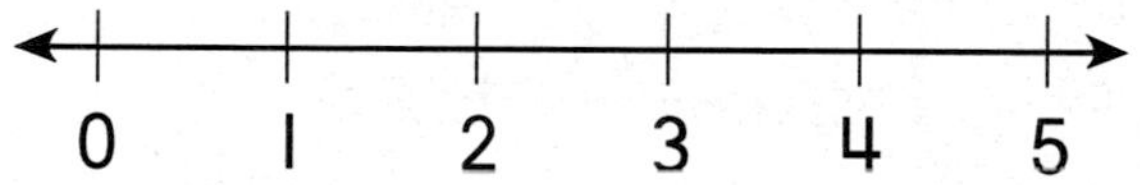

$5 - 3 =$ ____

2.

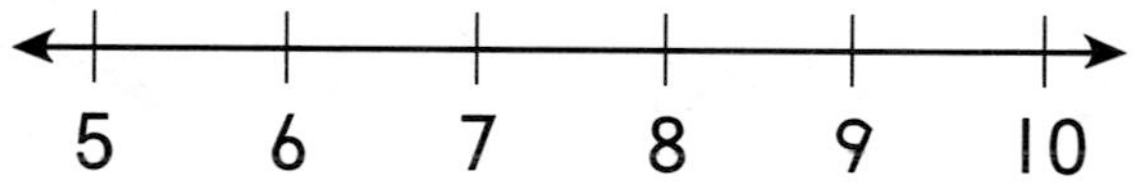

$8 - 3 =$ ____

3.

$10 - 3 =$ ____

4.

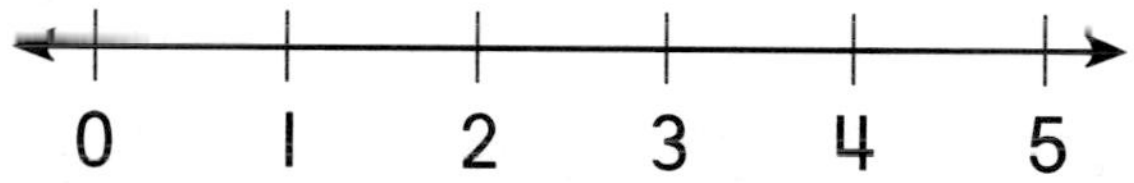

$4 - 3 =$ ____

5.

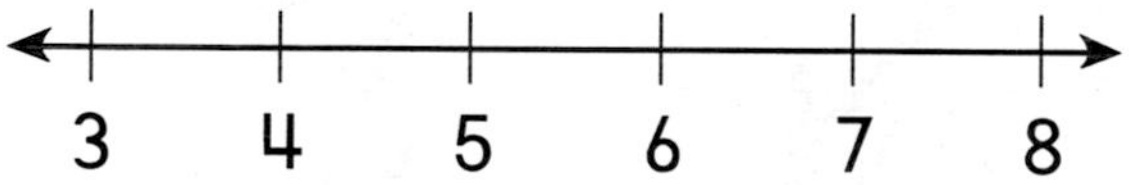

$7 - 3 =$ ____

6.

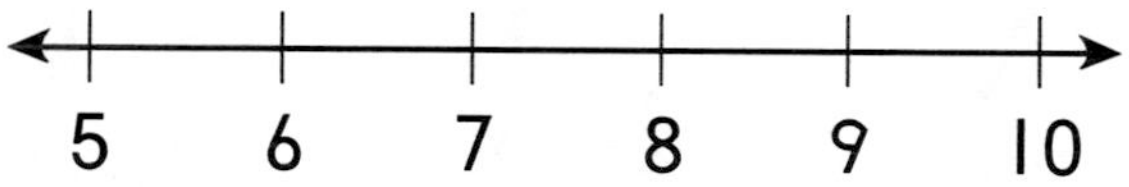

$9 - 3 =$ ____

7.

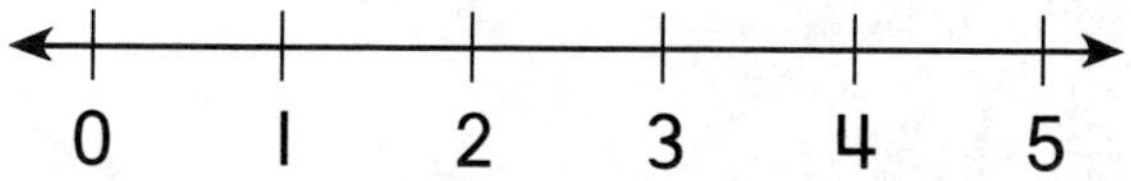

$3 - 2 =$ ____

8.

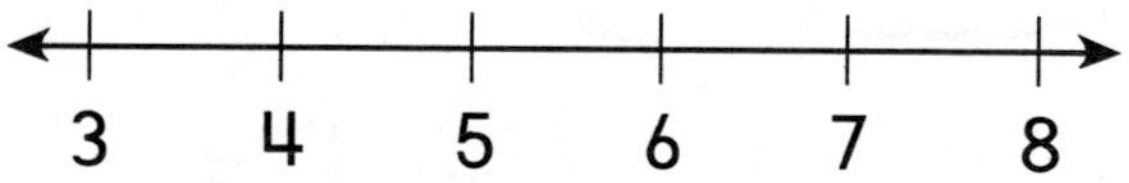

$6 - 3 =$ ____

Practice

Count back to subtract.
Color.

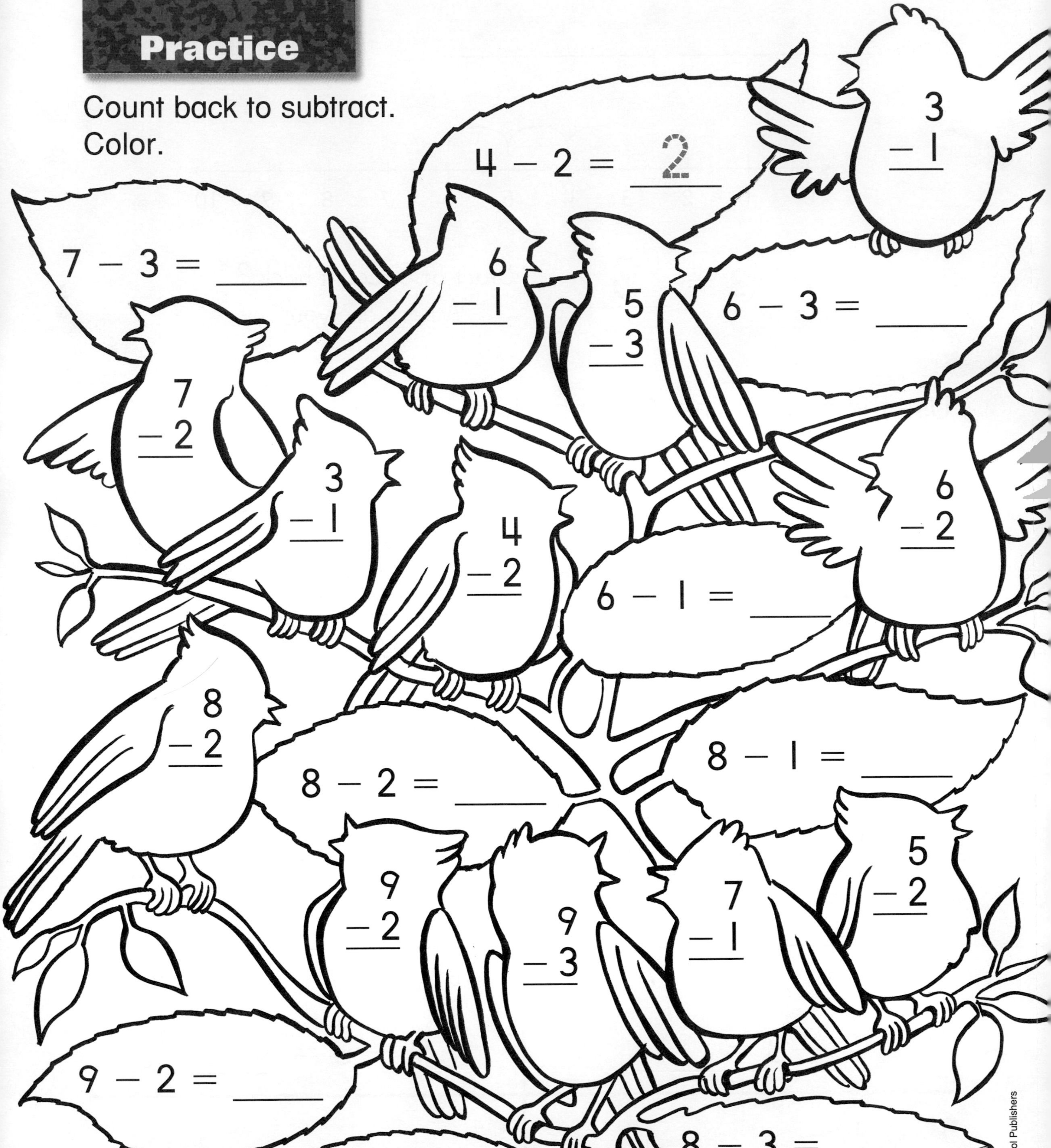

Home Note Your child counted back 3 to subtract from 10 or less.
ACTIVITY Have your child show you how he or she counted back to solve each problem.

Name ______________________________

Subtracting Zero

$\begin{array}{r} 6 \\ -\,6 \\ \hline 0 \end{array}$

I subtract all.
I have zero left.

$\begin{array}{r} 6 \\ -\,0 \\ \hline 6 \end{array}$

I subtract zero.
I have the same number left.

Cross out pictures to show how many are taken away. Subtract.

1. $\begin{array}{r} 7 \\ -\,0 \\ \hline \end{array}$

2. $\begin{array}{r} 3 \\ -\,3 \\ \hline \end{array}$

3. $\begin{array}{r} 6 \\ -\,0 \\ \hline \end{array}$

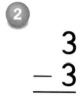

4. $\begin{array}{r} 5 \\ -\,0 \\ \hline \end{array}$

5. $\begin{array}{r} 4 \\ -\,4 \\ \hline \end{array}$

6. $\begin{array}{r} 2 \\ -\,2 \\ \hline \end{array}$

7. $\begin{array}{r} 3 \\ -\,0 \\ \hline \end{array}$

8. $\begin{array}{r} 2 \\ -\,0 \\ \hline \end{array}$

9. $\begin{array}{r} 1 \\ -\,1 \\ \hline \end{array}$

Talk About It • Critical Thinking

What happens when you subtract all that you have?

Practice

Subtract. Circle all the zero facts.

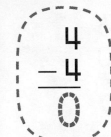

1.  $4 - 4 = 0$ (circled)  $4 - 0 = 4$ (circled)

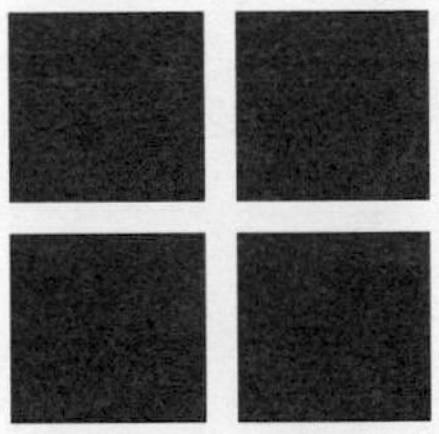

2. $5 - 3$ $6 - 0$ $7 - 3$ $4 - 4$ $8 - 3$ $9 - 9$

3. $3 - 3$ $9 - 7$ $5 - 5$ $4 - 0$ $7 - 2$ $5 - 0$

4. $6 - 1$ $5 - 5$ $3 - 0$ $3 - 3$ $8 - 3$ $5 - 5$

Mixed Review

Add or subtract.

5. $4 - 2 = \square$ $3 + 1 = \square$ $6 - 0 = \square$ $2 + 3 = \square$ $3 + 3 = \square$ $4 - 1 = \square$

Home Note Your child subtracted 0 or all from a number and found the difference.
ACTIVITY Show your child a group of small objects. Take away all of them and have your child tell you the subtraction fact. Show the objects again. This time take away none.

Name ______________________

Facts Practice

Fill in each table.

1

Add 0.	
9	9
4	
2	
6	
8	

2

Add 1.	
6	
1	
3	
0	
5	

3

Add 2.	
7	
3	
5	
2	
1	

4

Subtract 0.	
8	
1	
6	
7	
5	

5

Subtract 1.	
8	
1	
2	
5	
4	

6

Subtract 2.	
6	
4	
9	
7	
3	

Practice

Add or subtract. Fill in the numbers.

1 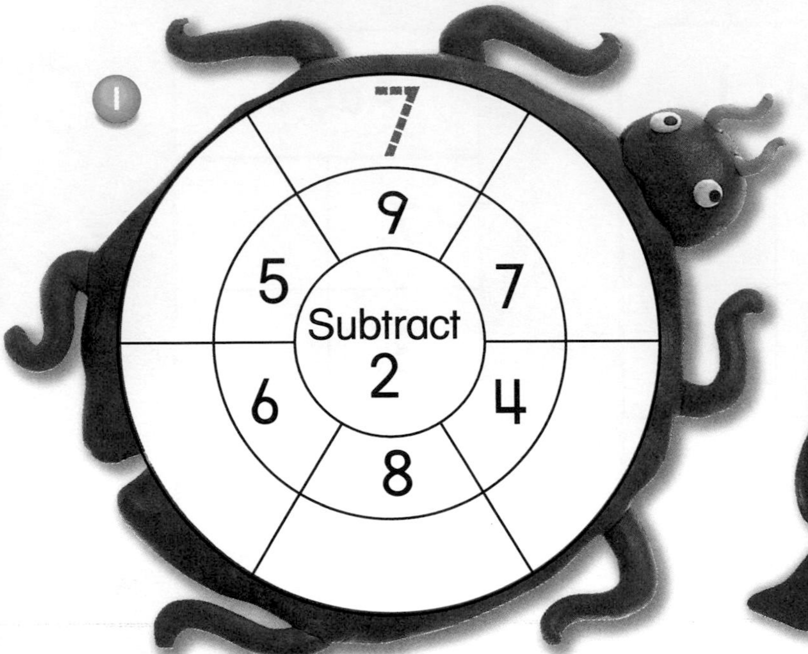

2 

3 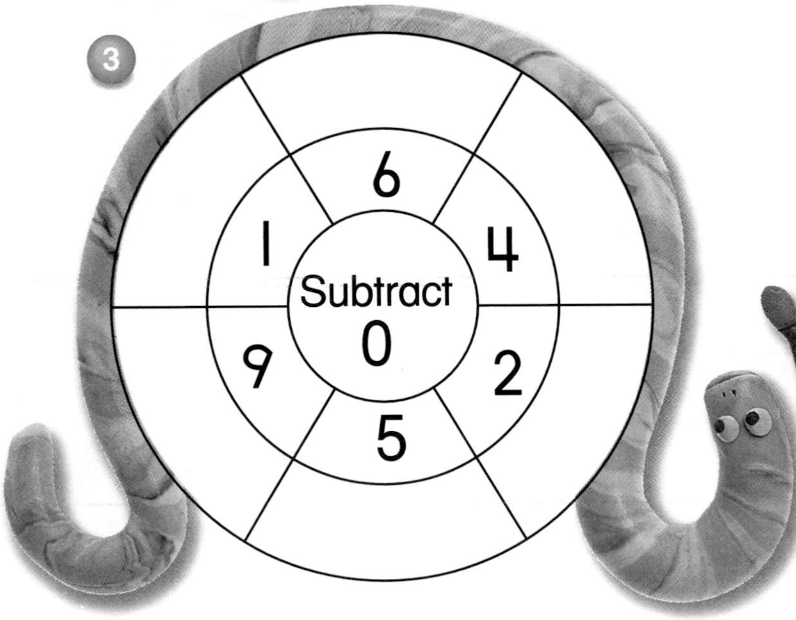

4 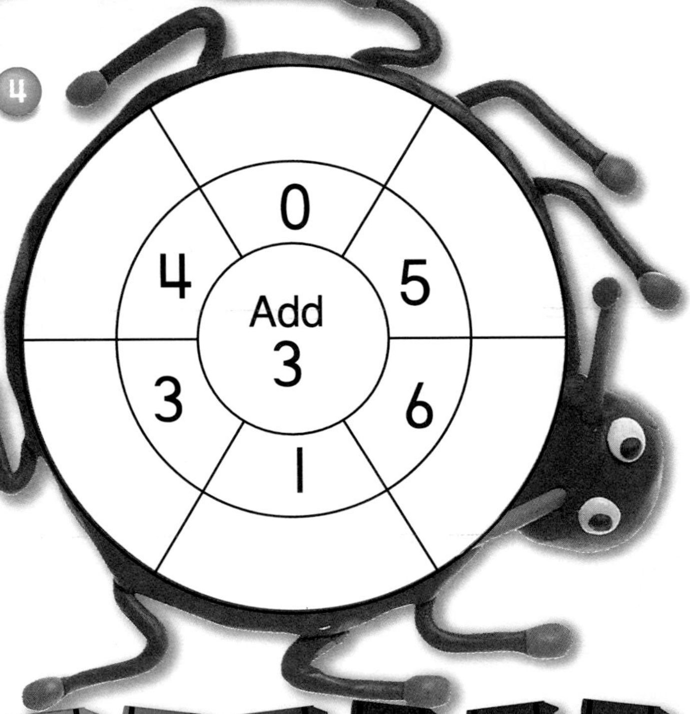

Write About It

5 Tell a friend a story about the picture.
Then write the subtraction sentence.

Home Note **Your child added and subtracted to find sums and differences.**
ACTIVITY **Have your child use small objects to show the problems on the number wheels.**

Name ______________________________

Understand • Plan • Solve • Look Back

Problem Solving
Draw a Picture

Add or subtract.
Draw more things, or cross things out.

1. 6 birds are on a branch.
2 fly away.
How many now? 4

2. 4 bugs are on a leaf.
3 more come.
How many now? ______

3. 3 frogs sing in a pond.
2 more come.
How many now? ______

4. 2 squirrels are in a tree.
2 run away.
How many now? ______

5. 5 ants walk in a row.
1 more comes.
How many now? ______

6. 4 fish play in the sea.
2 swim away.
How many now? ______

Practice

Add or subtract.
Draw more things, or cross things out.

1. 2 boys play here.
   2 girls come.
   How many now? 4

2. The house has 4 signs.
   Take 2 down.
   How many now? ______

3. The house has 3 windows.
   Make 2 more.
   How many now? ______

4. The house has 1 door.
   Make 1 more.
   How many now? ______

5. 4 rabbits are playing.
   1 runs away.
   How many now? ______

6. There is 1 ladder.
   Make 1 more.
   How many now? ______

Home Note Your child used the strategy *draw a picture* to solve problems.
ACTIVITY Give your child paper and crayons. Tell a story that has some addition and subtraction problems for him or her to draw.

Name ______________________________

Chapter 6 Review/Test

Concepts and Skills

Use the number line.
Count back to subtract.

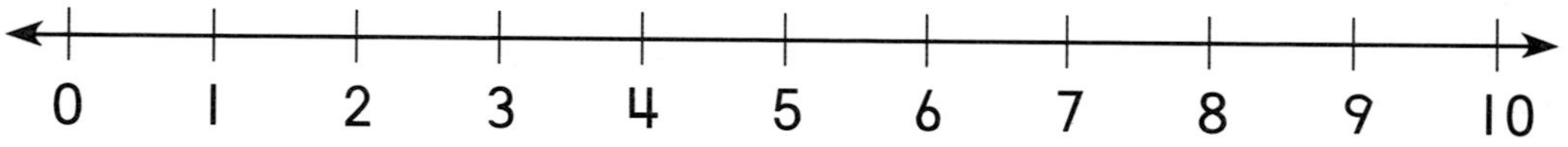

1. 7 − 1 = ____
2. 4 − 1 = ____
3. 8 − 2 = ____
4. 6 − 2 = ____
5. 4 − 3 = ____
6. 5 − 3 = ____

Cross out pictures to show how many are taken away. Subtract.

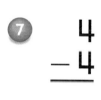

7. $\begin{array}{r} 4 \\ -\,4 \\ \hline \end{array}$

8. $\begin{array}{r} 6 \\ -\,0 \\ \hline \end{array}$

9. $\begin{array}{r} 2 \\ -\,2 \\ \hline \end{array}$

Problem Solving

Add or subtract.
Draw more things, or cross things out.

10. 6 clouds are in the sky.
2 float away.
How many now? ____

11. 2 leaves are on the tree.
Make 3 more.
How many now? ____

Name ______________________________

Cumulative Chapters 1–6

TAAS Prep

Mark the best answer.

1. Which number is missing?

   7, 8, _____, 10

   - ⬭ 6
   - ⬭ 9
   - ⬭ 11
   - ⬭ 15

2. Which number sentence tells how many are left?

   - ⬭ 6 − 1 = 5
   - ⬭ 7 + 2 = 9
   - ⬭ 7 − 2 = 5
   - ⬭ 7 − 5 = 2

3. Which picture shows how many are left?

   4 − 4 = 0

   - ⬭ 
   - ⬭ 
   - ⬭ 
   - ⬭ 

4. Which number shows the difference?

$$\begin{array}{r} 9 \\ -\ 3 \\ \hline \end{array}$$

   - ⬭ 4
   - ⬭ 5
   - ⬭ 6
   - ⬭ 7

5. Which fact has the same sum as 3 + 4?

   - ⬭ 4 + 3
   - ⬭ 3 + 5
   - ⬭ 8 + 1
   - ⬭ 2 + 6

6. 6 apples are in the bowl. Sue puts 3 more in the bowl. How many are now in the bowl?

   - ⬭ 6
   - ⬭ 7
   - ⬭ 8
   - ⬭ 9

Name ____________________

You will need:

Follow the Path to Add and Subtract

1. Spin the spinner, and move your game piece to the right space.
2. Find the the sum or difference.
3. Take turns with your partner.
4. The first one to reach END is the winner.

Home Note Your child has been learning addition and subtraction facts to 10.
ACTIVITY Play this game at home to help your child practice these facts.

Name ______________________________

Calculator | Computer

Use a [calculator].
Find the sums and differences.
Write which keys you press.
Write what you see.

1. 3 + 4 = ____

AC [3] [+] [4] = [7]

2. 7 − 2 = ____

AC [ ] [ ] [ ] = [ ]

3. 4 + 4 = ____

AC [ ] [ ] [ ] = [ ]

4. 9 − 1 = ____

AC [ ] [ ] [ ] = [ ]

5. 2 + 6 = ____

AC [ ] [ ] [ ] = [ ]

6. 7 − 2 = ____

AC [ ] [ ] [ ] [ ] [ ]

Mom Made Cookies

written by Shirley Frederick
illustrated by Karen Strelecki

This book will help me review addition and subtraction facts.

This book belongs to ______________________.

A

Mom made cookies.
Cookies on the plate.

$8 + 2 =$ ____

Cookies for me.
Cookies on the plate. 10 − 2 = ____

Cookies for Sister.
Cookies on the plate.

$8 - 2 = \_\_\_\_$

Cookies for Dad.
Cookies on the plate.

$6 - 3 = \_\_\_\_$

Cookies for Mom.
Cookies on the plate.

3 − 3 = ____

Mom made 6 more cookies.
Cookies on the plate.

Cookies for Dog!
Cookies on the plate. $6 - 6 = \_\_\_\_$

Name ______________________

Chapters 3–6

Review/Test

Concepts and Skills

Complete.

1. 

____ + ____ = ____

2. 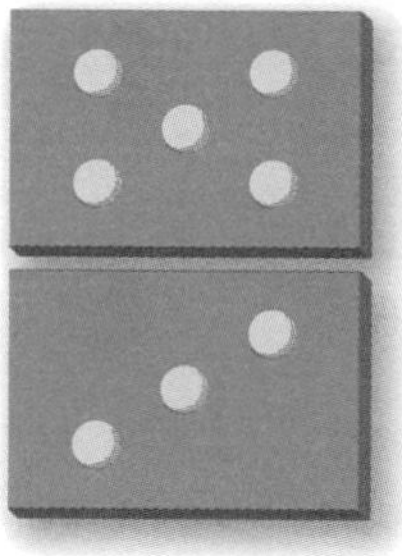

____
+ ____
____

Count on to add.

3. 

3 + 2 = ____

4. 

7 + 3 = ____

Count back to subtract.

5. 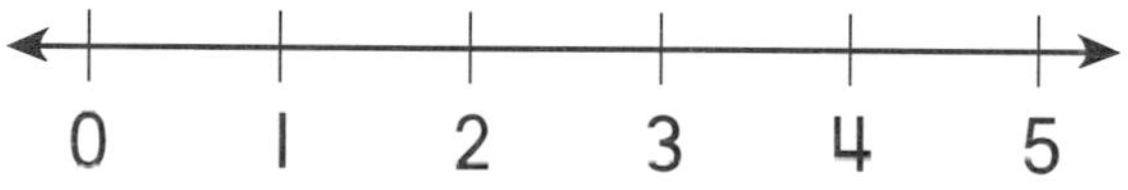

4 − 3 = ____

6. 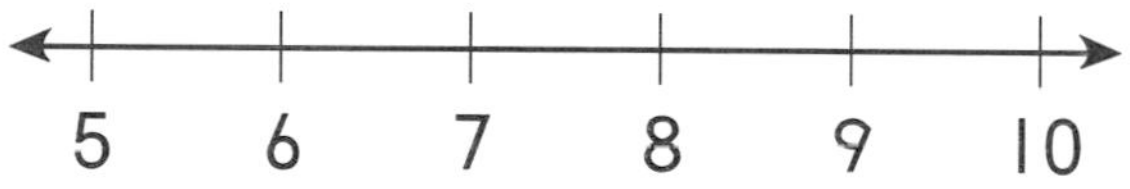

8 − 2 = ____

Add or subtract.
Write the numbers in the fact family.

7. 

5 + 4 = ____ 9 − 4 = ____

4 + 5 = ____ 9 − 5 = ____

☐ ☐ ☐

Write each sum. Which boxes in each row show the same sum? Color them yellow.

8 $3 + 0 =$ ____	9 $2 + 2 =$ ____	10 $0 + 3 =$ ____
11 $\begin{array}{r} 3 \\ +4 \\ \hline \end{array}$	12 $\begin{array}{r} 4 \\ +3 \\ \hline \end{array}$	13 $\begin{array}{r} 3 \\ +6 \\ \hline \end{array}$

Draw lines to match. Subtract. Write how many more.

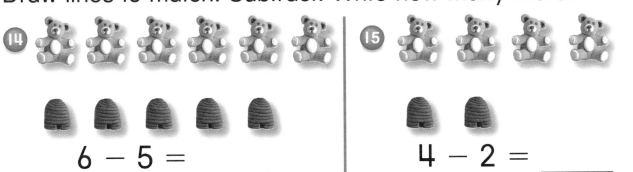

14 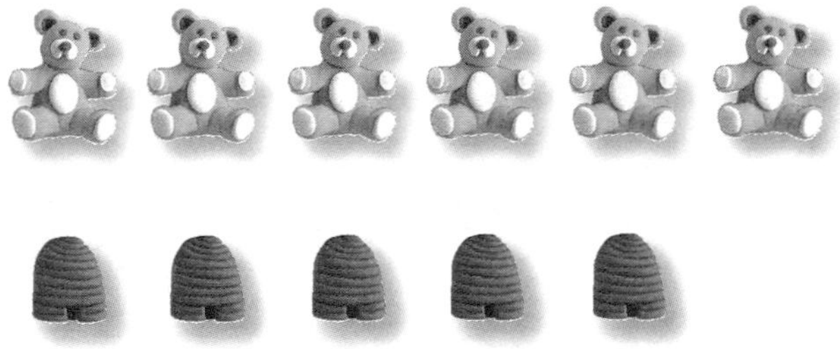

$6 - 5 =$ ____

____ more 

15 

$4 - 2 =$ ____

____ more 

Add or subtract.

16 $\begin{array}{r} 4 \\ +4 \\ \hline \end{array}$

17 $\begin{array}{r} 9 \\ -2 \\ \hline \end{array}$

18 $\begin{array}{r} 6 \\ -3 \\ \hline \end{array}$

19 $\begin{array}{r} 8 \\ +2 \\ \hline \end{array}$

Problem Solving

20 Draw  to show each price. Write the total amount.

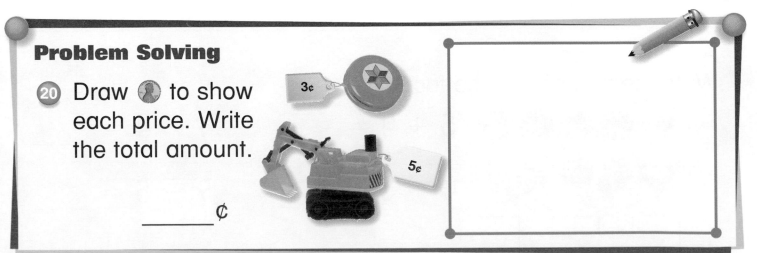

____¢

Name ______________________________

Performance Assessment

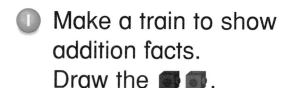

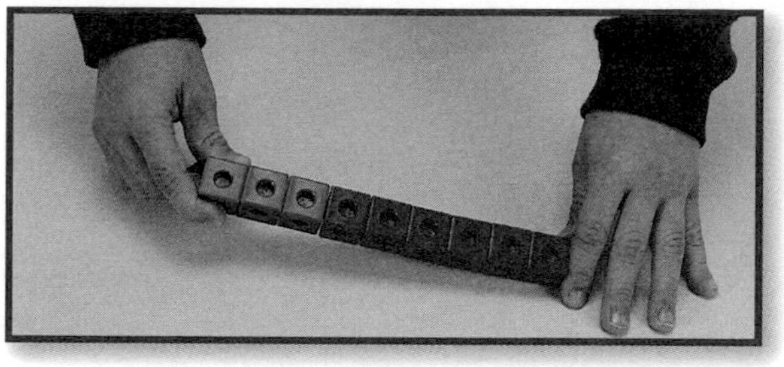

Use and .

Take some and some .
Take 9 in all.

1. Make a train to show addition facts.
Draw the .

Write the facts.

_____ + _____ = _____

_____ + _____ = _____

2. Break the train to show subtraction facts.
Draw the .

Write the facts.

_____ − _____ = _____

_____ − _____ = _____

3. Write the numbers in your fact family. _____, _____, _____

Write About It

4. Write one of your facts.

_____ ○ _____ = _____

Draw a picture showing the fact.

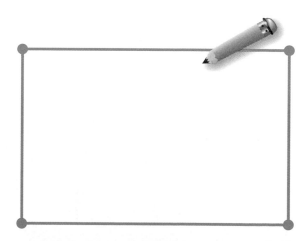

Name ______________________________

Chapters 1–6 Cumulative Review

Fill in the ⬭ for the correct answer.

1. Which number is missing?

2, 3, ____, 5

- ⬭ 4
- ⬭ 6
- ⬭ 7
- ⬭ Not Here

2. 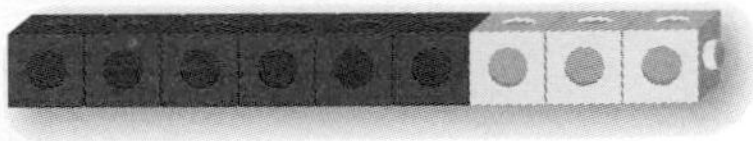

$6 + 3 =$ ____

- ⬭ 9
- ⬭ 7
- ⬭ 6
- ⬭ Not Here

3. Which number sentence tells the story?

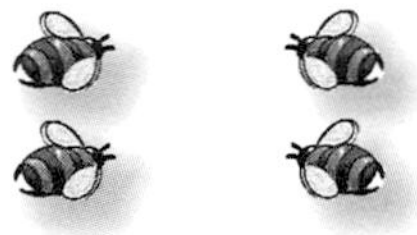

- ⬭ $2 + 2 = 4$
- ⬭ $4 - 1 = 3$
- ⬭ $4 - 2 = 2$
- ⬭ $2 - 2 = 0$

4. 

$7 - 3 =$ ____

- ⬭ 2
- ⬭ 3
- ⬭ 5
- ⬭ Not Here

5. Which one has the same sum as

$5 + 1$?

- ⬭ $2 + 3$
- ⬭ $1 + 5$
- ⬭ $5 + 2$
- ⬭ $0 + 8$

6. Which one has the same sum as

$4 + 3$?

- ⬭ $4 + 6$
- ⬭ $7 + 2$
- ⬭ $8 + 0$
- ⬭ $3 + 4$

7. How many more  ?

$3 - 1 =$ ____

____ more 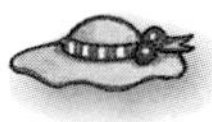

- ⬭ 1
- ⬭ 2
- ⬭ 3
- ⬭ Not Here

8. Add or subtract.

Tom gets a book and a ball. What does he pay in all?

- ⬭ 3¢
- ⬭ 5¢
- ⬭ 7¢
- ⬭ 10¢

9. $\begin{array}{r} 4 \\ +\,2 \\ \hline \end{array}$

- ⬭ 6
- ⬭ 5
- ⬭ 2
- ⬭ 1

10. $\begin{array}{r} 6 \\ -\,2 \\ \hline \end{array}$

- ⬭ 2
- ⬭ 3
- ⬭ 4
- ⬭ 5

11. $\begin{array}{r} 9 \\ -\,0 \\ \hline \end{array}$

- ⬭ 9
- ⬭ 5
- ⬭ 1
- ⬭ 0

CHAPTER 7

Solid Figures

Home Note In this chapter, your child will learn about solid figures.
ACTIVITY Have your child find and identify the solid shapes in the picture on this page.

SCHOOL-HOME CONNECTION

Dear Family,

Today we started Chapter 7. We will learn about solid figures. We will name the figures and sort them. Here are the new vocabulary words and an activity for us to do together at home.

Love,

____

Vocabulary

rectangular prism

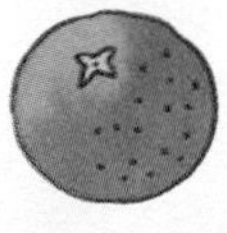
sphere

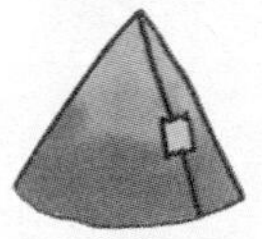
cone

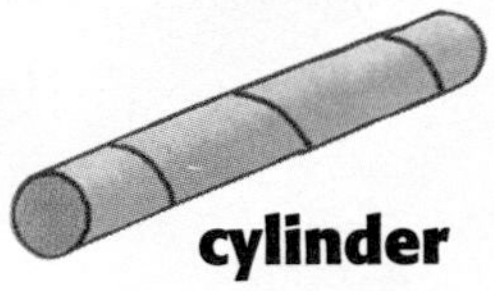
cylinder

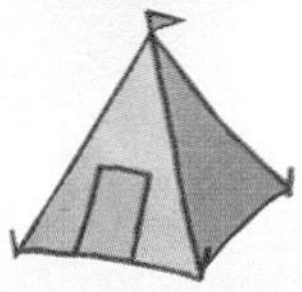
pyramid

cube

ACTIVITY

Have your child sort toys and other objects into groups by shape. Ask your child to name the shape each group shows. Then have your child find other ways to sort objects, for example, by size and by color.

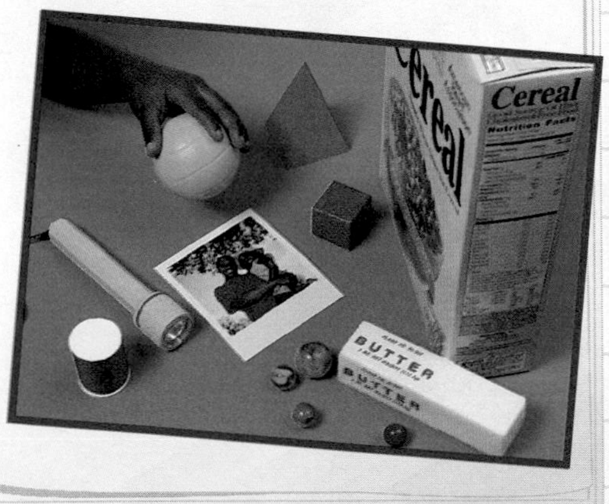

Visit our Web site for additional activities and ideas.
http://www.hbschool.com

Name ____________________

Solid Figures

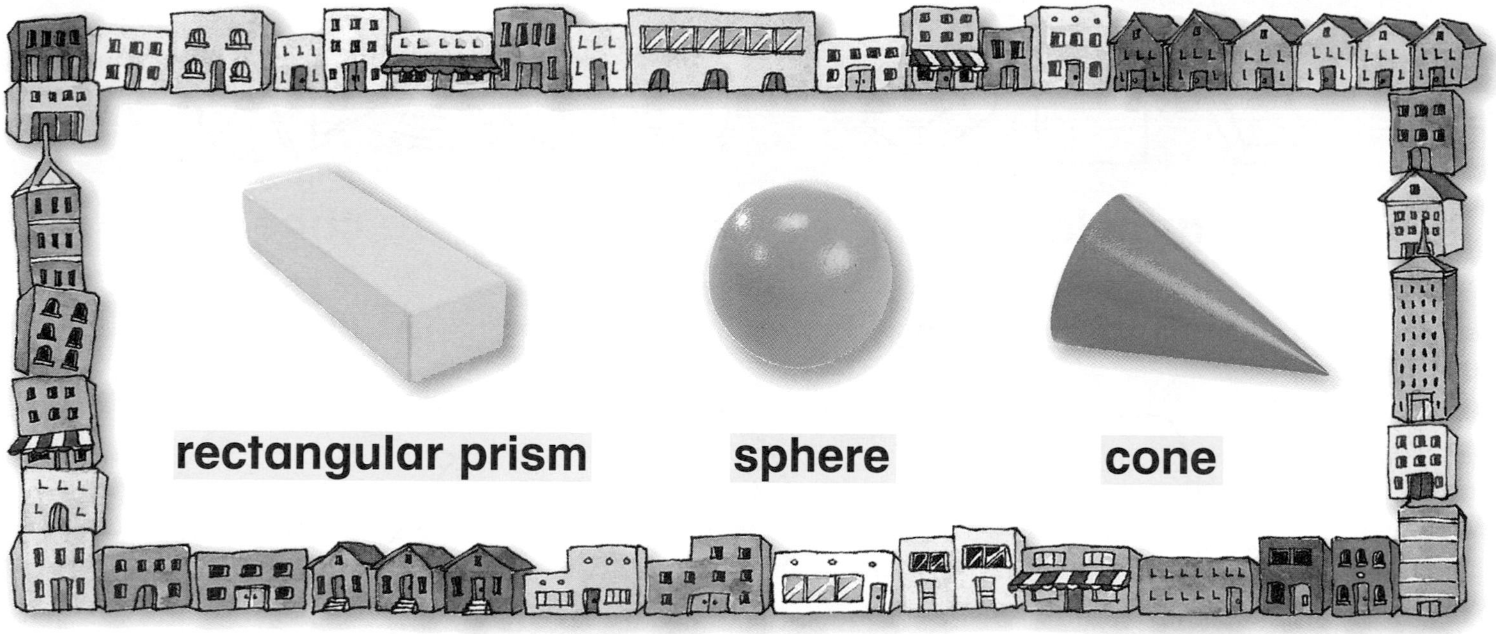

Color the objects to match the figures.

Talk About It • Critical Thinking

How are these figures the same?
How are they different?

Practice

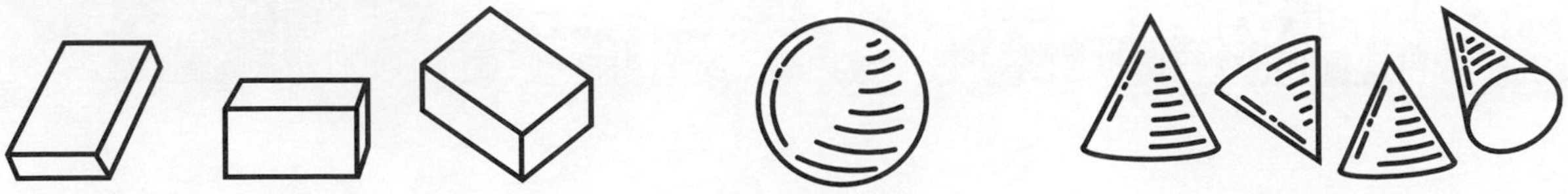

rectangular prism **sphere** **cone**

Color the figures that have the same shape.

1 

2 

3 

Mixed Review

Write each sum.

4. 4 + 1 = ____
5. 4 + 2 = ____
6. 4 + 3 = ____
7. 6 + 1 = ____
8. 6 + 2 = ____
9. 6 + 3 = ____

Home Note **Your child identified solid figures and related them to everyday objects.**
ACTIVITY **Have your child point out objects that are cones, spheres, and rectangular prisms.**

Name ______________________________

More Solid Figures

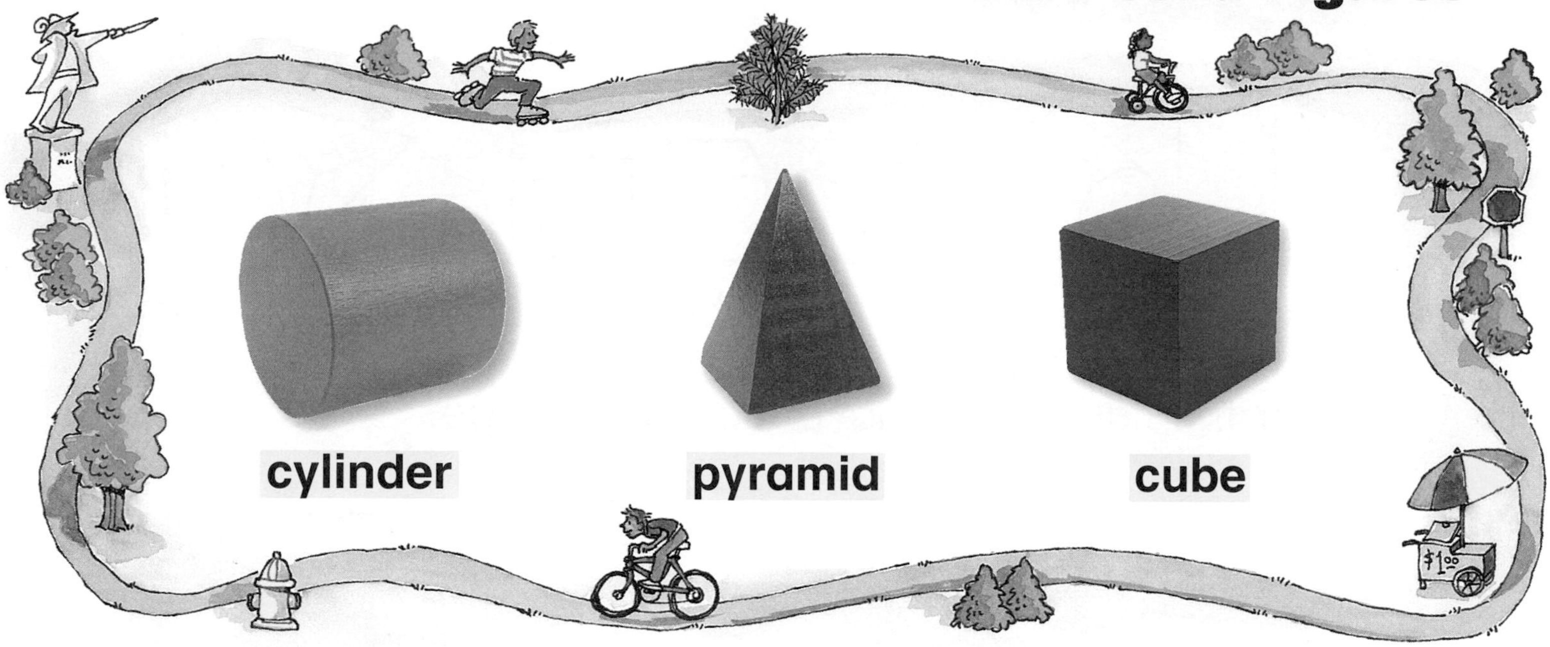

Color the objects to match the figures.

Talk About It • Critical Thinking

How are these figures the same?
How are they different?

Practice

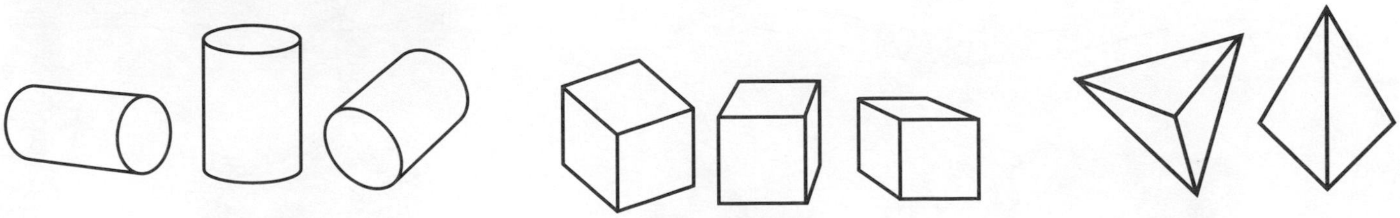

cylinder **cube** **pyramid**

Color the figures that have the same shape.

1

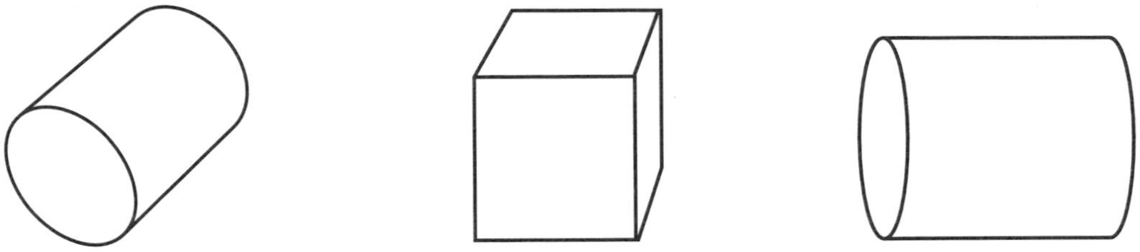

2

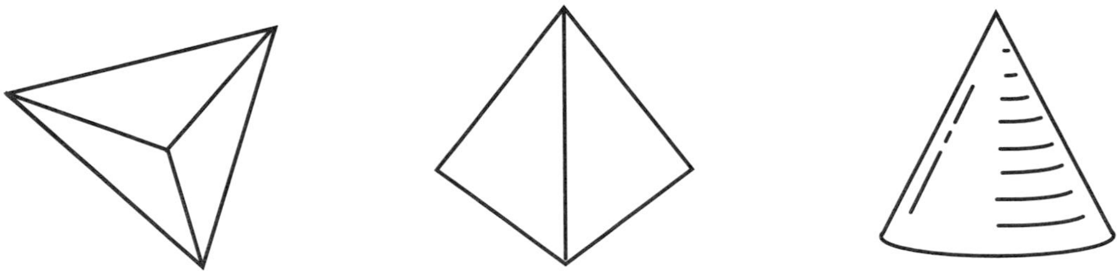

3

Write About It

4 Draw a picture of something that has this shape. Write a sentence about your picture.

Home Note **Your child identified solids from different views.**
ACTIVITY **Have your child point out objects that are cylinders, pyramids, and cubes.**

Name ______________________________

Sorting Solid Figures

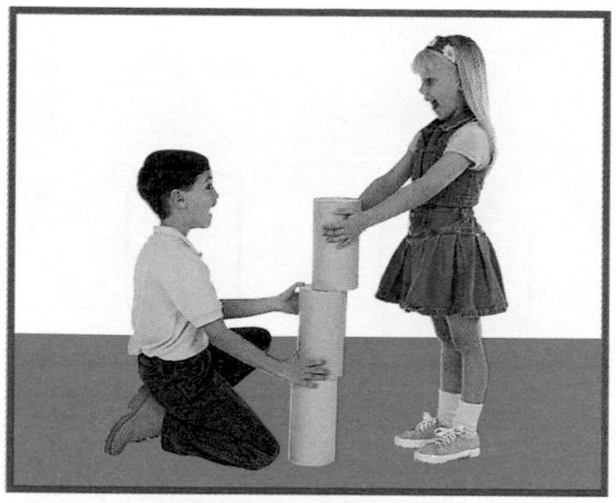

stack

roll

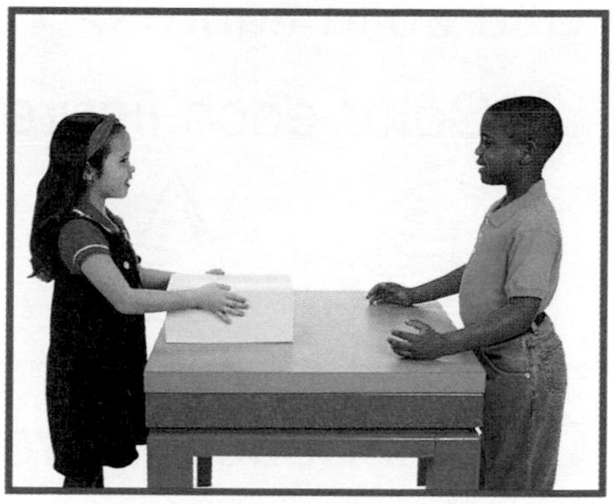

slide

Use solid figures.
Try to stack, roll, and slide them.
Write **yes** or **no**.

	stack	roll	slide
1 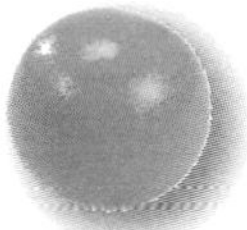 **sphere**	no		
2  **cone**			
3  **cube**			
4  **cylinder**			
5  **pyramid**			
6  **rectangular prism**			

Practice

Use solid figures.

1. Color each figure that will stack.

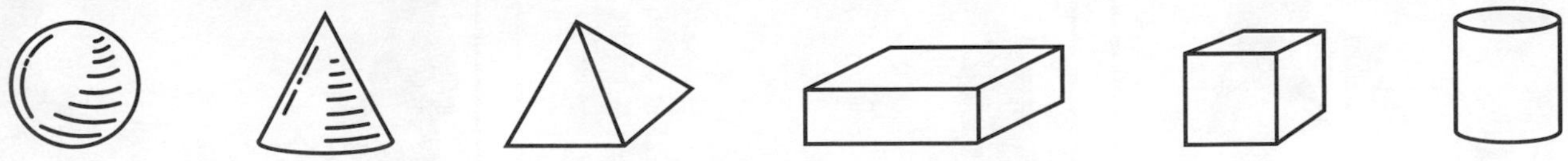

2. Color each figure that will roll.

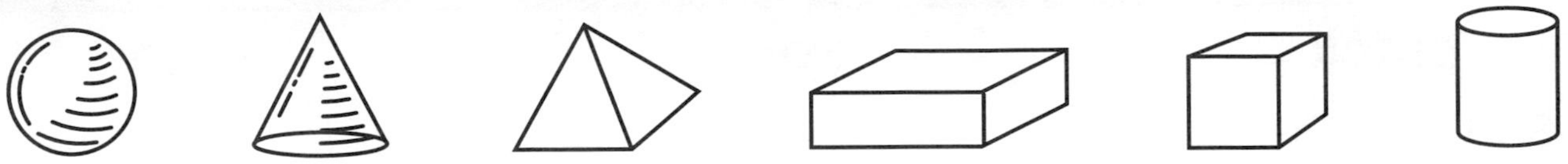

3. Color each figure that will slide.

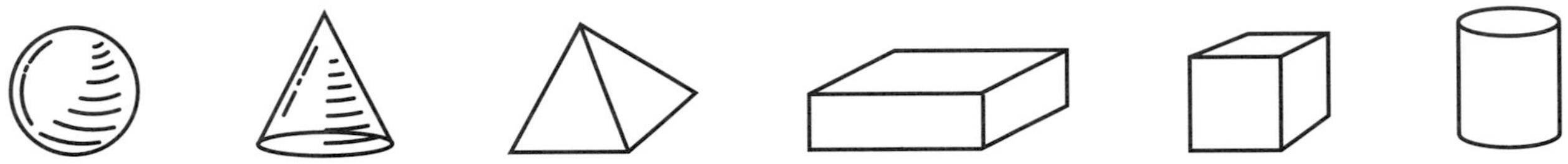

4. Color each figure that will stack and roll.

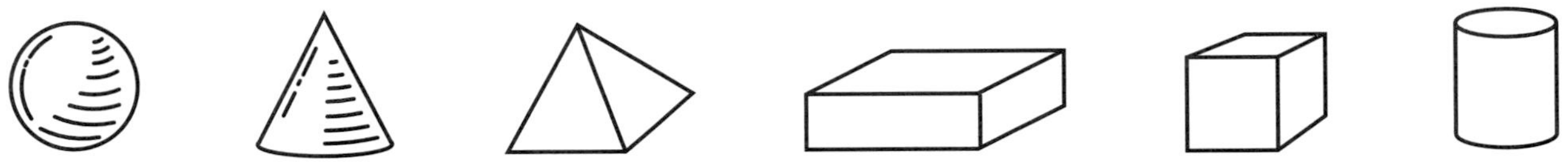

5. Color each figure that will slide and stack.

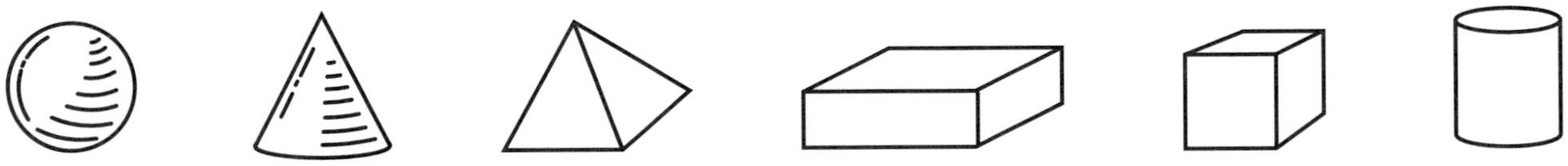

Problem Solving • Reasoning

Cross out the shape that does not belong.
Underline the sentence that tells why.

6. 

It will not stack. It will not roll.

Home Note Your child classified solid figures by what they will do.
ACTIVITY Find objects that are solid figures. Work with your child to find out which ones will stack, roll, and slide.

Name ____________________

More Sorting Solid Figures

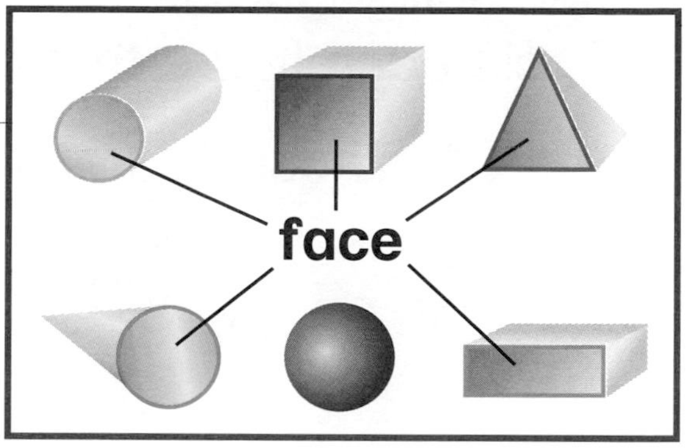

Use solid figures.
Sort them by the number of faces.
Color the figures with the number of faces.

1. only 1 face

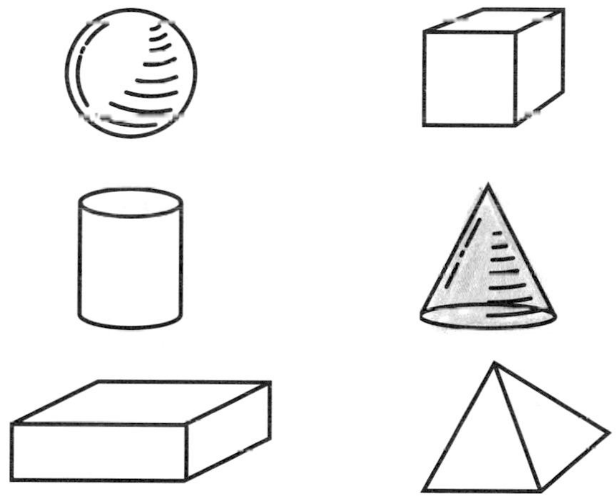

2. only 2 faces

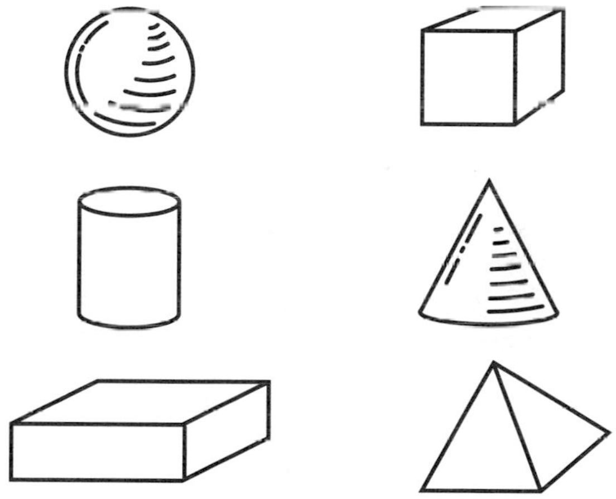

3. 6 faces

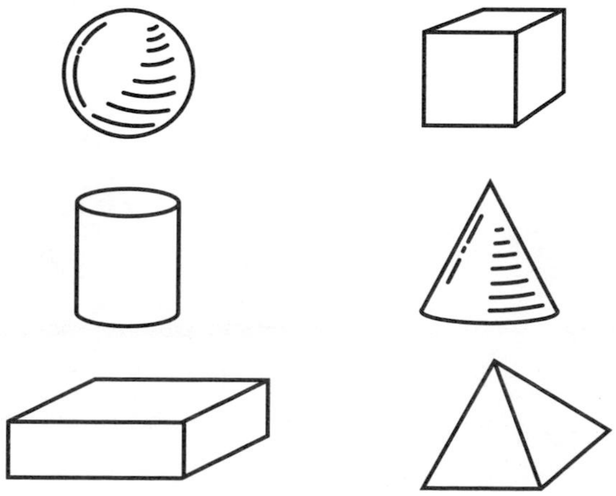

4. only 5 faces

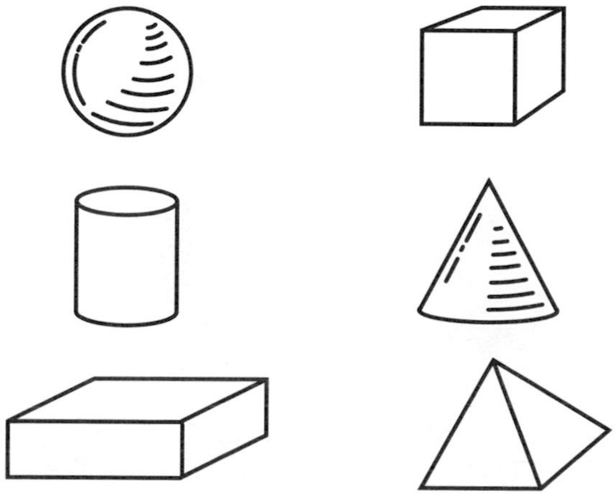

Practice

Use solid figures.
Circle each figure that goes with the sentence.

1. These figures have 6 faces.

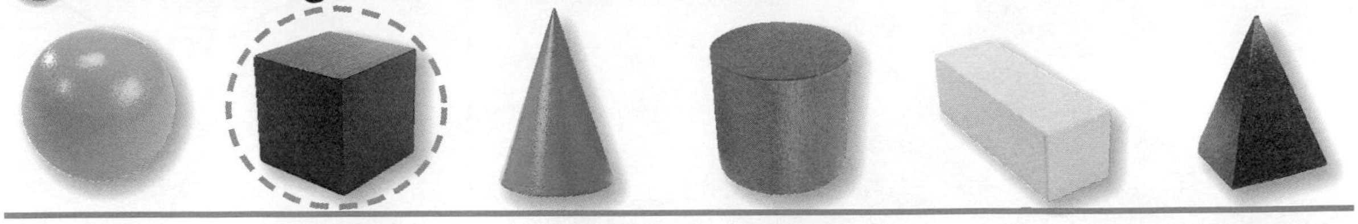

2. This figure has only 2 faces.

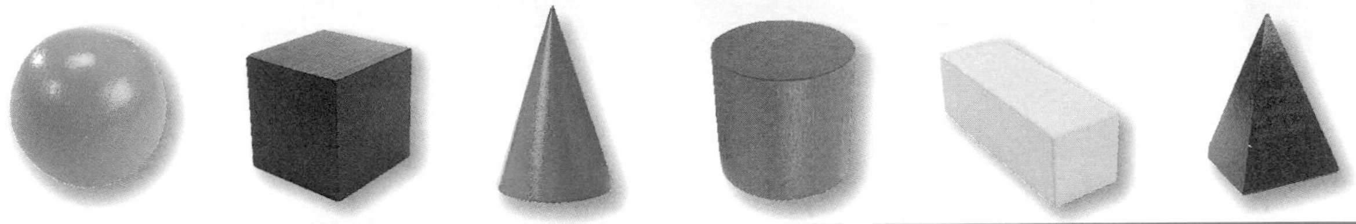

3. This figure has 0 faces.

4. These figures have only 1 or 2 faces.

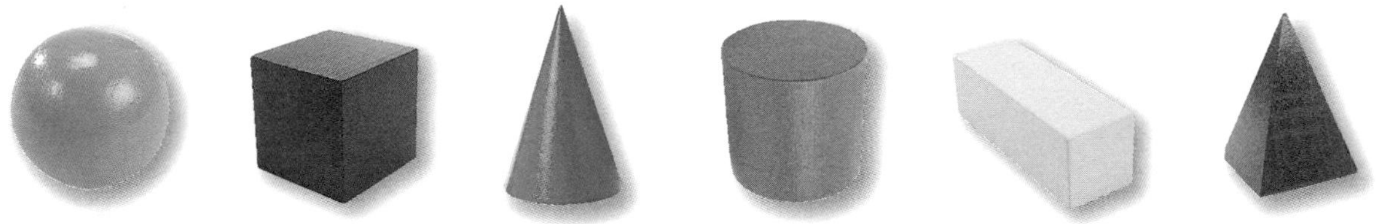

Problem Solving • **Reasoning**

Sort the solid figures into 2 groups.
Circle each group.
Explain to a classmate how you sorted the solid figures.

5.

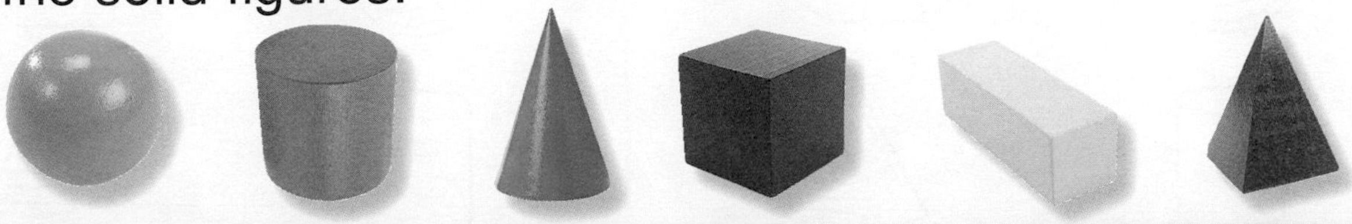

Home Note Your child sorted solid figures by the number of faces.
ACTIVITY Find objects that have the faces that are shown in this lesson. Then have your child sort the objects into the four groups shown above.

Name ______________________________

Understand • Plan • Solve • Look Back

Problem Solving
Make a Model

6 cubes

Use .
Build the model.
Write how many  you used.

1 

______ cubes

2 

______ cubes

3 

______ cubes

4 

______ cubes

5 

______ cubes

6 

______ cubes

Practice

Use .
Build the model.
Write how many  you used.

1 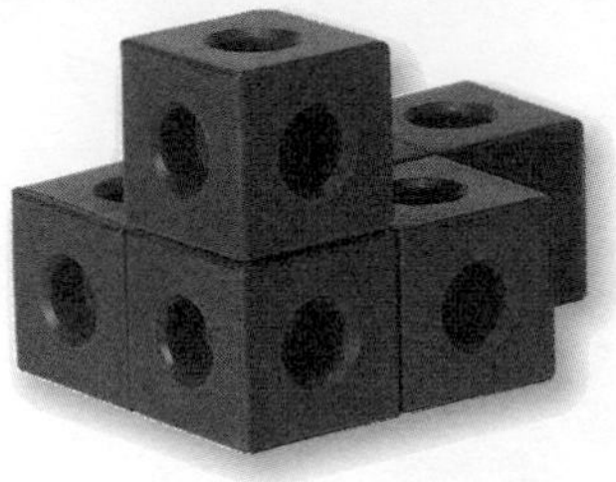

6 cubes

2 

______ cubes

3 

______ cubes

4 

______ cubes

5 

______ cubes

6 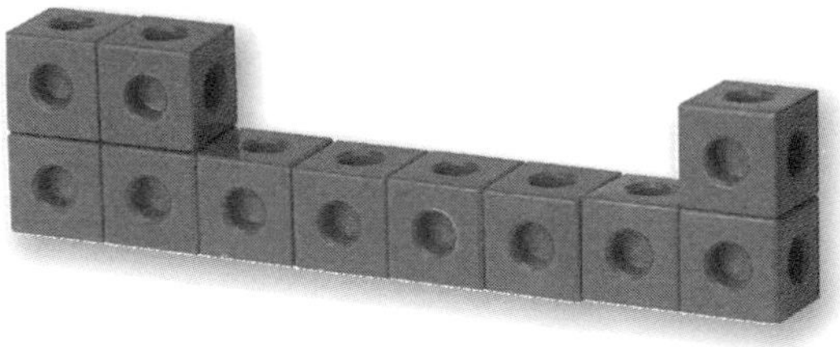

______ cubes

7 

______ cubes

8 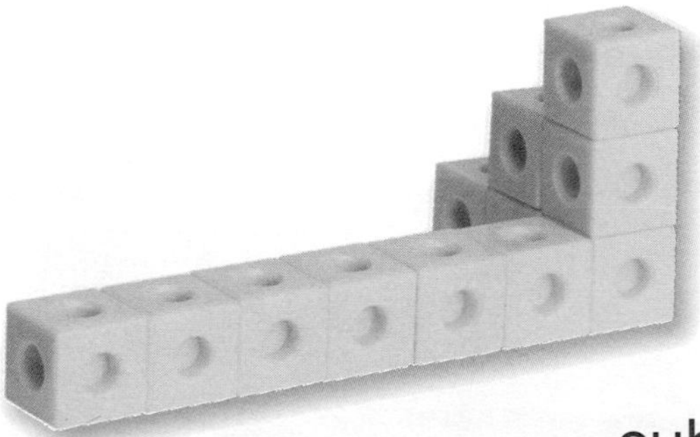

______ cubes

Home Note Your child used cubes to make a model.
ACTIVITY Use toy blocks or sugar cubes to build models like the ones in this lesson. Have your child look at the models and copy them.

Name ______________________

Chapter 7
Review/Test

Concepts and Skills

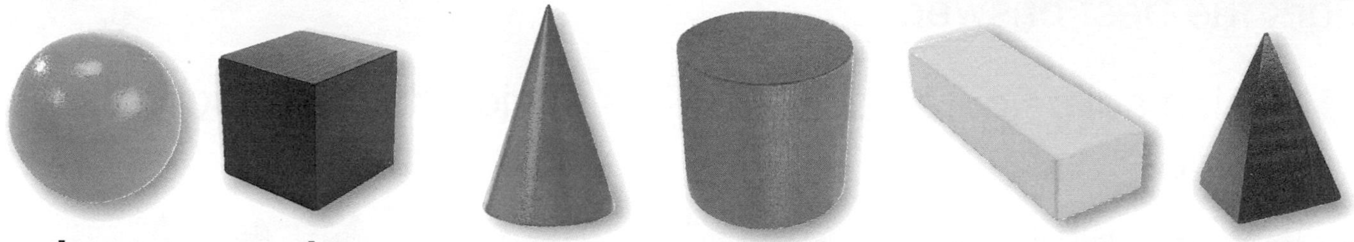

sphere | cube | cone | cylinder | rectangular prism | pyramid

1 Color the objects to match the figures.

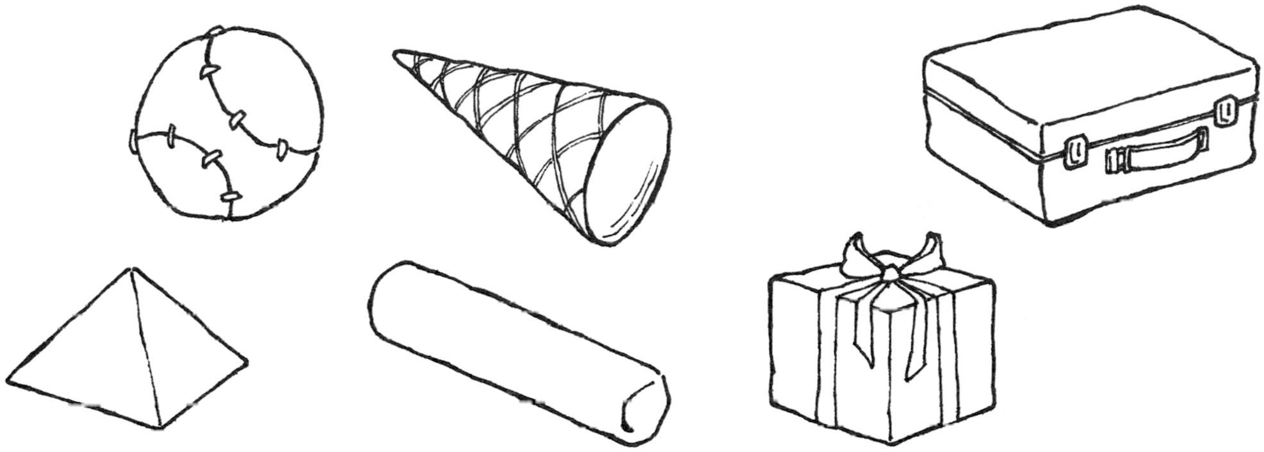

Color each figure that goes with the sentence.

2 These figures have 6 faces.

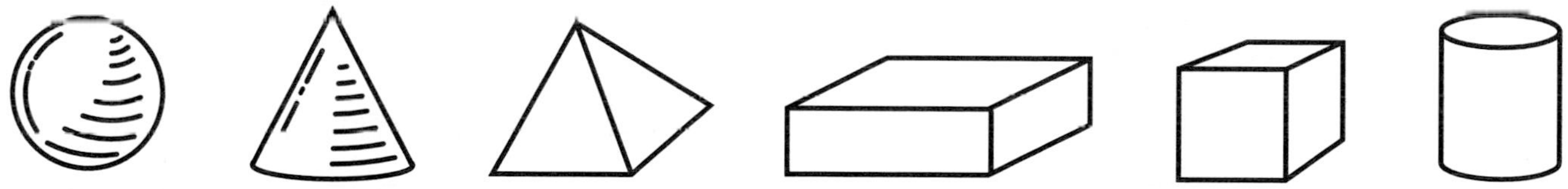

Problem Solving

3 Use .
Build this model.
Write how many you used.

____ cubes

Name ____________________

Cumulative Chapters 1–7

TAAS Prep

Mark the best answer.

1. Which solid figure has 5 faces?

- ⬭ 
- ⬭ 
- ⬭ 
- ⬭ 

2. Which has the same shape?

- ⬭ 
- ⬭ 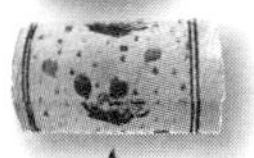
- ⬭ 
- ⬭ 

3. Which has the same sum as 2 + 3?

- ⬭ 1 + 3
- ⬭ 4 + 5
- ⬭ 3 + 2
- ⬭ 2 + 5

4. Which doubles fact tells how many in all?

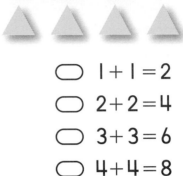

- ⬭ 1 + 1 = 2
- ⬭ 2 + 2 = 4
- ⬭ 3 + 3 = 6
- ⬭ 4 + 4 = 8

5. Which number is missing?

4, 5, ____, 7

- ⬭ 6
- ⬭ 7
- ⬭ 8
- ⬭ 9

6. Which subtraction sentence tells how many are left?

- ⬭ 5 − 3 = 2
- ⬭ 5 − 2 = 3
- ⬭ 3 − 2 = 1
- ⬭ 3 − 1 = 2

CHAPTER 8

Plane Figures

Tell about all the shapes you see in the picture.

Home Note **In this chapter, your child will learn about plane figures.**
ACTIVITY Have your child find and name the plane shapes in the picture on this page.

SCHOOL-HOME CONNECTION

Dear Family,

Today we started Chapter 8. We will learn about plane shapes and about symmetry. Here are the new vocabulary words and an activity for us to do together at home.

Love,

____

Vocabulary

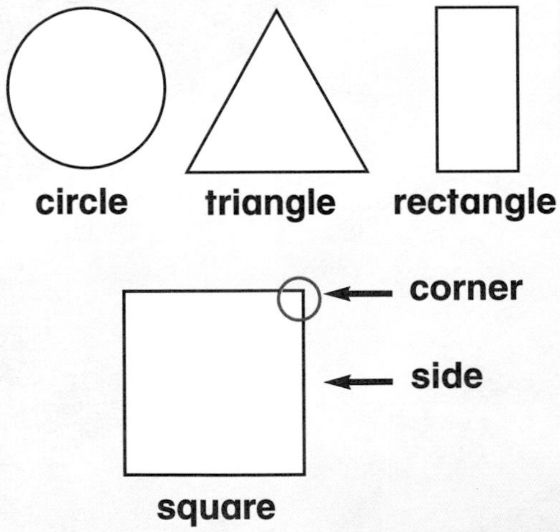

ACTIVITY

Gather objects that show the shapes of cones, cylinders, cubes, and rectangular prisms. Have your child trace the face of each solid shape and name the plane shape.

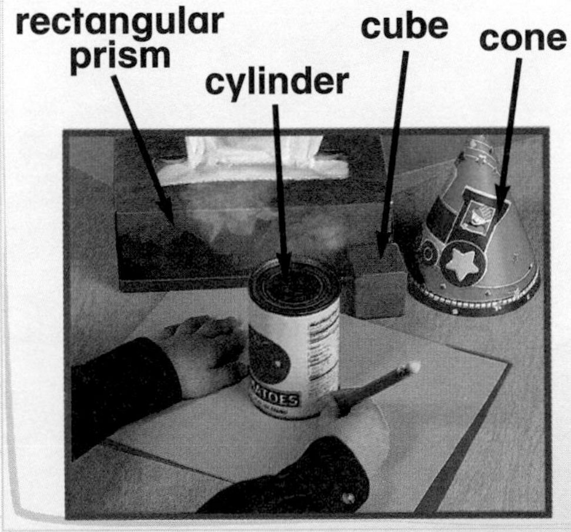

Visit our Web site for additional ideas and activities.
http://www.hbschool.com

Name ____________________

Plane Figures

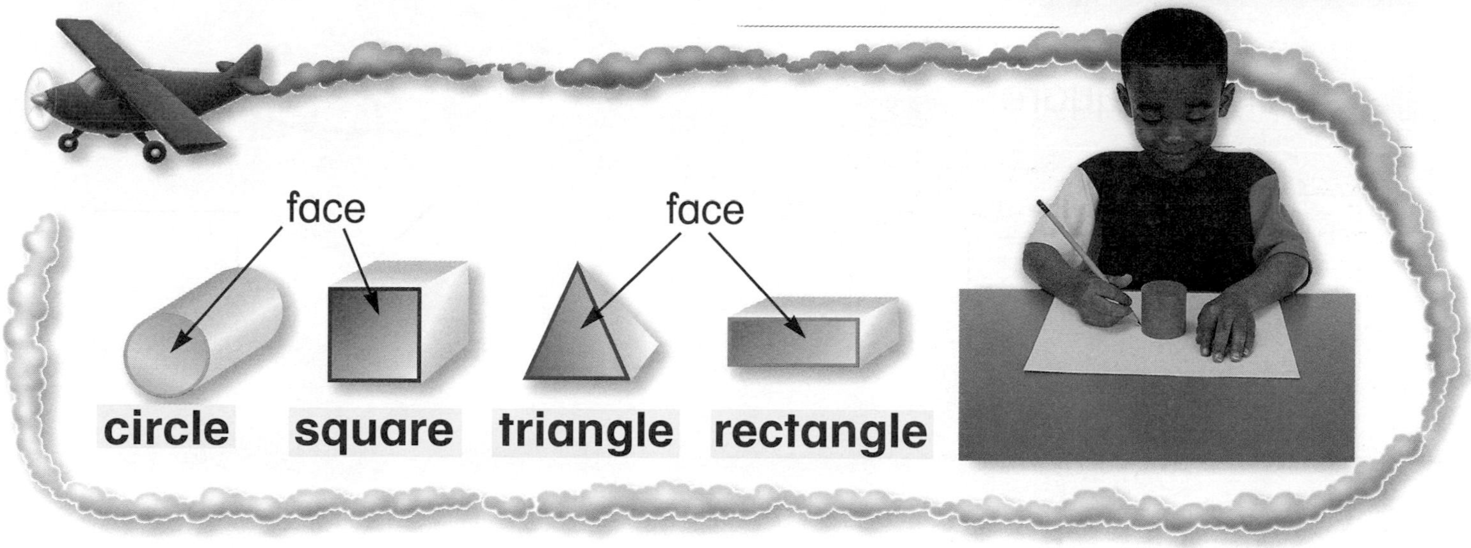

Use solid figures. Draw around a face.
Write the name of the figure you drew.

1 

____

2 

____

3 

____

4 

____

Talk About It • Critical Thinking

How can you tell which parts of a solid figure are faces?

Practice

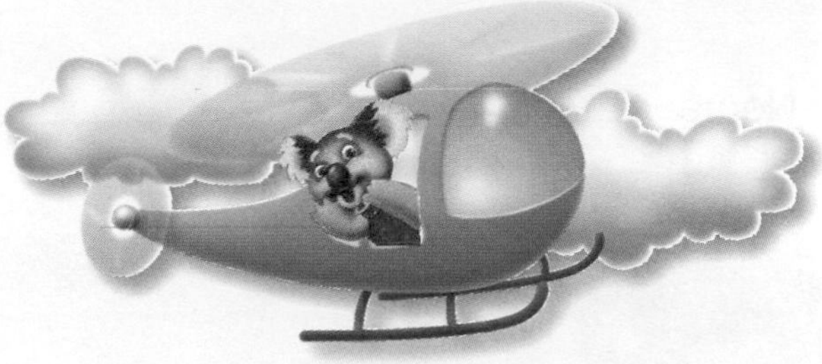

1. Color the squares.

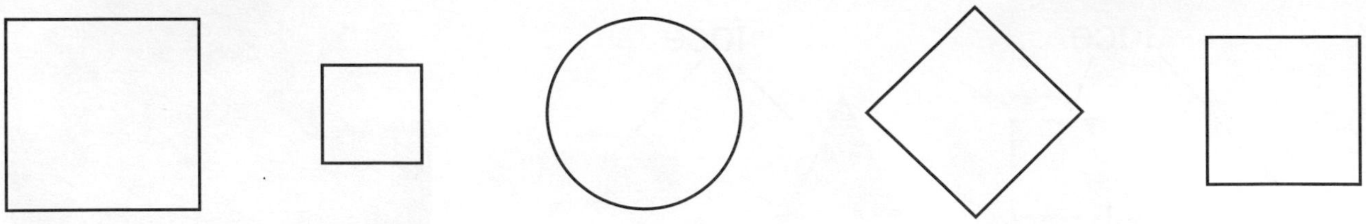

2. Color the triangles.

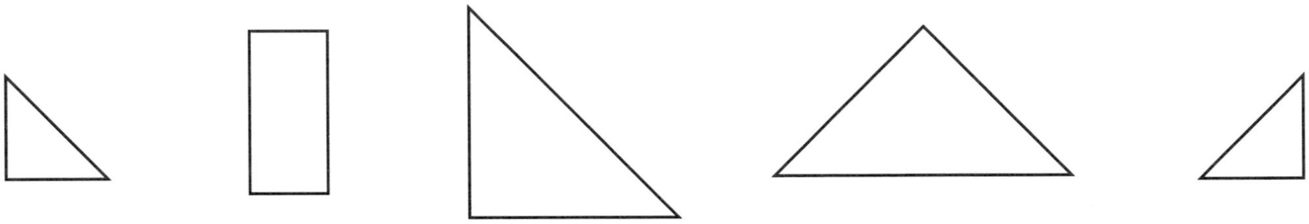

3. Color the circles.

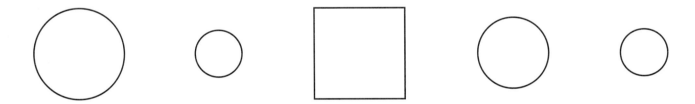

Problem Solving • **Visual Thinking**

4. Draw a large square.

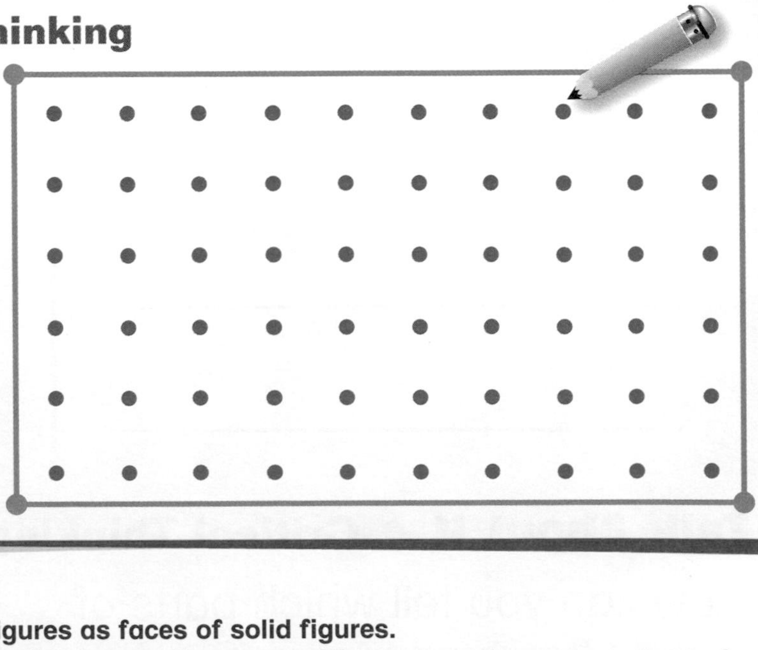

Home Note Your child identified plane figures as faces of solid figures.
ACTIVITY Gather some objects that are solid shapes. Have your child place each one on a sheet of paper, trace one face, and name the shape he or she drew.

Name ______________________________

Sorting Plane Figures

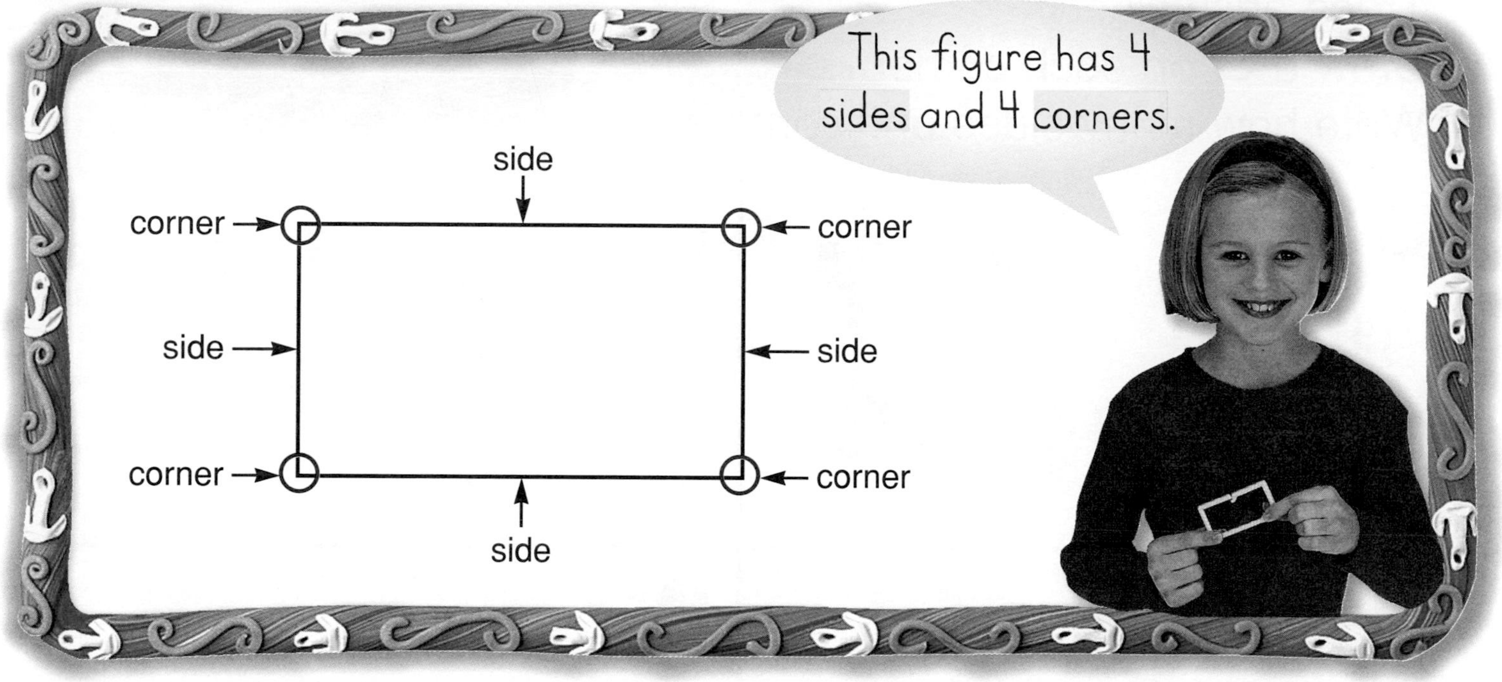

Use small Attrilinks.™ Sort them by the number of sides and corners. Draw them.

1 3 sides 3 corners	2 4 sides 4 corners
3 0 sides 0 corners	4 6 sides 6 corners

Practice

Trace each side.
Draw a O on each corner.
Write how many sides and corners.

1. 4 sides
4 corners

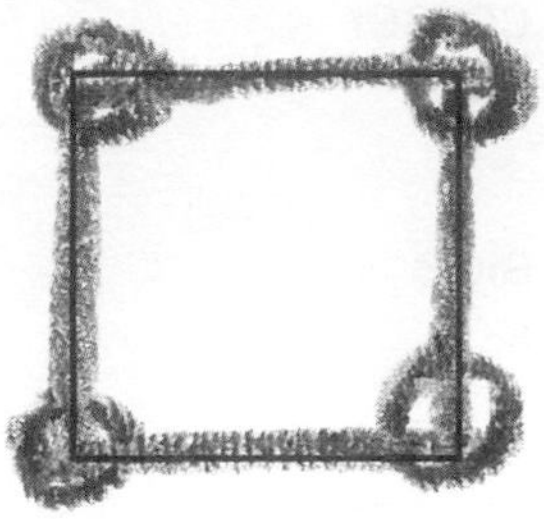

2. ____ sides
____ corners

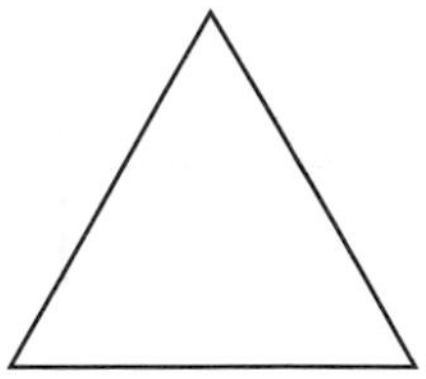

3. ____ sides
____ corners

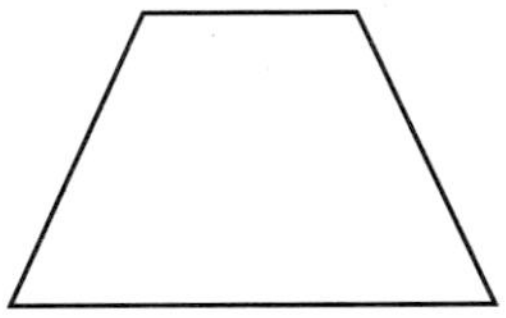

4. ____ sides
____ corners

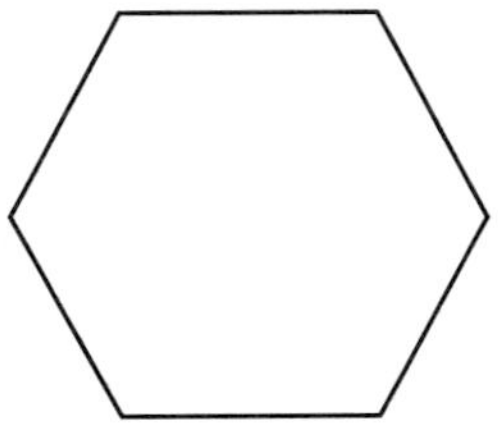

5. ____ sides
____ corners

6. ____ sides
____ corners

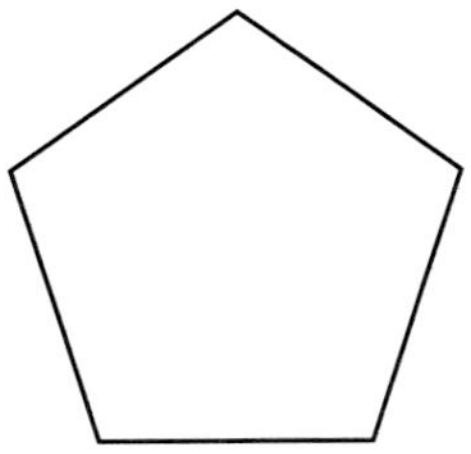

Write About It

7. Draw a figure that has 4 sides and 4 corners. Write a sentence that tells about what you drew.

Home Note Your child sorted and described plane figures by the number of sides and corners.
ACTIVITY Give your child a matching number of sides and corners, such as 4 sides and 4 corners. Have him or her draw the figure.

Name ______________________________

Draw a figure that is the same size and shape.

1. 

2. 

3.  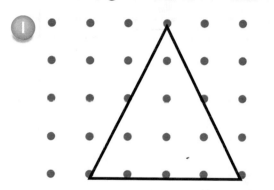

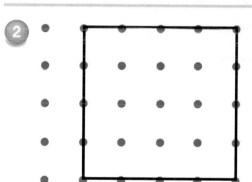

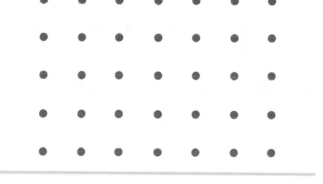

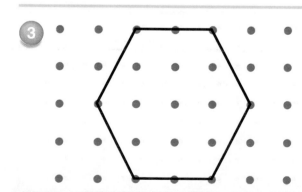

Talk About It • Critical Thinking

How can you tell if two figures are the same size and shape?

Practice

Color the figures that are the same size and shape.

1
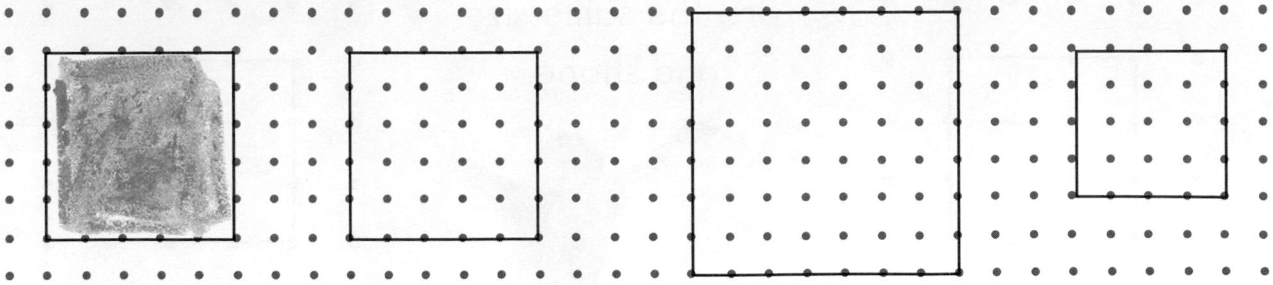

2
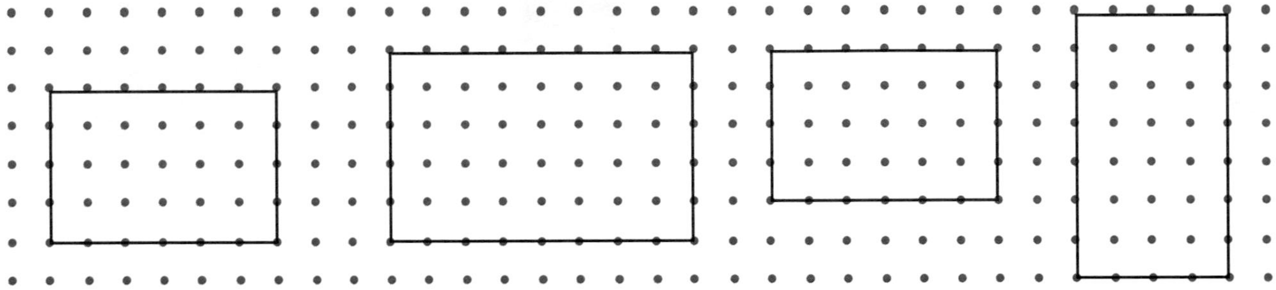

3
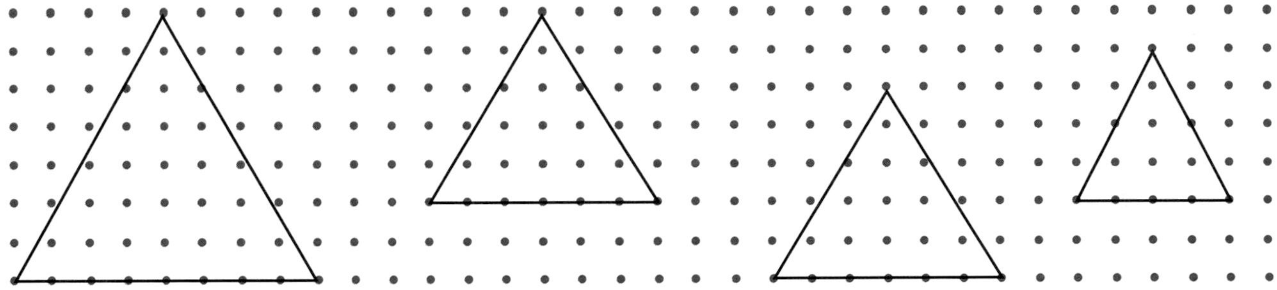

Problem Solving • Visual Thinking

4 Work with a partner. Draw a shape. Ask your partner to draw one that is the same size and shape.

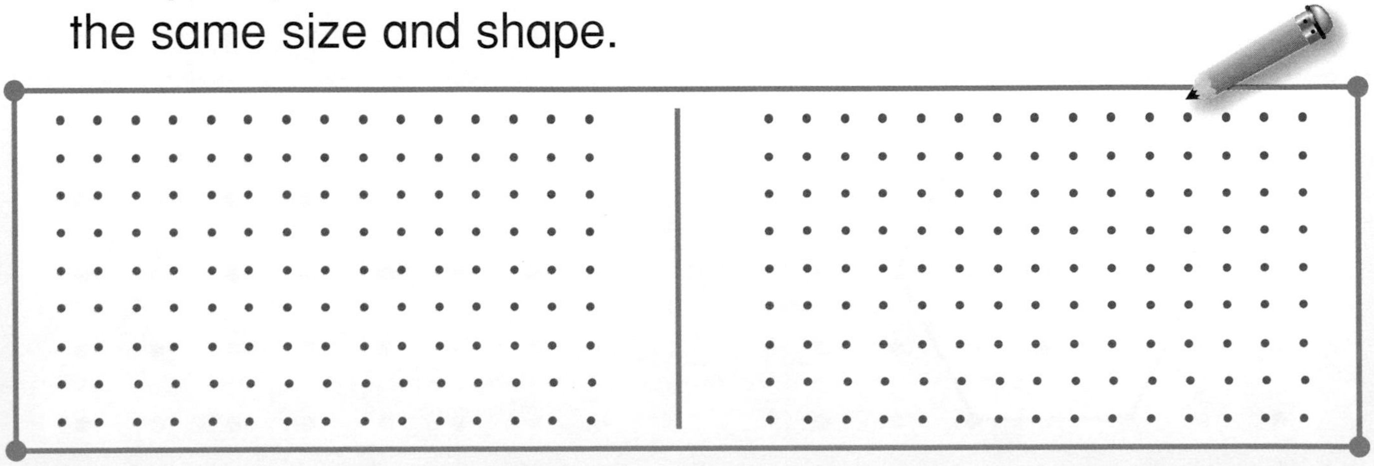

Home Note Your child identified and drew congruent figures.
ACTIVITY Draw and cut out a shape. Have your child use your shape to draw and cut out a shape that is the same size and shape.

Name ______________________

Symmetry

Use 

1. Fold your paper.

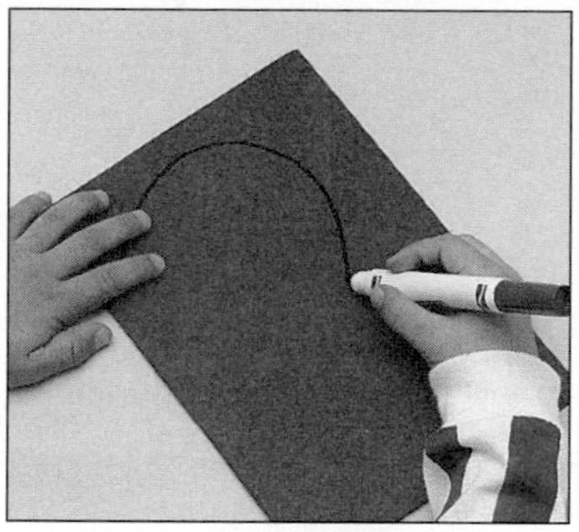

2. Start at the fold. Draw a shape.

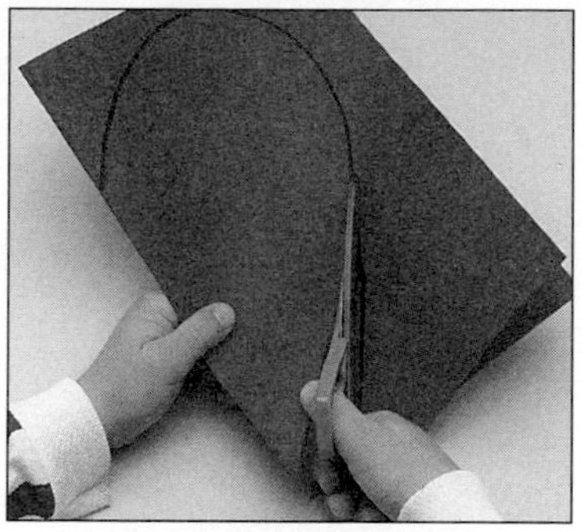

3. Cut along the line.

4. Open your shape.

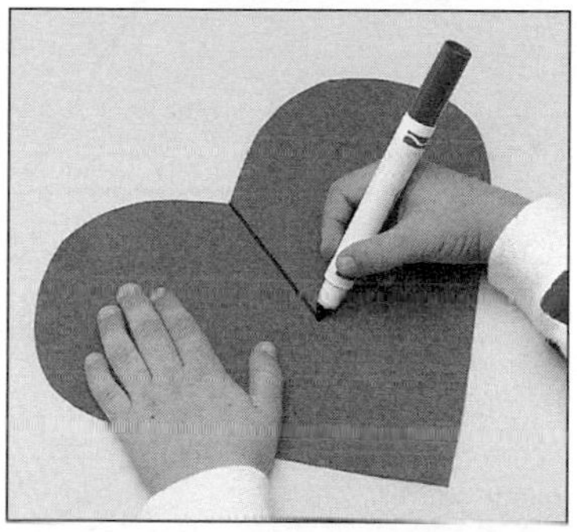

5. Draw a line down the middle.

Look at your shape. The two sides should match.

Draw a line to make two sides that match.

1 

2 

Practice

Draw a line to make two sides that match.

1
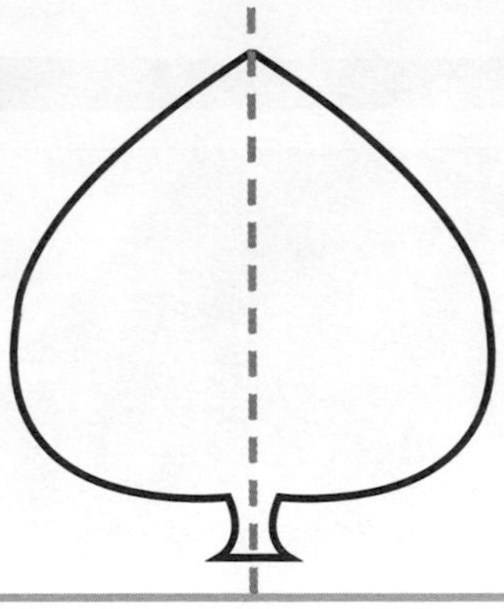

2
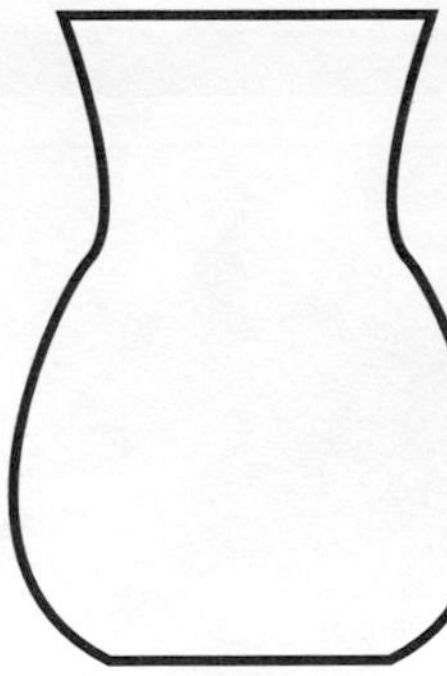

3
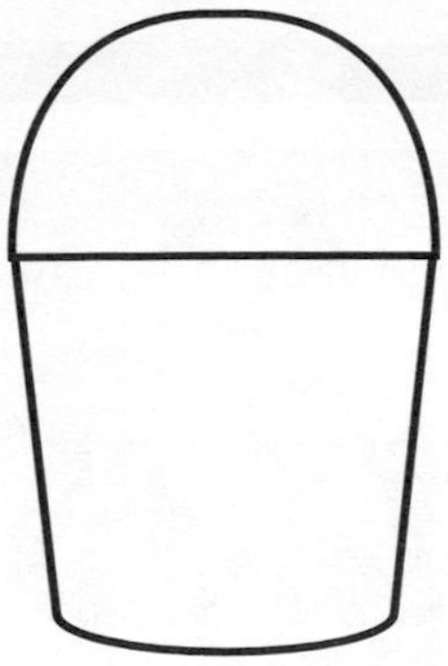

4

5
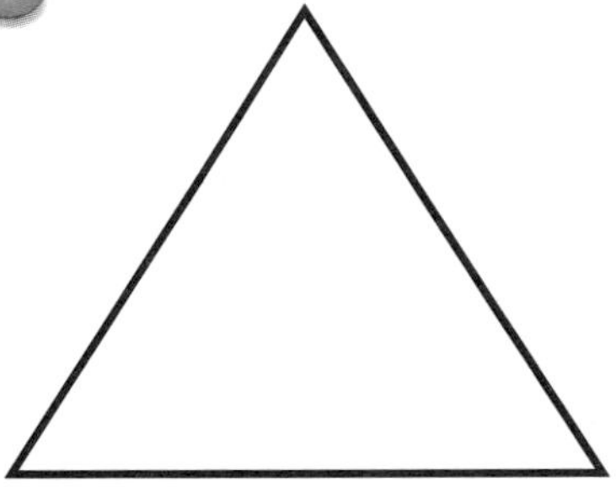

6
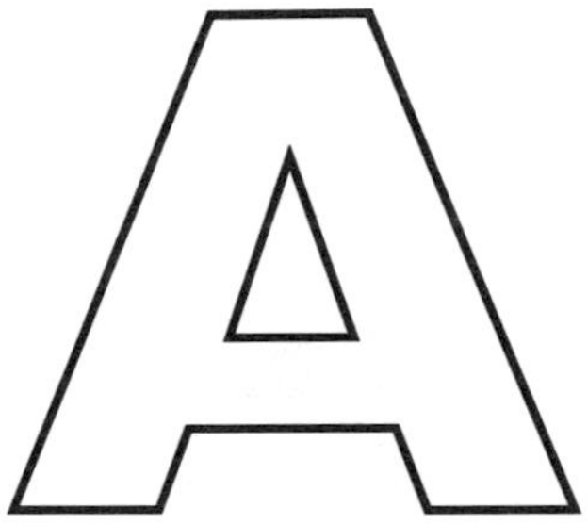

7
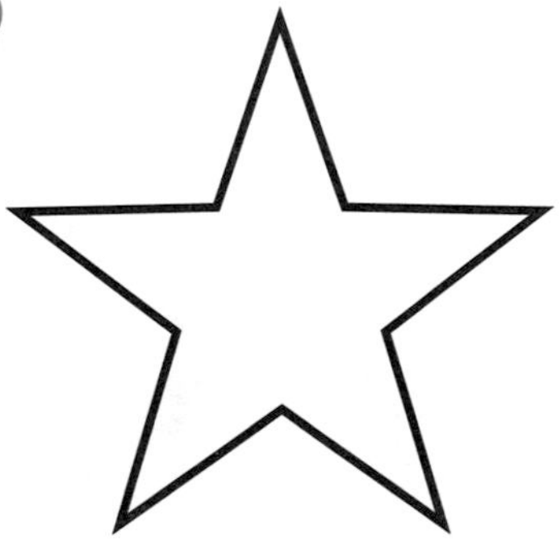

8
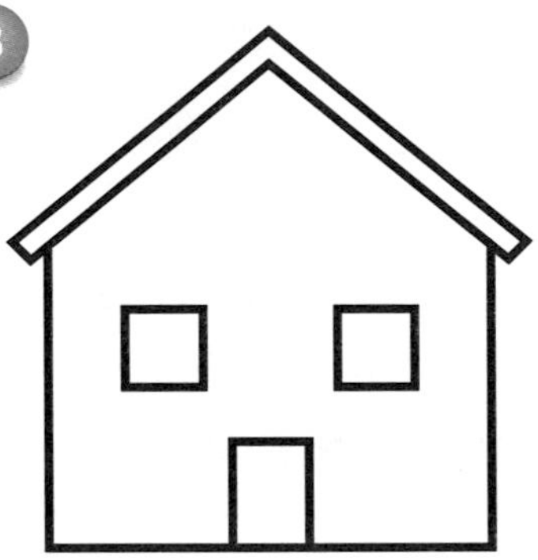

9
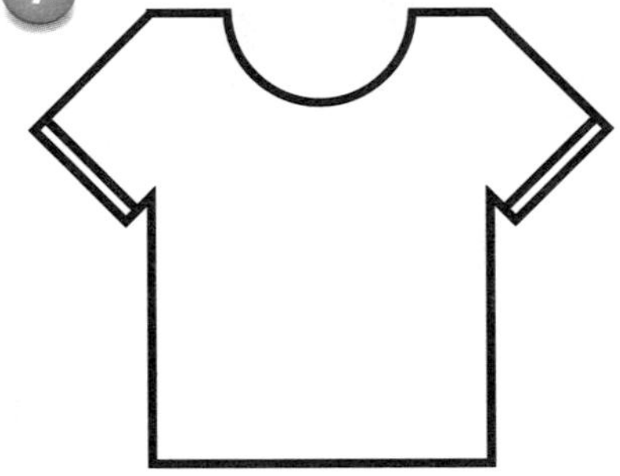

Mixed Review

Count back to subtract.

10. 9 − 3 = ____
11. 4 − 2 = ____
12. 8 − 2 = ____
13. 7 − 1 = ____
14. 5 − 3 = ____
15. 6 − 1 = ____

Home Note Your child identified lines of symmetry.
ACTIVITY Find an object or draw an object that is symmetrical. Have your child trace the line of symmetry with a finger or draw the line.

Name ______________________________

Concepts and Skills

Color the circle.
Color the rectangle.
Color the square.
Color the triangle.

1
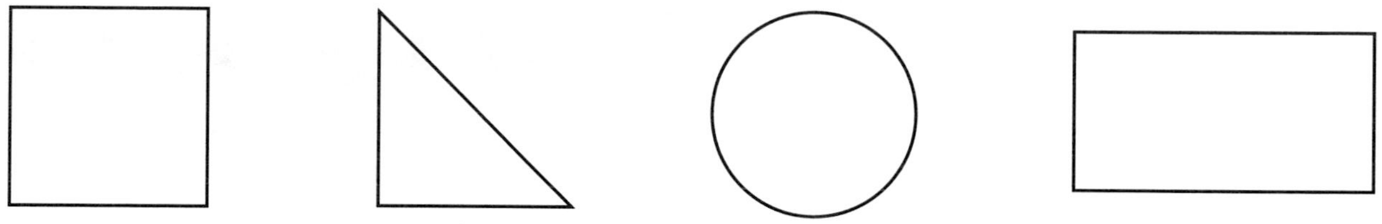

Write how many sides and corners.

2
____ sides
____ corners
 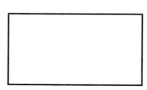

3
____ sides
____ corners
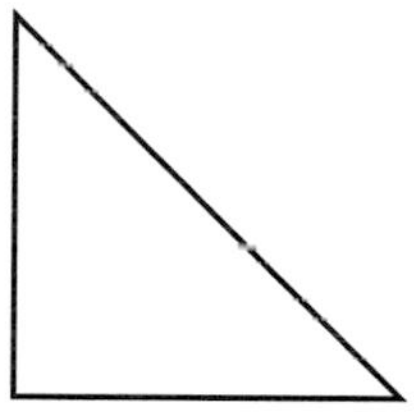

Color the figures that are the same size and shape.

4
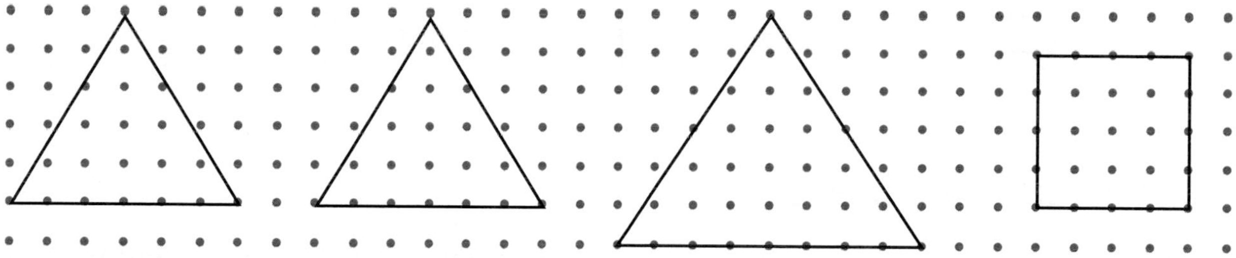

Draw a line to make two sides that match.

5

6

Name ______________________

Cumulative Chapters 1–8

TAAS Prep

Mark the best answer.

1. Which number is missing?

3, 4, ____, 6

- ⬭ 2
- ⬭ 5
- ⬭ 7
- ⬭ 8

2. Which number is missing?

6, ____, 8, 9

- ⬭ 5
- ⬭ 6
- ⬭ 7
- ⬭ 8

3. How many are there?

- ⬭ 5
- ⬭ 6
- ⬭ 7
- ⬭ 8

4. How many are there?

- ⬭ 5
- ⬭ 6
- ⬭ 7
- ⬭ 8

5. Which has the same sum as 6 + 2?

- ⬭ 5 + 2
- ⬭ 2 + 6
- ⬭ 6 + 3
- ⬭ 5 + 5

6. Which has the same sum as 4 + 3?

- ⬭ 6 + 4
- ⬭ 4 + 5
- ⬭ 1 + 7
- ⬭ 3 + 4

CHAPTER 9

Location and Movement

Choose an animal and describe to a classmate how to find it.

Home Note **In this chapter, your child will learn about open and closed figures and will learn the terms used to describe location.**
ACTIVITY **Have your child describe the location of objects in this picture.**

SCHOOL-HOME CONNECTION

Dear Family,

Today we started Chapter 9. We will look for open and closed figures. We will also learn ways to find objects on plane figures, in pictures, on grids, and on maps. Here are the new vocabulary words and an activity for us to do together at home.

Love,

____

Vocabulary

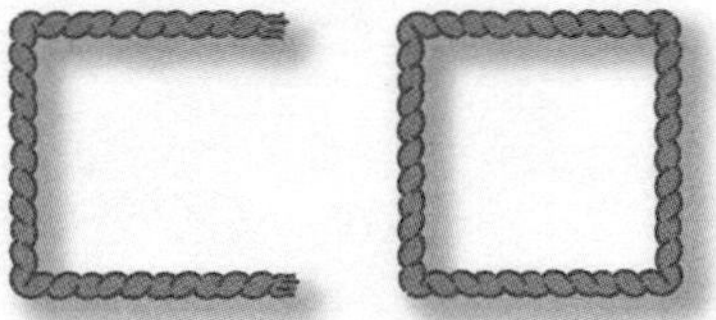

ACTIVITY

Draw a square. Have your child draw objects to show the meanings of *inside*, *outside*, and *on*. For example, say, "Draw a ball inside the square."

Visit our Web site for additional ideas and activities.
http://www.hbschool.com

Name ______________________________

Open and Closed

Color each closed figure.
Circle each open figure.

1

2

3

4

Talk About It • Critical Thinking

How can you tell if a figure is open or closed?

Practice

Color each closed figure.
Circle each open figure.

1

2

3

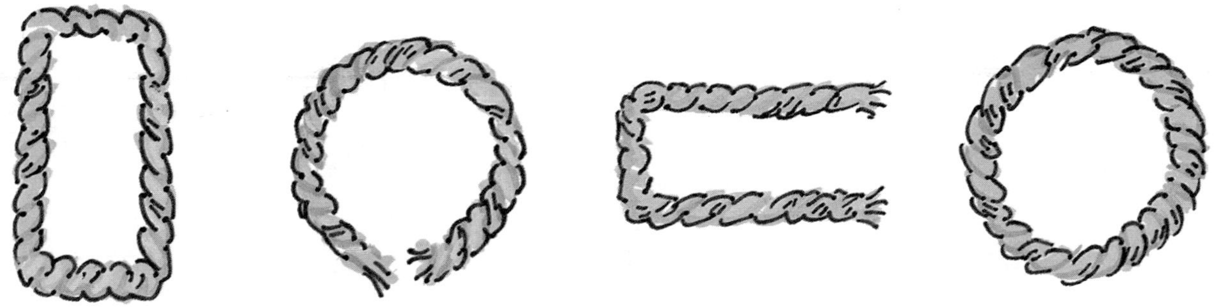

Mixed Review

Write the subtraction sentence.

4

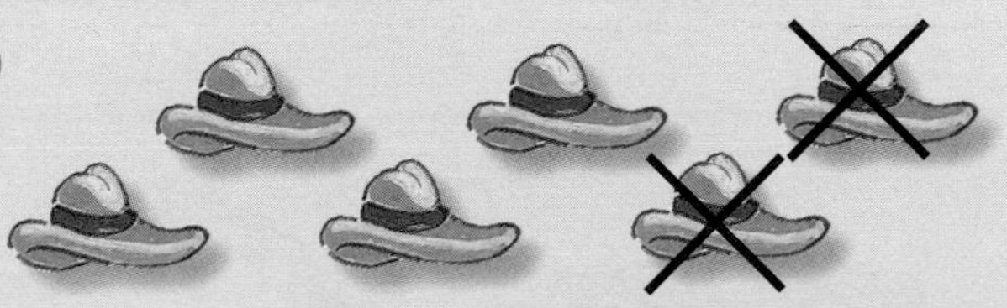

____ − ____ = ____

5

____ − ____ = ____

Home Note Your child identified open and closed figures.
ACTIVITY Have your child use a piece of string or yarn to make open and closed figures.

Name ______________________________ **Inside, Outside, On**

Color each animal inside the fence.
Color each animal outside the fence.
Color each animal on the fence.

1

2

Talk About It • Critical Thinking

How did you decide which animals to color?

Practice

inside 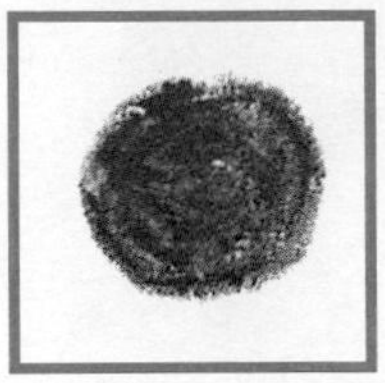 outside 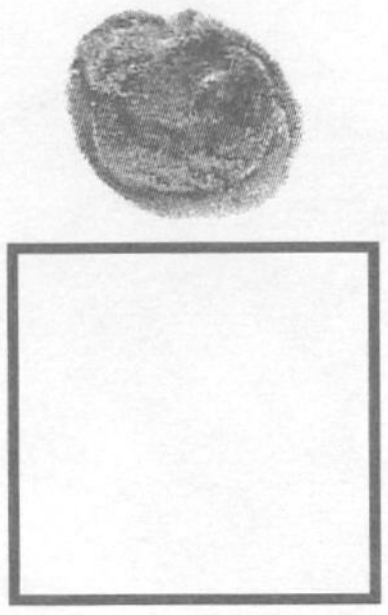 on 

Color the circles.

1 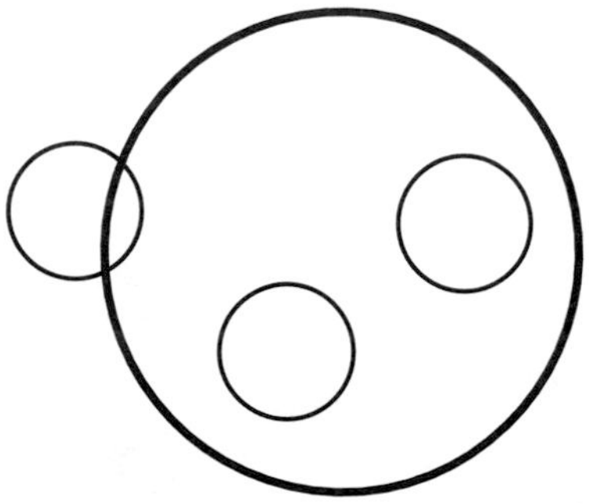

2 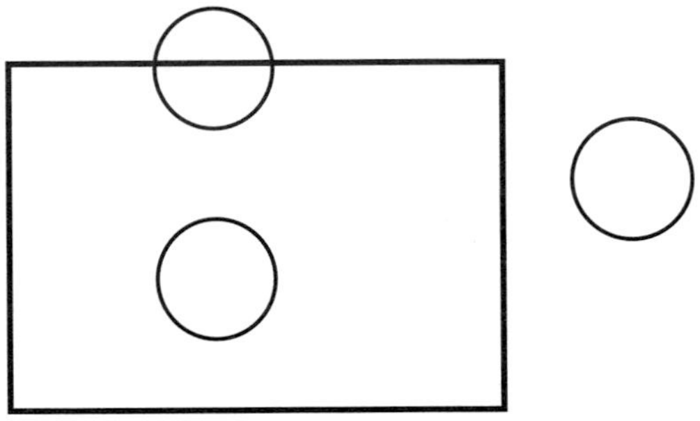

3 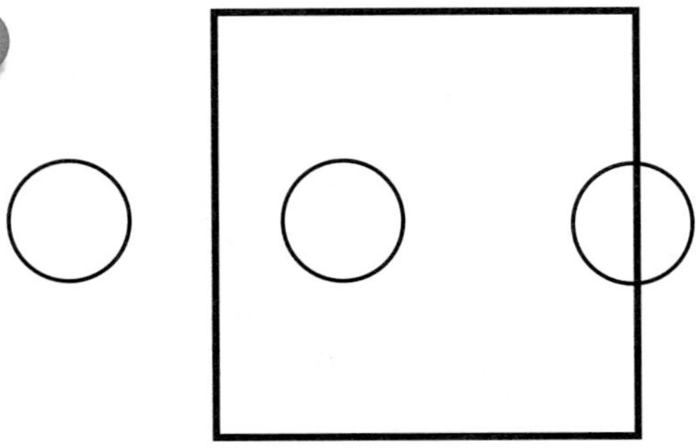

4 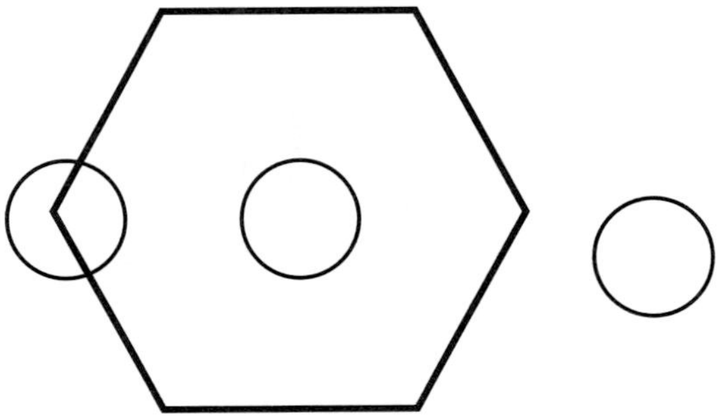

Problem Solving • Visual Thinking

5 Draw a triangle that is inside both circles.

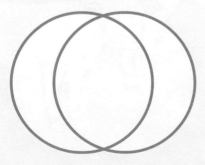

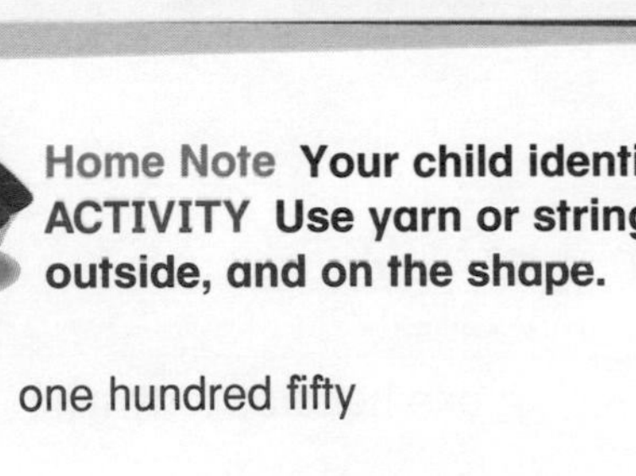

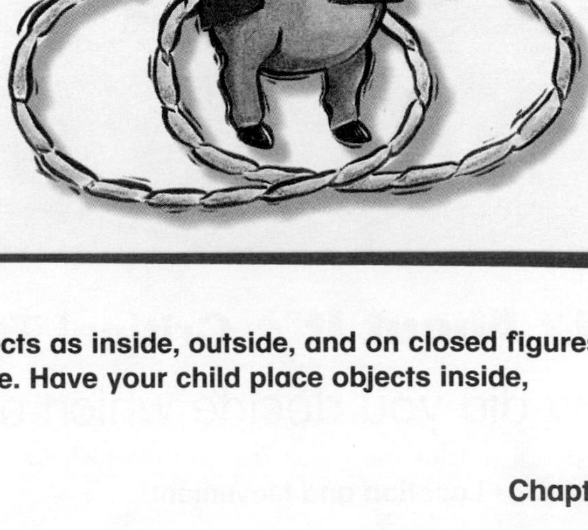

Home Note Your child identified the position of objects as inside, outside, and on closed figures.
ACTIVITY Use yarn or string to make a closed figure. Have your child place objects inside, outside, and on the shape.

Name ______________________________

Understand • Plan • Solve • Look Back

Problem Solving
Draw a Picture

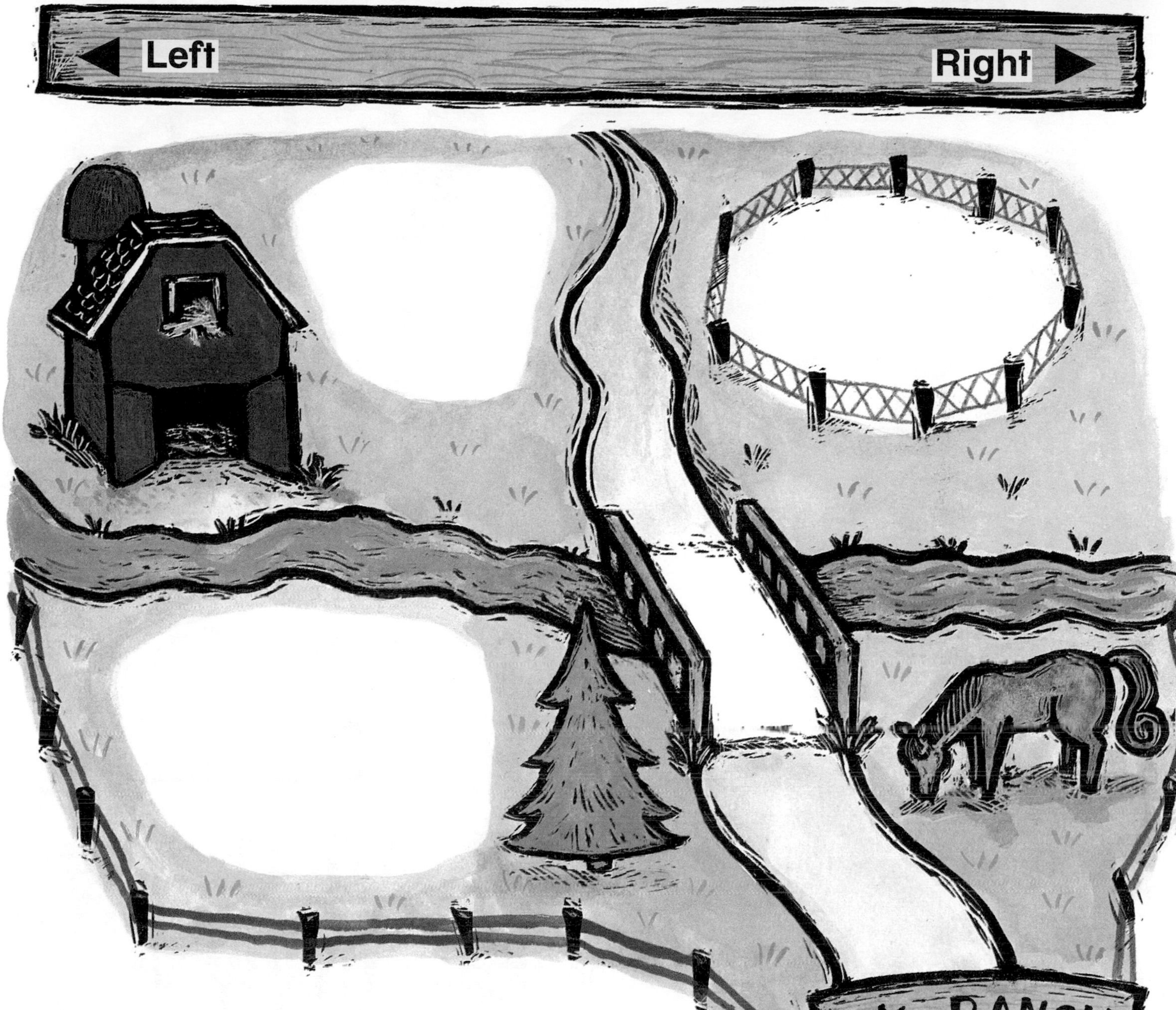

Draw to complete the picture.

1. Draw a  to the left of the .
2. Draw a  to the right of the .
3. Draw a  inside the .
4. Draw yourself on the .
5. Draw a friend outside the .

Practice

Draw to complete the picture.

1. Draw a  to the left of the .
2. Draw a  to the right of the .
3. Draw a  inside the .
4. Draw yourself outside the .
5. Draw a  on the .

Home Note Your child used direction words to describe location.
ACTIVITY Place an object in the middle of a room. Have your child name things that are to the left and right of the object.

Name ______________________________________

Positions on a Grid

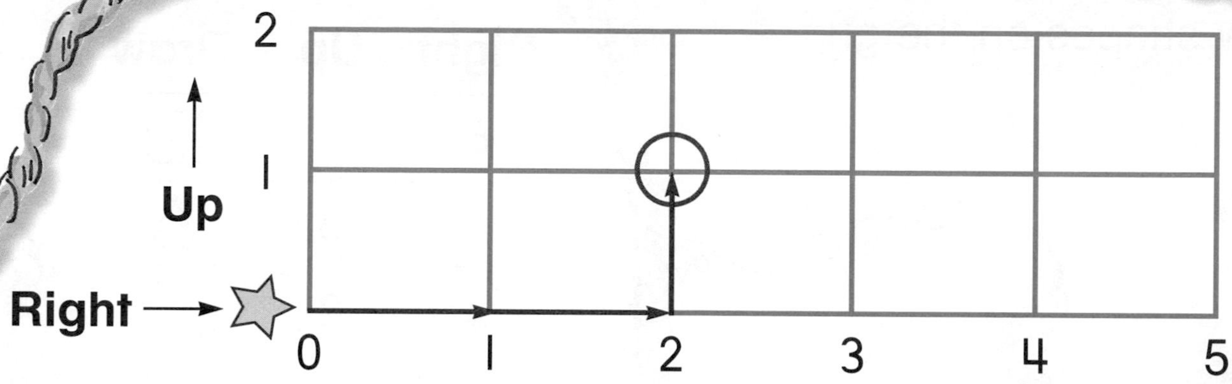

The circle is 2 spaces to the right and then 1 space up.

Start at ☆.
Follow directions to draw circles on the grid.

Right	Up	Draw
4	2	○
3	3	○
2	4	○
4	3	○

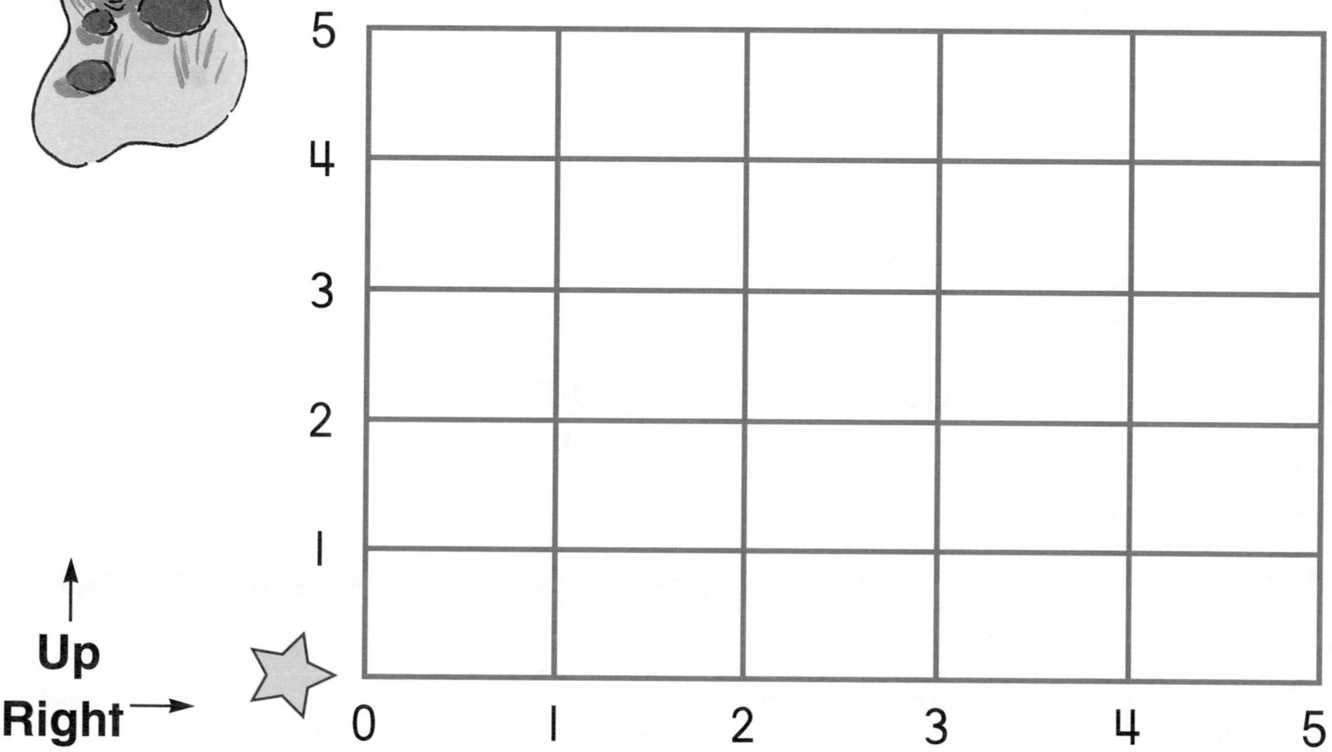

Practice

Start at ☆. Follow directions to draw shapes on the grid.

Right	Up	Draw
6	1	□
1	3	△
3	2	▭
5	4	○

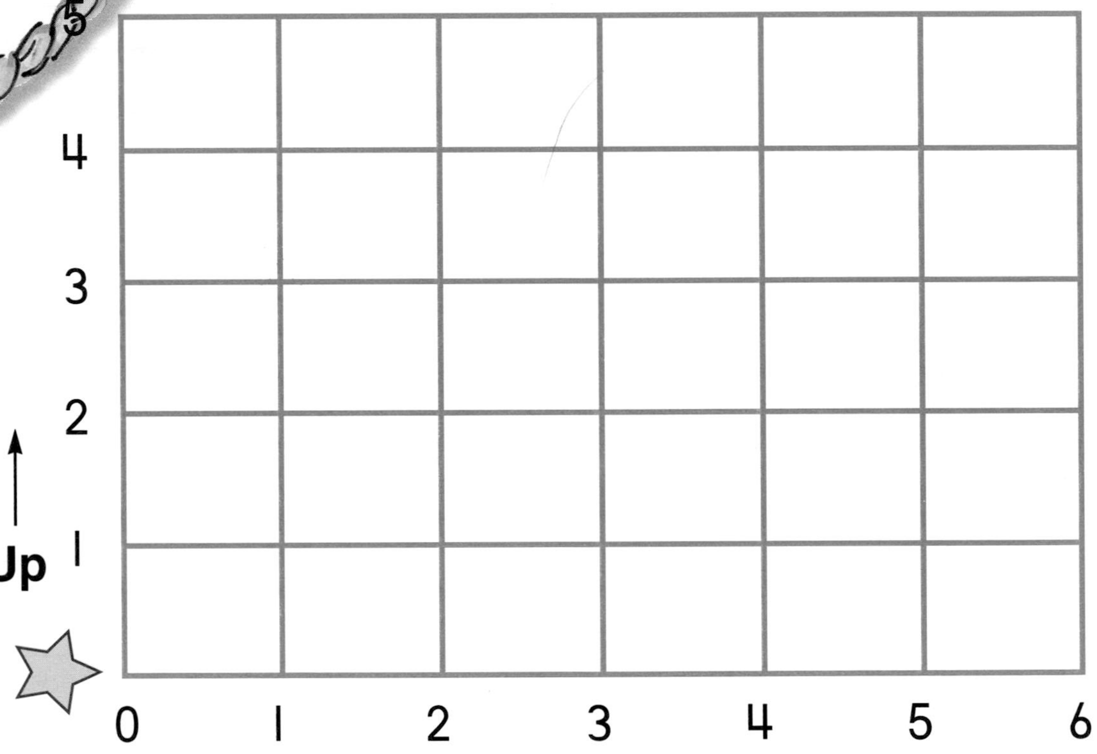

Write About It

Draw a picture of yourself inside or outside your school. Write a sentence that tells where you are.

Home Note Your child identified positions on a grid.
ACTIVITY Have your child follow your directions to place small objects on the grid.

Name ______________________________

Concepts and Skills

Color each closed figure. Circle each open figure.

1 

Color the circle inside . Color the circle outside .
Color the circle on .

2 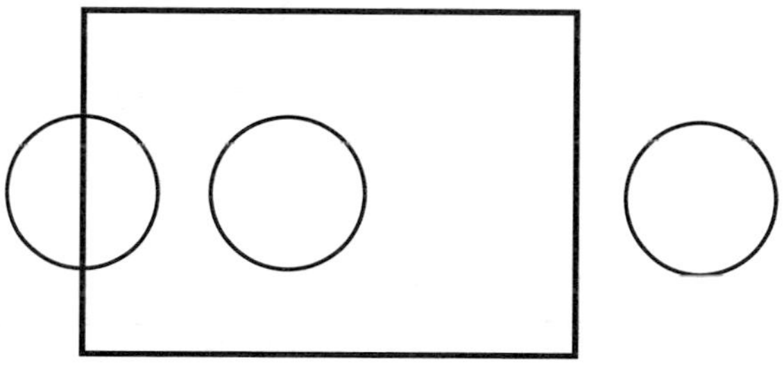

3 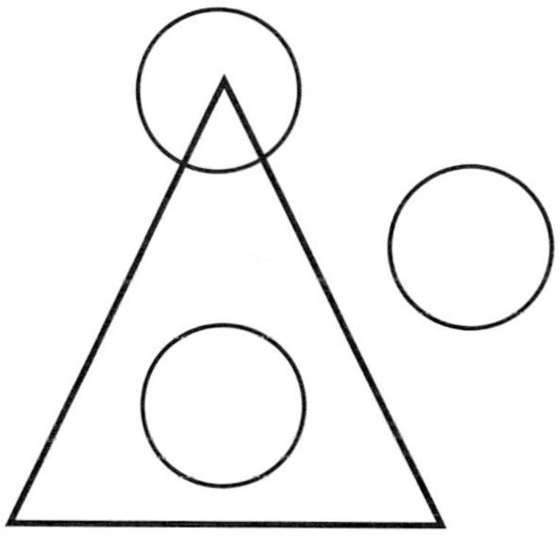

4 Start at . Follow directions to draw shapes on the grid.

Right	Up	Draw
5	2	□
3	1	△

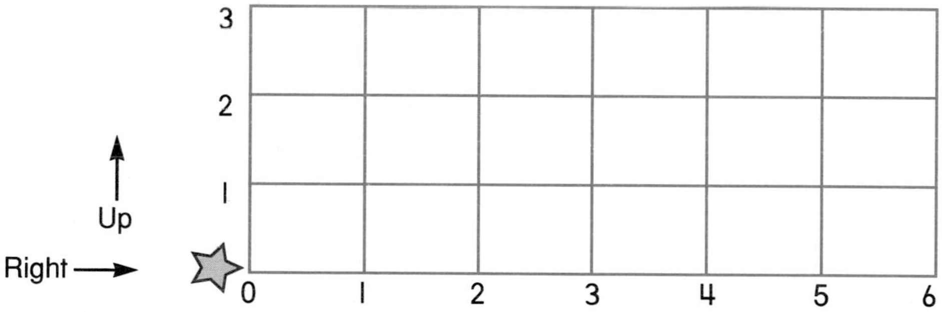

Problem Solving

Complete the picture.

5 Draw a red ball
to the left of the fence.
Draw a blue ball
to the right of the fence.

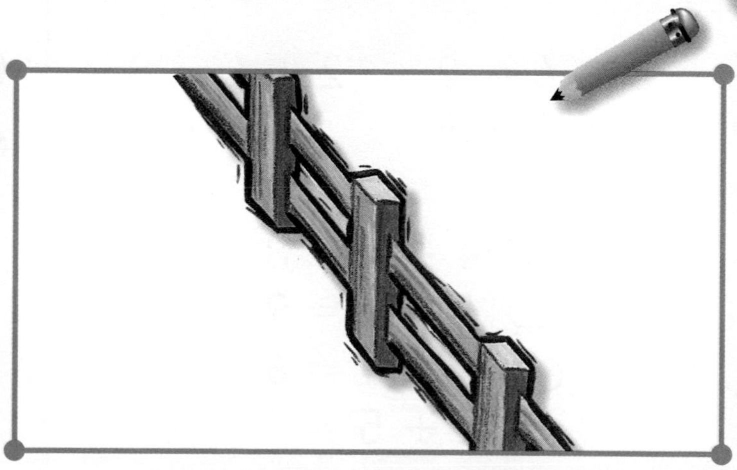

Name ______________________________

Cumulative Chapters 1–9

TAAS Prep

Mark the best answer.

1. How many ○ do you see?

- ⬭ 1
- ⬭ 2
- ⬭ 3
- ⬭ 4

2. How many?

- ⬭ 3
- ⬭ 4
- ⬭ 5
- ⬭ 6

3. Which is the missing fact?

$8 - 6 = 2$ $2 + 6 = 8$

$6 + 2 = 8$

- ⬭ $4 + 4 = 8$
- ⬭ $6 + 1 = 9$
- ⬭ $8 - 2 = 6$
- ⬭ $2 + 5 = 7$

4. Which names the shape?

- ⬭ circle
- ⬭ triangle
- ⬭ rectangle
- ⬭ square

5. Which is the missing number?

2, ____, 4, 5

- ⬭ 1
- ⬭ 2
- ⬭ 3
- ⬭ 4

6. Which number is greater?

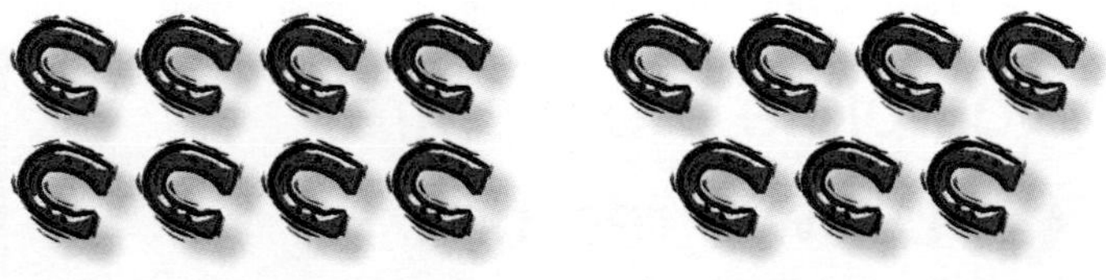

8 7

- ⬭ 7
- ⬭ 8

CHAPTER 10

Patterns

What patterns do you see in the picture?

Harcourt Brace School Publishers

Home Note In this chapter, your child is learning about patterns.
ACTIVITY Have your child identify and describe the patterns in the picture on this page.

SCHOOL-HOME CONNECTION

Dear Family,

Today we started Chapter 10. We will learn about patterns. We will read, make, and continue patterns. Here is the new vocabulary and an activity for us to do together at home.

Love,

____

Vocabulary

This is a pattern.

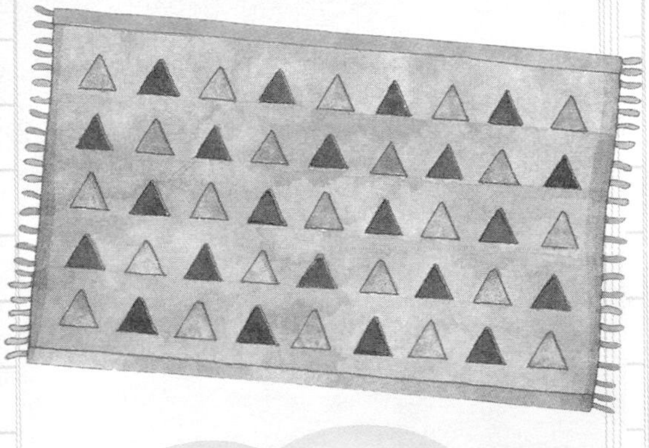

Blue triangle,
red triangle, blue triangle,
red triangle, blue triangle,
red triangle

ACTIVITY

Arrange forks and spoons in a pattern. Have your child continue the pattern. Then have your child make a pattern for you to continue.

Visit our Web site for additional ideas and activities.
http://www.hbschool.com

Name ______________________________

Identifying Patterns

I can read the pattern on my shirt. blue, white, blue, white, blue
I can read the pattern on my shirt. red, blue, yellow, red, blue, yellow

Read the pattern. Then color to continue it.

1. 
2. 
3. 
4. 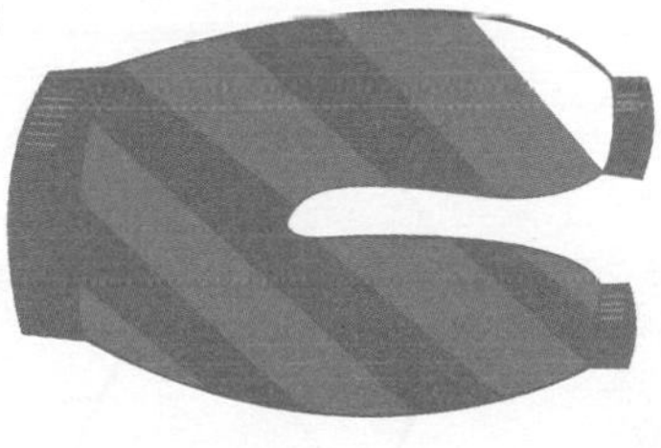
5. 
6. 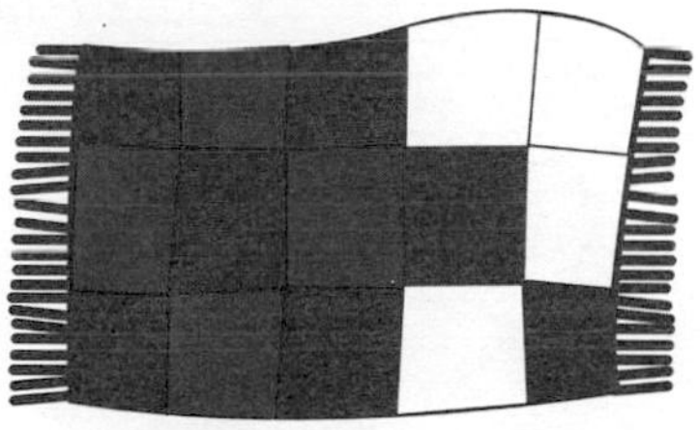
7. 
8. 
9. 

Talk About It • Critical Thinking

How can you tell what the pattern is?

Practice

Read the pattern. Then color to continue it.

1 

2 

3 

4 

5 

Problem Solving • Visual Thinking

Draw and color to continue the pattern.

6 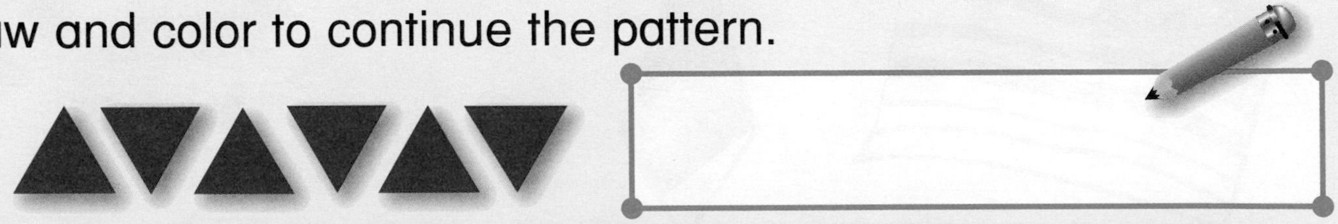

Home Note Your child identified, read, and extended patterns.
ACTIVITY With your child, look for patterns in clothing, wallpaper, and other things. Have him or her read the patterns to you.

Name ______________________________

Reproducing and Extending Patterns

Use shapes to copy and continue the pattern. Draw your shapes.

1 

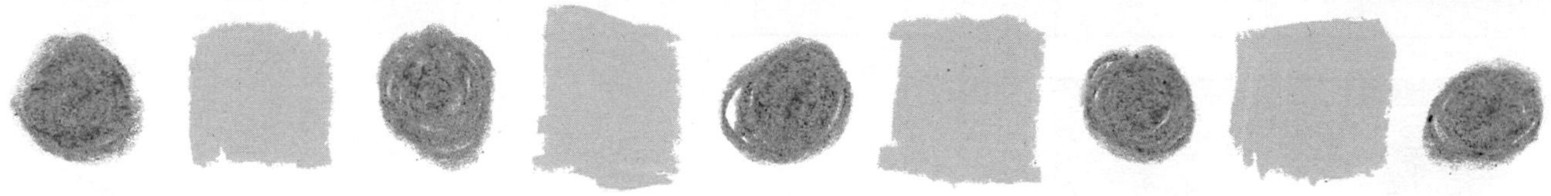

2 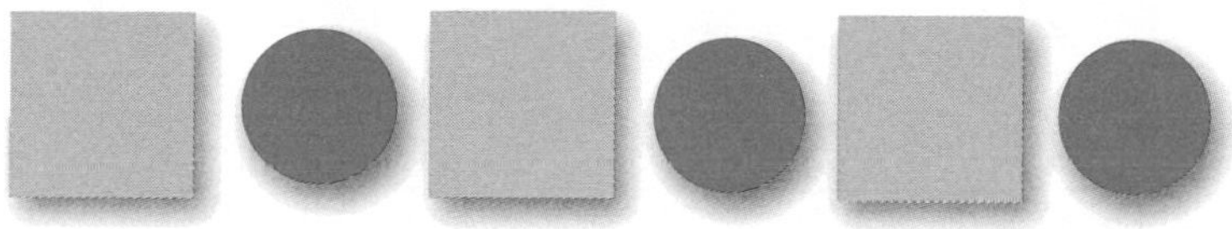

3 

4 

Practice

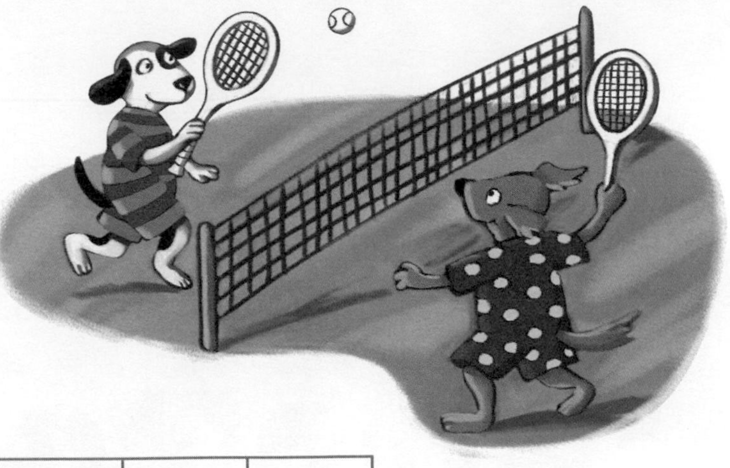

Color the squares to copy and continue the pattern.

1 

2 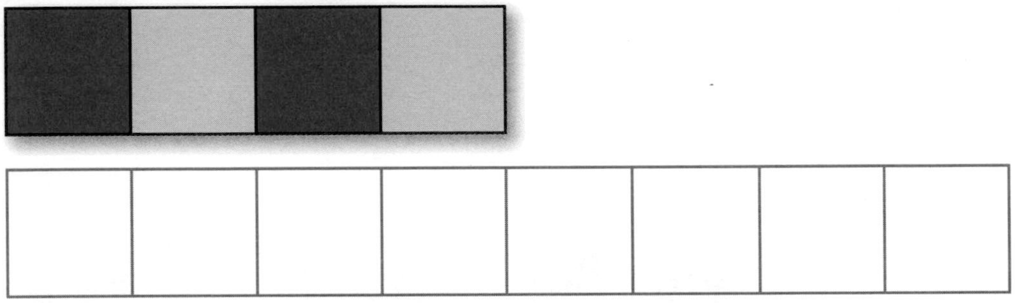

3 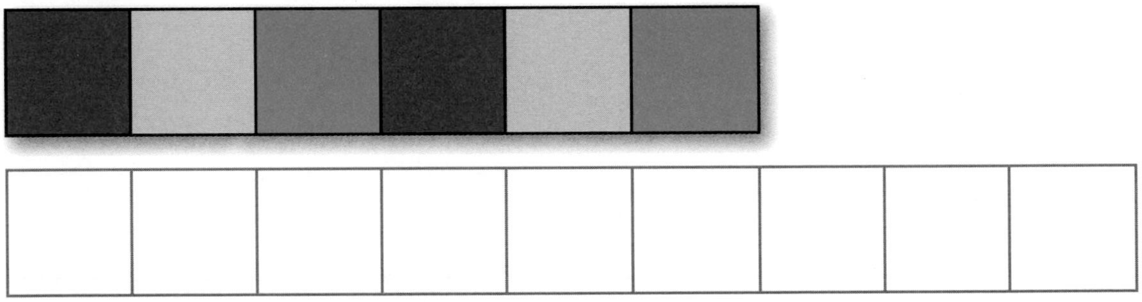

4 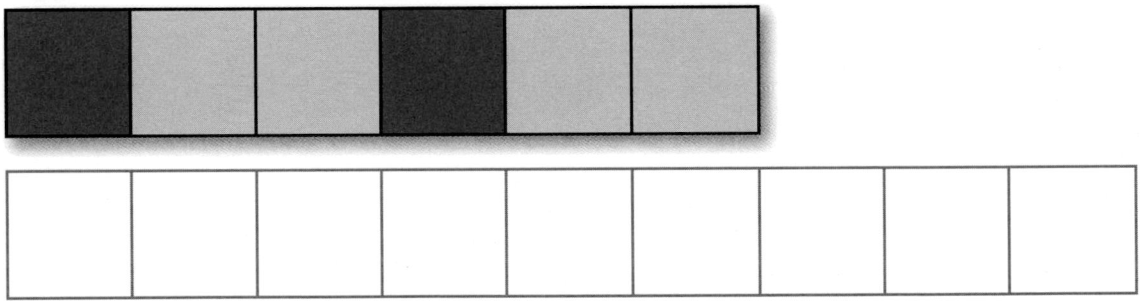

Write About It

5 Draw a pattern.
Write a sentence telling what the pattern is.

Home Note **Your child copied and extended patterns**
ACTIVITY Arrange some objects in a pattern. Have your child read the pattern.

Name ______________________________

Making and Extending Patterns

Use shapes.
Draw and color the shapes to continue the pattern.
Then use the same shapes to make a different pattern.
Draw and color the shapes to show your new pattern.

1 

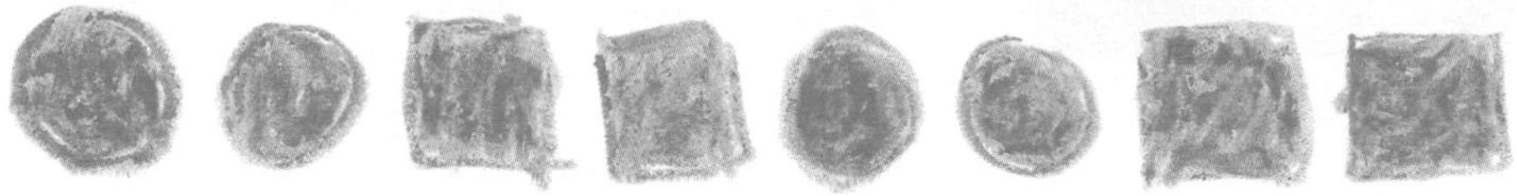

2 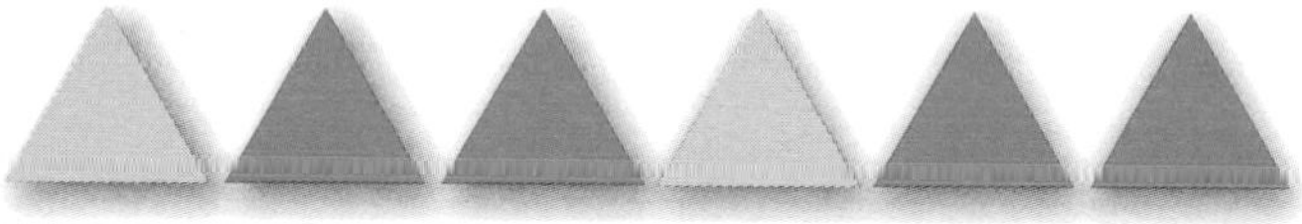

3 

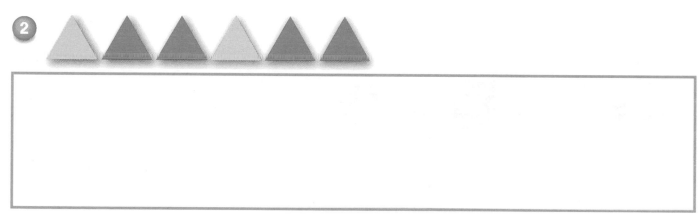

Talk About It • Critical Thinking

How are the patterns the same?
How are the patterns different?

Practice

Use shapes. Draw and color the shapes to continue the pattern. Then use the same shapes to make a different pattern. Draw and color the shapes to show your new pattern.

1 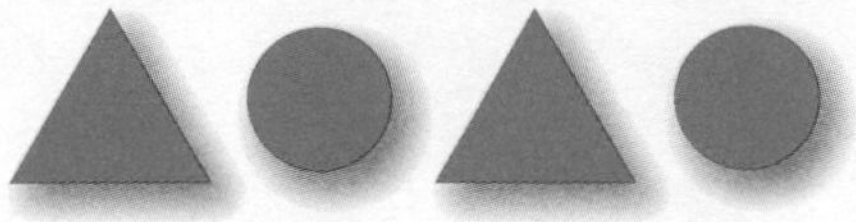

2 

3 

Mixed Review

Add or subtract.

4

7	5	9	6	8	10
+ 2	+ 4	− 3	− 4	+ 2	− 4

Home Note Your child made and extended patterns.
ACTIVITY Have your child arrange objects in a pattern and then read the pattern to you.

Name ______________________________

Understand • Plan • Solve • Look Back

Problem Solving
Look For A Pattern

Look for a pattern.
Circle the mistake in the pattern.
Then use shapes to show the correct pattern. Draw and color the shapes.

1 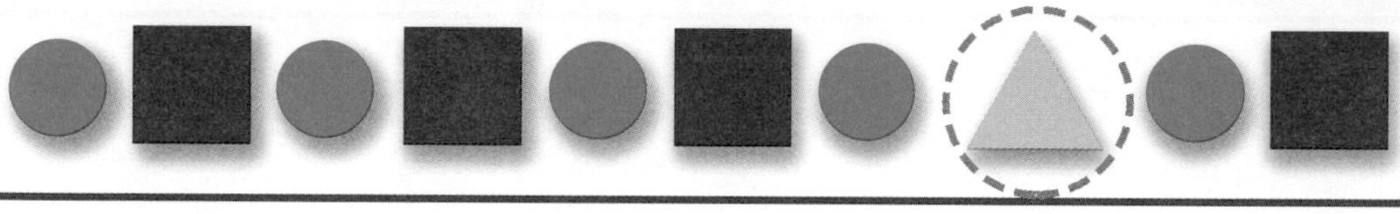

2 

3 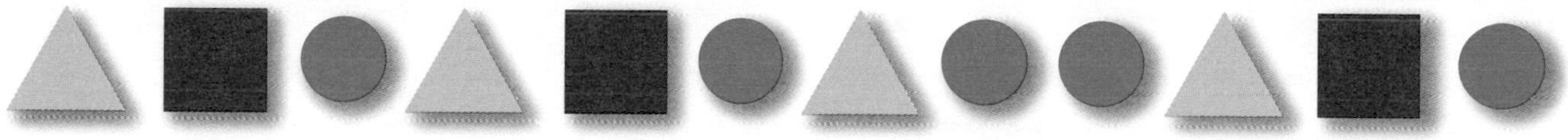

4 

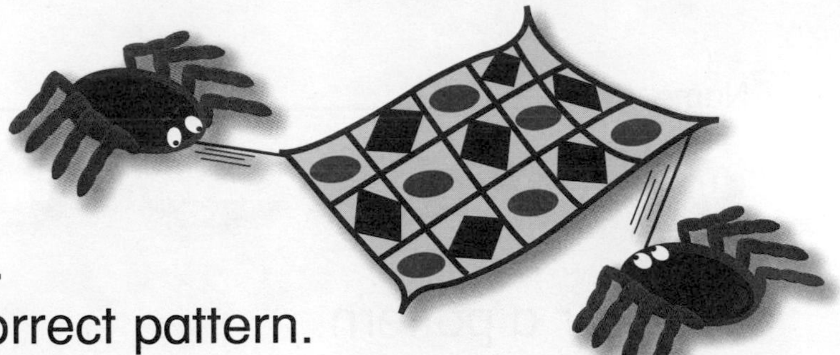

Practice

Look for a pattern.
Circle the mistake in the pattern.
Then use shapes to show the correct pattern.
Draw and color the shapes.

1 

2 

3 

4 

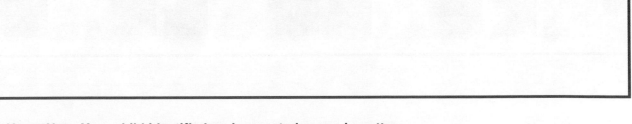

Home Note **Your child identified and corrected errors in patterns.**
ACTIVITY Arrange objects in a pattern. Change one of the objects so that there is a mistake in the pattern. Have your child find and correct the mistake.

Name ______________________________

Chapter 10
Review/Test

Concepts and Skills

1. Read the pattern.
Then color to continue it.

2. Use shapes to copy and continue
the pattern. Draw your shapes.

3. Use shapes. Draw and color the shapes to
continue the pattern. Then use the same
shapes to make a different pattern. Draw and
color the shapes to show your new pattern.

Problem Solving

4. Look for a pattern.
Circle the mistake in the pattern. Then use shapes to
show the correct pattern. Draw and color the shapes.

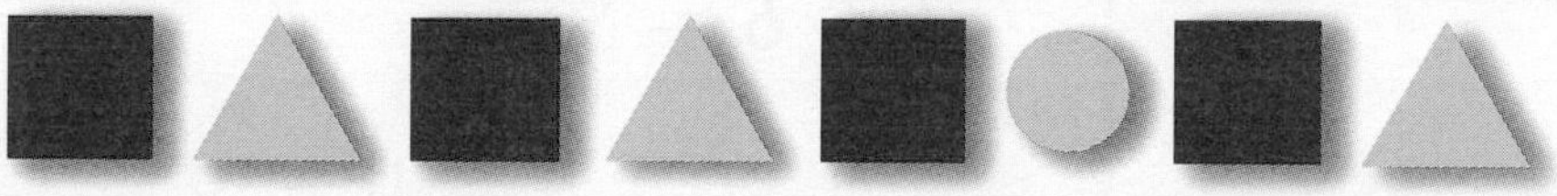

Name ______________________________

Cumulative Chapters 1–10

TAAS Prep

Mark the best answer.

1. Which shape comes next?

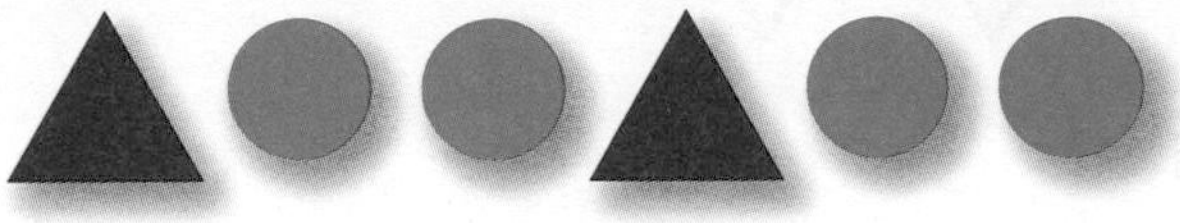

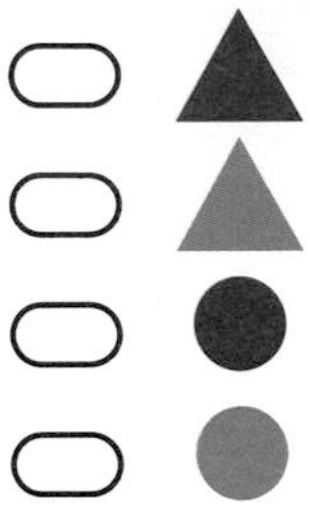

2. Which shape is a mistake in the pattern?

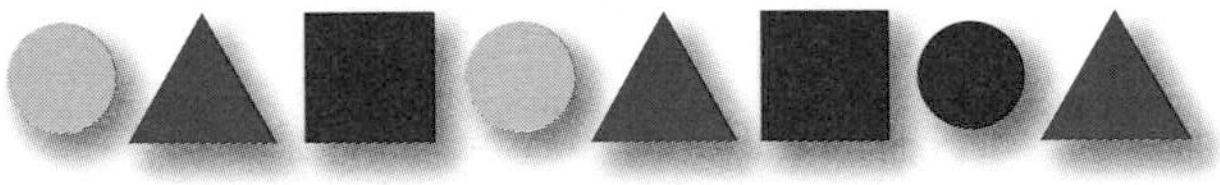

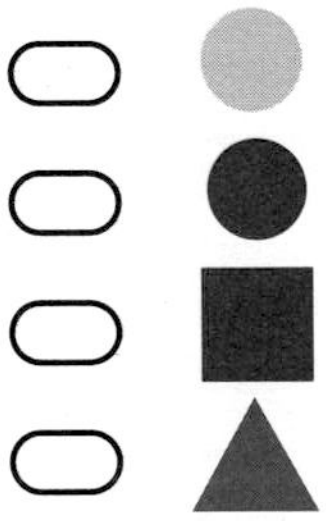

3. How many circles?

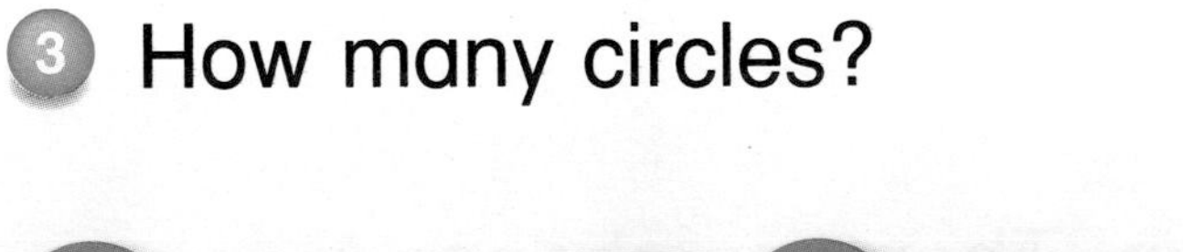

- 3
- 4
- 5
- 6

4. How many sides and corners?

- 2 sides, 3 corners
- 3 sides, 2 corners
- 3 sides, 3 corners
- 4 sides, 4 corners

5. Which figure has 6 sides and 6 corners?

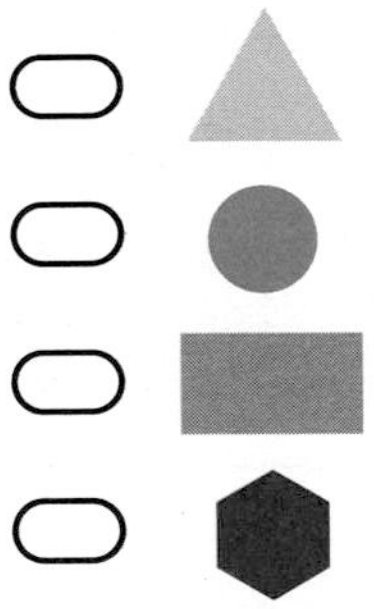

6. Which number sentence completes the fact family?

$6 + 3 = 9$ $\quad$ $9 - 6 = 3$

$3 + 6 = 9$

- $5 + 4 = 9$
- $6 + 2 = 8$
- $9 - 3 = 6$
- $4 + 2 = 6$

Name ______________________

MATH FUN

You will need:
A [marker] and a [number cube].

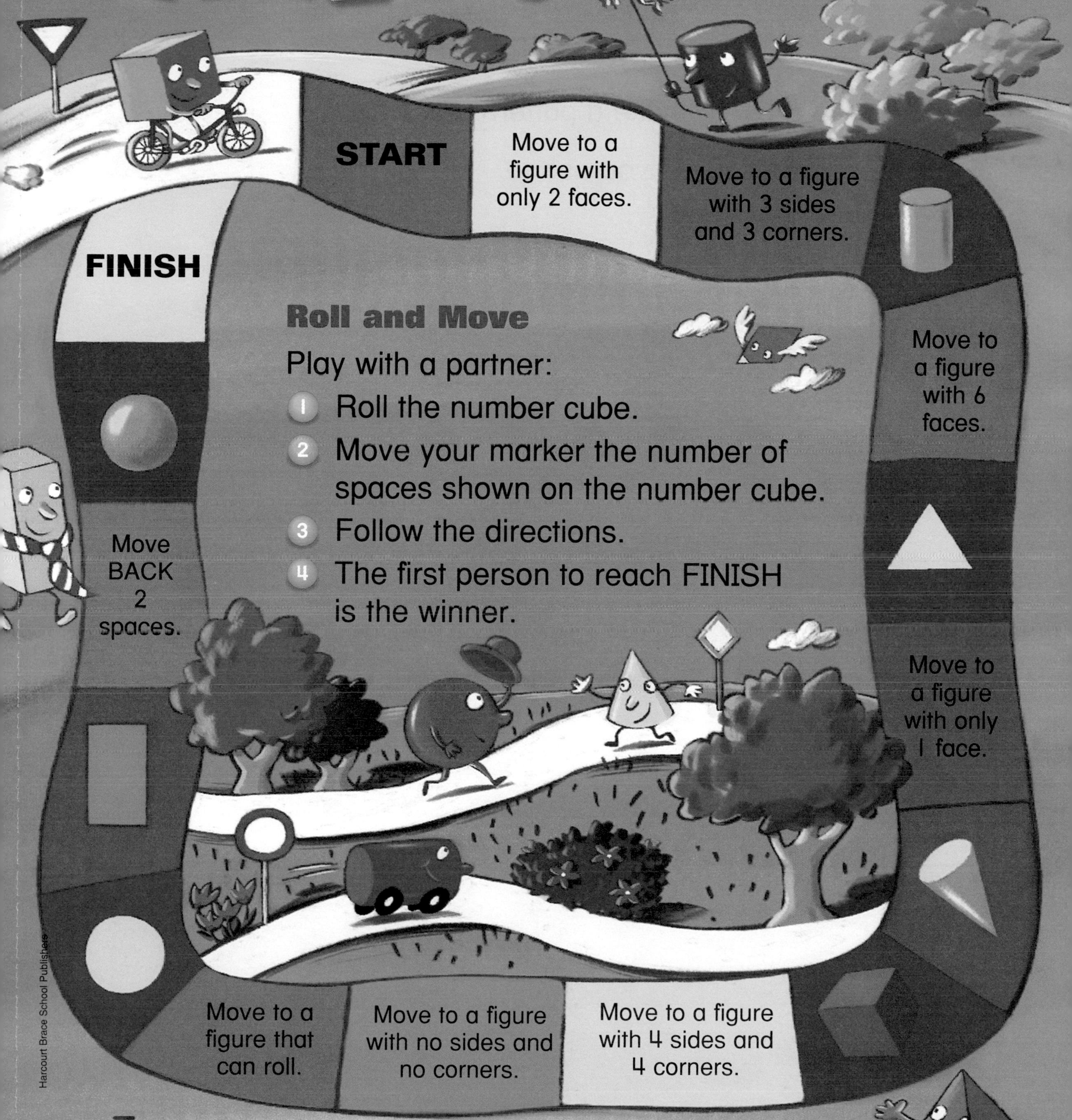
START
Move to a figure with only 2 faces.
Move to a figure with 3 sides and 3 corners.
Move to a figure with 6 faces.
Move to a figure with only 1 face.
Move to a figure with 4 sides and 4 corners.
Move to a figure with no sides and no corners.
Move to a figure that can roll.
Move BACK 2 spaces.
FINISH

Roll and Move

Play with a partner:

1. Roll the number cube.
2. Move your marker the number of spaces shown on the number cube.
3. Follow the directions.
4. The first person to reach FINISH is the winner.

Home Note Your child learned about solid figures and plane figures.
ACTIVITY As you play the game with your child, encourage him or her to talk about the figures.

Name ______________________

Calculator | Computer

You can use a  to make a pattern.

1. Which shapes could you use to continue the pattern?
Draw the shapes.

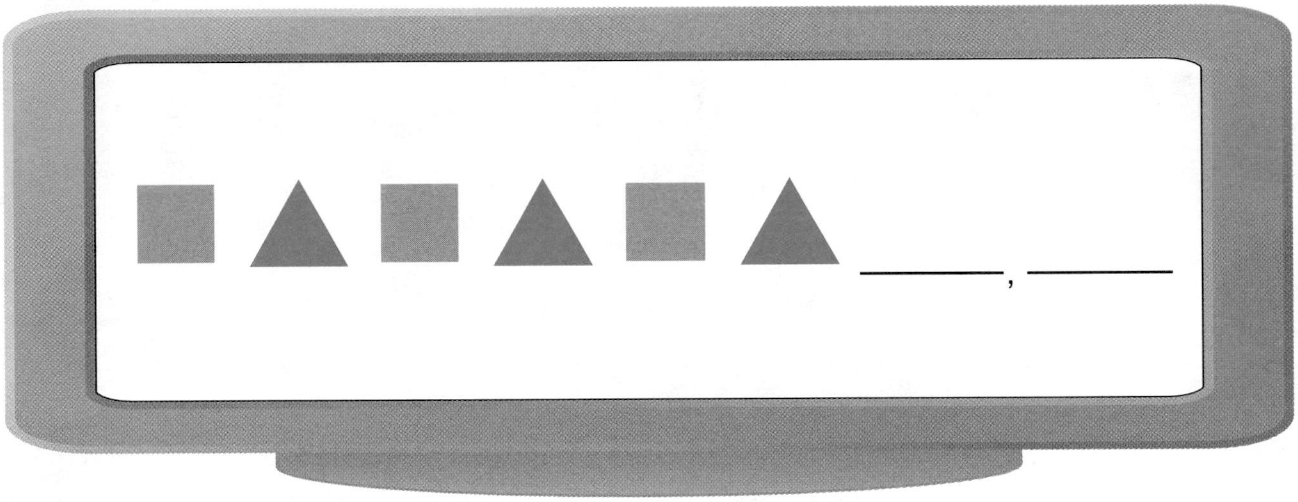

2. Use , or a  to make a pattern.
Draw your pattern.

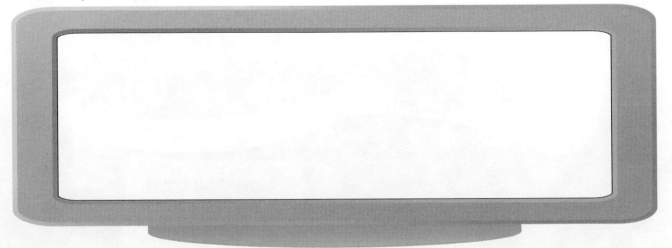

3. Have a classmate continue the pattern.

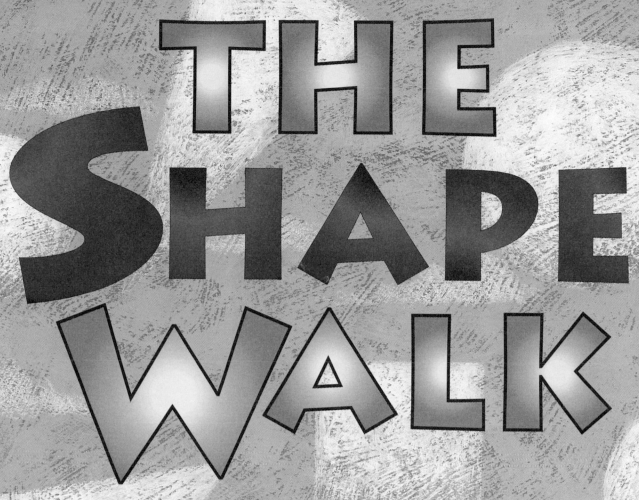

THE SHAPE WALK

WRITTEN BY ROZANNE LANCZAK WILLIAMS ILLUSTRATED BY TIM BOWERS

This book will help me review solid and plane figures.

This book belongs to ________________.

Come on a shape walk.
Come with me.
Point to the circles that you see.

B

Come on a shape walk.
Come with me.
Point to the squares that you see.

Come on a shape walk.
Come with me.
Point to the triangles that you see.

D

Come on a shape walk.
Come with me.
Point to the spheres that you see.

Come on a shape walk.
Come with me.
Point to the cones that you see.

F

Come on a shape walk.
Come with me.
Point to the cubes that you see.

G

Circles on cones and cubes with squares.
Look for the shapes—they're everywhere!

H

Name ______________________________

Concepts and Skills

cube

cylinder

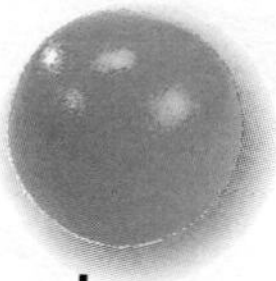
sphere

cone

Color the objects to match the figures.

1 

2 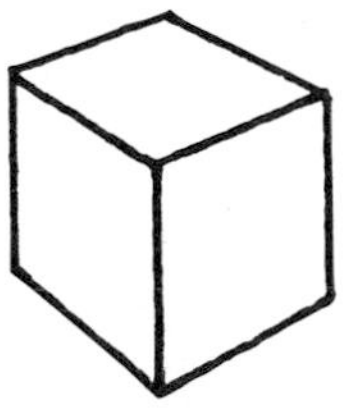

3 

4 

Circle each figure that goes with the sentence.
All faces are flat.

5 

Write how many sides and corners.

6 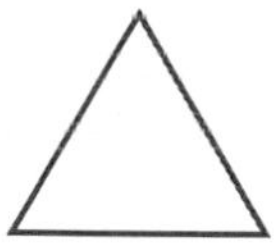

_____ sides

_____ corners

7 

_____ sides

_____ corners

Draw a line to make two sides that match.

8 

9 

Color each closed figure. Circle each open figure.

10

11
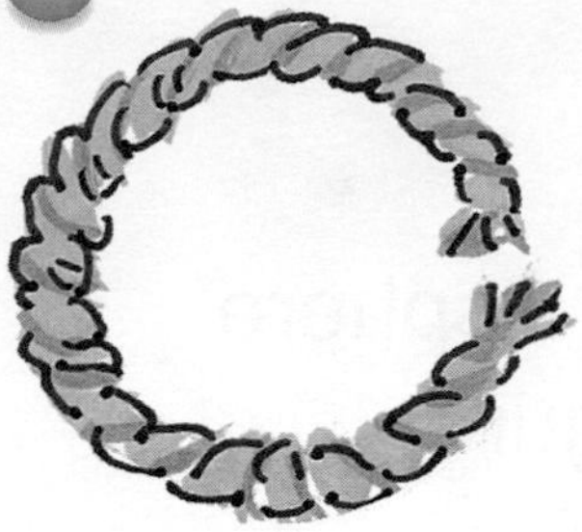

12

13

Color the circles inside . Color the circles outside .

14
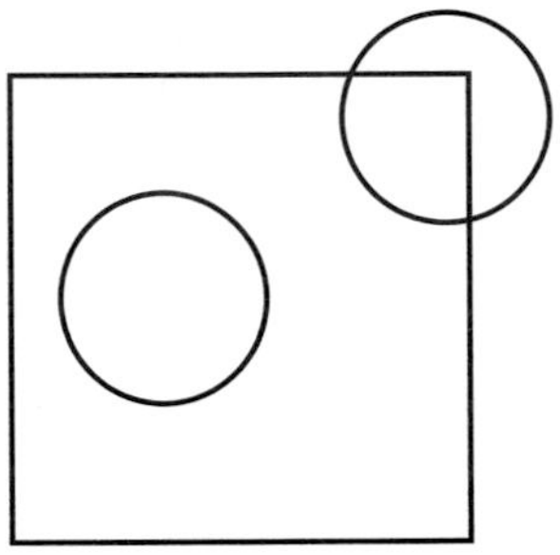

15

Start at ☆. Follow directions to draw shapes on the grid.

16

Right	Up	Draw
4	1	△
1	2	□

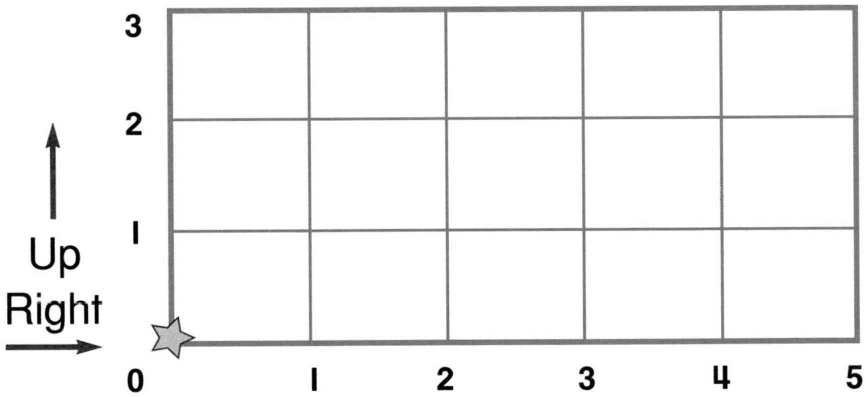

Problem Solving

Look for a pattern.
Circle the mistake in the pattern.
Then use shapes to show the correct pattern.
Draw and color the shapes.

17

Name ______________________

Performance Assessment

Use Workmat 1 and solid figures.

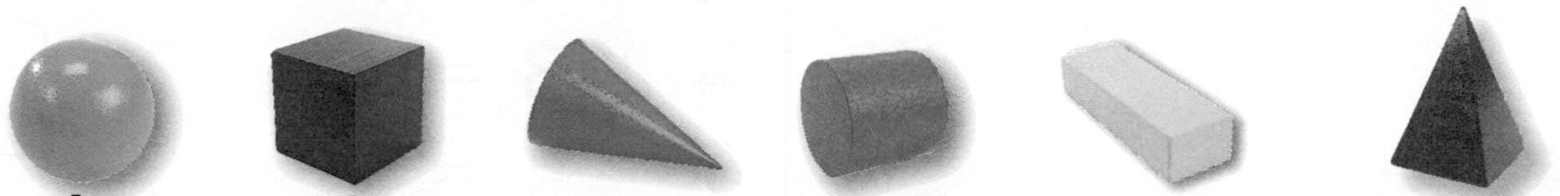

1. Put figures that are alike in some way on the Workmat. Circle the names of the figures.

sphere cube cone cylinder rectangular prism pyramid

2. How are these figures alike?

____

3. Choose a figure that has 1 or more faces. Trace around all the faces. How many faces does the figure have? _____

Write About It

4. Use shapes to make a pattern. Draw and color the shapes to show the pattern.

Name ______________________

Chapters 1–10 Cumulative Review

Fill in the ⬭ for the correct answer.

1. Find the missing sum.

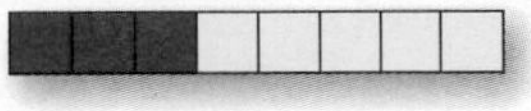

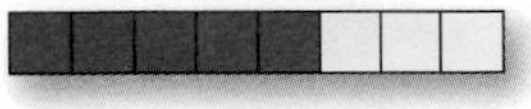

5 + 3 = ____

⬭ 3
⬭ 8
⬭ 5
⬭ not here

2. Complete the fact family.

3 + 4 = 7
4 + 3 = 7
7 − 4 = 3
7 − 3 = ____

⬭ 4
⬭ 8
⬭ 7
⬭ 10

3. 6 birds are resting.
3 birds are flying.
How many birds in all?

⬭ 3 birds ⬭ 6 birds
⬭ 8 birds ⬭ 9 birds

4. 7 bugs are on a rock.
2 bugs walk away.
How many are left?

⬭ 3 bugs ⬭ 4 bugs
⬭ 5 bugs ⬭ 6 bugs

5. $\begin{array}{r} 3 \\ +1 \\ \hline \end{array}$

⬭ 2
⬭ 3
⬭ 4
⬭ 5

6. $\begin{array}{r} 5 \\ -1 \\ \hline \end{array}$

⬭ 3
⬭ 4
⬭ 5
⬭ 6

7. $\begin{array}{r} 4 \\ +5 \\ \hline \end{array}$

⬭ 6
⬭ 7
⬭ 8
⬭ 9

8. $\begin{array}{r} 7 \\ -2 \\ \hline \end{array}$

⬭ 4
⬭ 5
⬭ 6
⬭ 7

9. Which is the same shape as this hat?

⬭ 
⬭ 
⬭ 
⬭ 

10. Which pattern has a mistake?

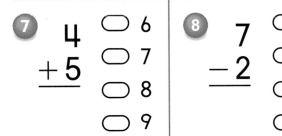

⬭ 

⬭ 

CHAPTER 11

Addition Facts to 12

Tell some addition stories about the pets.

Home Note In this chapter, your child will learn addition facts to 12.
ACTIVITY Have your child make up addition stories about the animals on this page.

SCHOOL-HOME CONNECTION

Dear Family,

Today we started Chapter 11. We will learn addition facts to 12 and ways to make adding numbers easier. Here are the new vocabulary words and an activity for us to do together at home.

Love,

____

Vocabulary

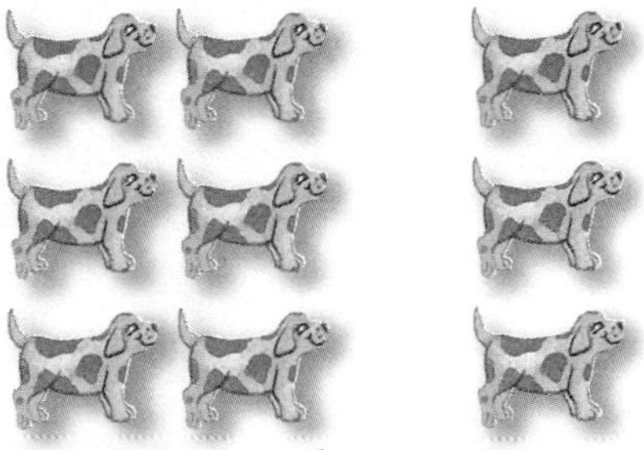

6 is **greater** than 3.

4 + 7 = 11

This is a **number sentence.**

ACTIVITY

Give your child doubles problems using calendar dates from 1 to 6. For example: If today's date is 2, what date will it be in 2 days? If today's date is 5, what date will it be in 5 days?

Visit our Web site for additional ideas and activities.
http://www.hbschool.com

Name ______________________________

Counting On to 12

Start with the greater number.
Count on to add.

(9)	Say 9.	2	Say 7.	(5)	Say 5.
+1	Count on 1.	+(7)	Count on 2.	+3	Count on 3.
10	10	9	8, 9	8	6, 7, 8

Circle the greater number.
Count on to add.

1	5	2	3	9	3	7
	+3	+7	+8	+1	+9	+3
2	8	2	9	8	1	6
	+1	+9	+2	+2	+7	+3
3	1	8	3	8	9	2
	+8	+2	+6	+3	+3	+8

Talk About It • Critical Thinking

How do you know which number is greater?

Practice

Circle the greater number.
Count on to add.

1. (8) + 3 = 11	2. 3 + 6 = ___	3. 9 + 1 = ___
4. 2 + 6 = ___	5. 9 + 3 = ___	6. 1 + 8 = ___
7. 6 + 3 = ___	8. 9 + 2 = ___	9. 2 + 7 = ___
10. 2 + 8 = ___	11. 3 + 9 = ___	12. 7 + 3 = ___
13. 1 + 7 = ___	14. 3 + 8 = ___	15. 2 + 9 = ___

Problem Solving

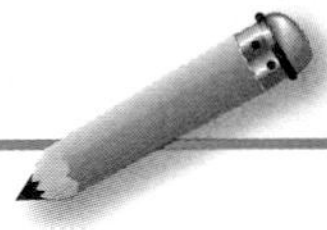

Solve. Draw a picture.

16. Leslie spent 9¢.
Lin spent 3¢.
How much did they spend in all? ___¢

17. Ann Lee saved 8¢.
Carol saved 2¢.
How much did they save in all? ___¢

Home Note Your child counted on to add.
ACTIVITY Make flash cards for any facts your child needs help with. Practice those facts together.

Name ________________________

Doubles to 12

$\underline{6} + \underline{6} = \underline{12}$

Write the doubles facts.

1

___ + ___ = ___

2

___ + ___ = ___

3
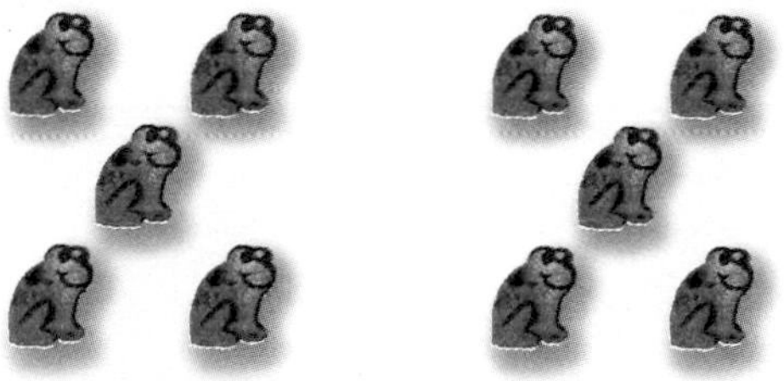

___ + ___ = ___

4

___ + ___ = ___

5

___ + ___ = ___

6

___ + ___ = ___

Practice

Write the sums.
Circle the doubles.

1. 2 + 2 = 4
2. 5 + 2 = ____
3. 7 + 3 = ____

4. 4 + 2 = ____
5. 6 + 6 = ____
6. 5 + 5 = ____

7. 3 + 3 = ____
8. 4 + 4 = ____
9. 6 + 4 = ____

10.

$$\begin{array}{r} 8 \\ +2 \\ \hline \end{array} \quad \begin{array}{r} 5 \\ +5 \\ \hline \end{array} \quad \begin{array}{r} 8 \\ +3 \\ \hline \end{array} \quad \begin{array}{r} 7 \\ +1 \\ \hline \end{array} \quad \begin{array}{r} 4 \\ +4 \\ \hline \end{array}$$

11.

$$\begin{array}{r} 3 \\ +3 \\ \hline \end{array} \quad \begin{array}{r} 5 \\ +3 \\ \hline \end{array} \quad \begin{array}{r} 7 \\ +2 \\ \hline \end{array} \quad \begin{array}{r} 6 \\ +2 \\ \hline \end{array} \quad \begin{array}{r} 1 \\ +1 \\ \hline \end{array}$$

Mixed Review

Color the squares .
Color the triangles .

12. 

Home Note Your child added doubles.
ACTIVITY Name the doubles. Have your child give the sums.

Name ______________________________

Three Addends

$2 + 2 + 6 =$ 10

4

2 + 2 = 4
4 + 6 = 10

$2 + 2 + 6 =$ 10

8

2 + 6 = 8
2 + 8 = 10

Use ▪.
Add the blue numbers first.
Write the sum.

1. $3 + 2 + 5 =$ ____ $3 + 2 + 5 =$ ____

2. $5 + 2 + 2 =$ ____ $5 + 2 + 2 =$ ____

3. $8 + 0 + 2 =$ ____ $8 + 0 + 2 =$ ____

4. $3 + 3 + 2 =$ ____ $3 + 3 + 2 =$ ____

Talk About It • Critical Thinking

Why are both sums the same?

Practice

Write the sum.

1.

2	6	4	4	5
3	6	5	2	3
+4	+0	+2	+2	+2
9				

2.

2	3	1	3	2
5	4	1	3	1
+1	+3	+6	+4	+6

3.

6	3	2	3	4
1	5	5	2	5
+4	+2	+5	+3	+0

Problem Solving • **Mental Math**

Solve.

4. Tim has 2 dogs.
Steve has 3 cats.
Greg has 6 fish.
How many animals do they have in all?

_____ animals

Home Note **Your child added with three addends.**
ACTIVITY Have your child use small objects to show how he or she found the sums in rows 1 and 2.

Name ______________________________

Practice the Facts

Add.
Complete each table.

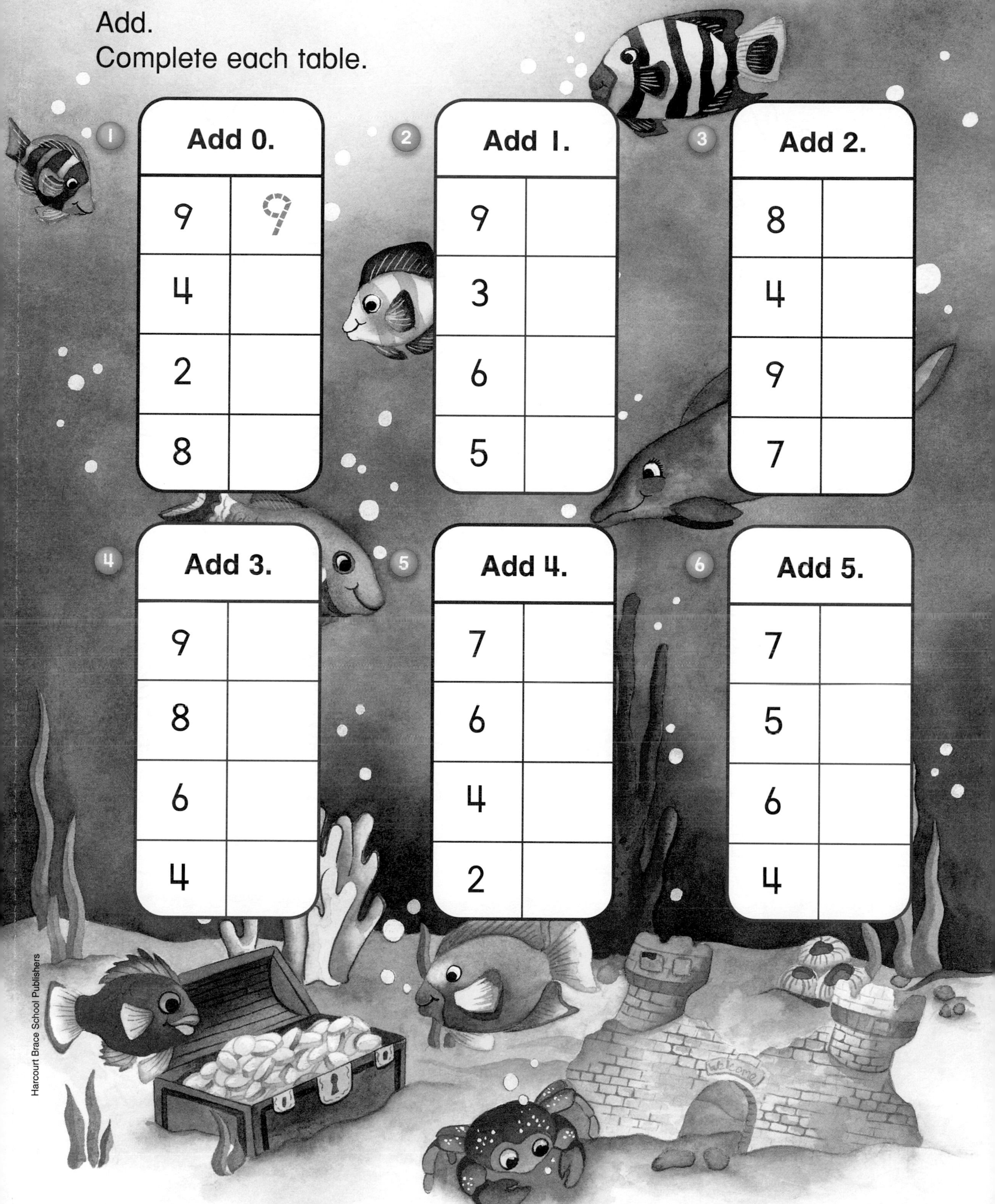

1

Add 0.	
9	9
4	
2	
8	

2

Add 1.	
9	
3	
6	
5	

3

Add 2.	
8	
4	
9	
7	

4

Add 3.	
9	
8	
6	
4	

5

Add 4.	
7	
6	
4	
2	

6

Add 5.	
7	
5	
6	
4	

Practice

Add. Color yellow the animals that have a sum of 10, 11, or 12.

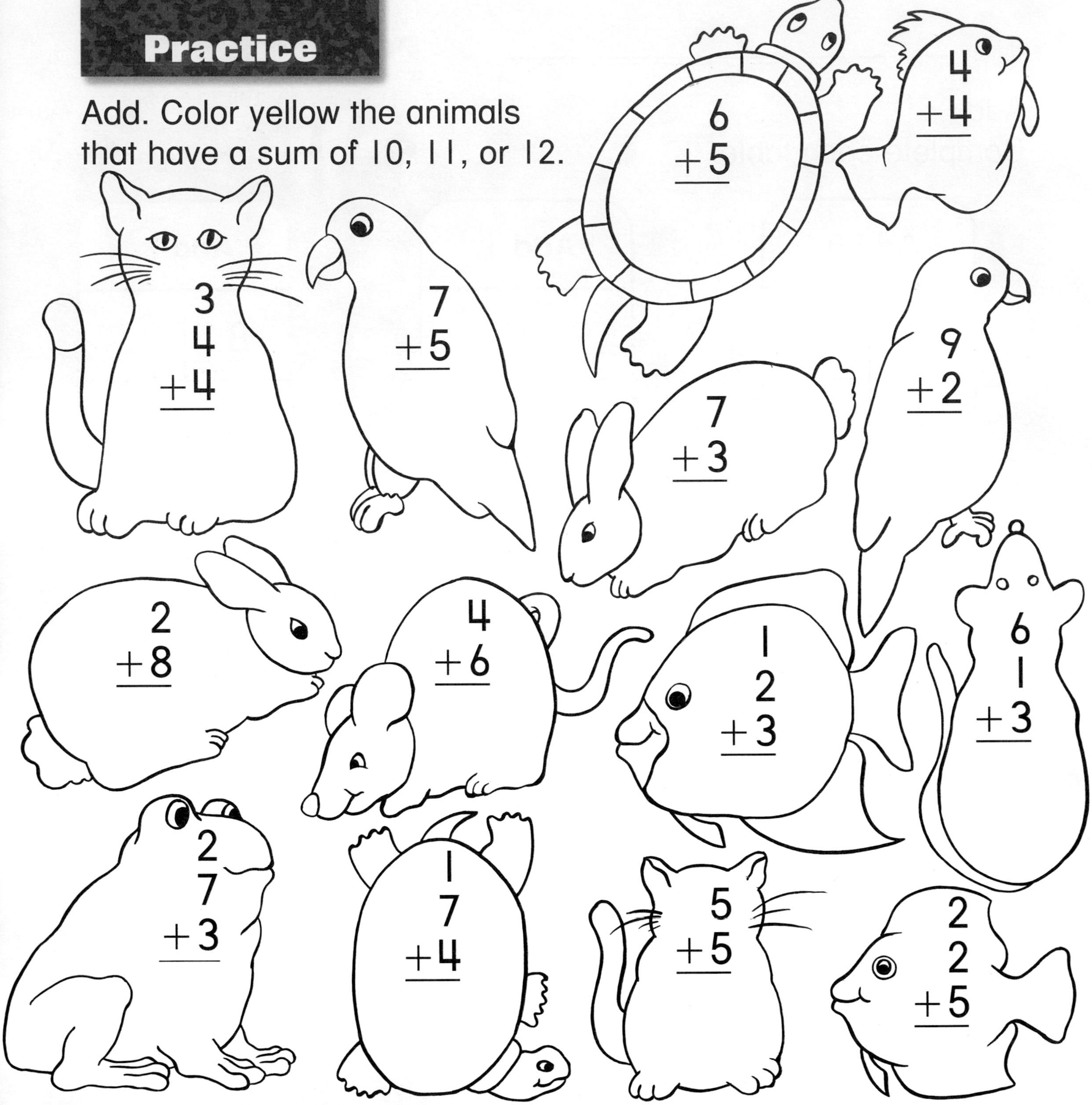

Write About It

Tell a classmate a story about some of the pets on the page. Then write an addition sentence.

Home Note Your child practiced adding numbers with sums to 12.
ACTIVITY Have your child use small objects to show you all the ways to make 12.

Name ______________________

Understand • Plan • Solve • Look Back

Problem Solving
Write a Number Sentence

2 boys feed the rabbit.
1 girl feeds the rabbit.
How many children feed the rabbit?

2 + 1 = 3 is a number sentence.

2 + 1 = 3 children

Write the number sentence.

1. 3 children swing.
   1 more comes to swing.
   How many are swinging?

   ___ + ___ = ___ children

2. 7 girls play ball.
   2 more come to play ball.
   How many are playing ball?

   ___ + ___ = ___ girls

3. 6 boys jump.
   4 more come to jump.
   How many are jumping?

   ___ + ___ = ___ boys

Practice

Write the number sentence.

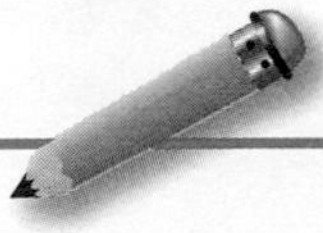

1. 8 cats play in the yard.
3 more come to play.
How many cats are playing?

___ + ___ = ___ cats

2. 6 bugs are on the rock.
6 more come.
How many bugs are on the rock?

___ + ___ = ___ bugs

3. 9 birds are flying.
Then 1 more flies.
How many birds are flying?

___ + ___ = ___ birds

4. 4 dogs sit in the sun.
2 dogs sit in the shade.
How many dogs are sitting?

___ + ___ = ___ dogs

Home Note Your child used the strategy. Write a Number Sentence to solve addition problems.
ACTIVITY Read the addition stories. Have your child point out the number sentence that goes with each.

Name ______________________________

Chapter 11

Review/Test

Concepts and Skills

Circle the greater number.
Count on to add.

1. $\begin{array}{r} 4 \\ +3 \\ \hline \end{array}$ $\quad \begin{array}{r} 5 \\ +2 \\ \hline \end{array}$ $\quad \begin{array}{r} 3 \\ +6 \\ \hline \end{array}$ $\quad \begin{array}{r} 9 \\ +3 \\ \hline \end{array}$ $\quad \begin{array}{r} 2 \\ +7 \\ \hline \end{array}$ $\quad \begin{array}{r} 9 \\ +1 \\ \hline \end{array}$

Write the sums.
Circle the doubles.

2. $4 + 4 =$ ____ $\quad 6 + 4 =$ ____ $\quad 3 + 3 =$ ____

3. $7 + 2 =$ ____ $\quad 6 + 6 =$ ____ $\quad 5 + 5 =$ ____

Write the sum.

4. $\begin{array}{r} 5 \\ 5 \\ +2 \\ \hline \end{array}$ $\quad \begin{array}{r} 2 \\ 3 \\ +3 \\ \hline \end{array}$ $\quad \begin{array}{r} 4 \\ 4 \\ +2 \\ \hline \end{array}$ $\quad \begin{array}{r} 3 \\ 2 \\ +1 \\ \hline \end{array}$ $\quad \begin{array}{r} 7 \\ 1 \\ +2 \\ \hline \end{array}$

Problem Solving

Write the number sentence.

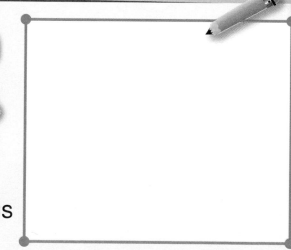

5. 7 dogs were running.
3 dogs were sitting.
How many dogs were there?

____ + ____ = ____ dogs

Name ______________________________

Cumulative Chapters 1-11

TAAS Prep

Mark the best answer.

1. Which number sentence does the story show?

   Eric had 6 pens.
   He gave away 3.
   How many pens does he have left?

   - ⬭ $6 + 3 = 9$
   - ⬭ $6 - 3 = 3$
   - ⬭ $6 - 6 = 0$
   - ⬭ not here

2. How many are there?

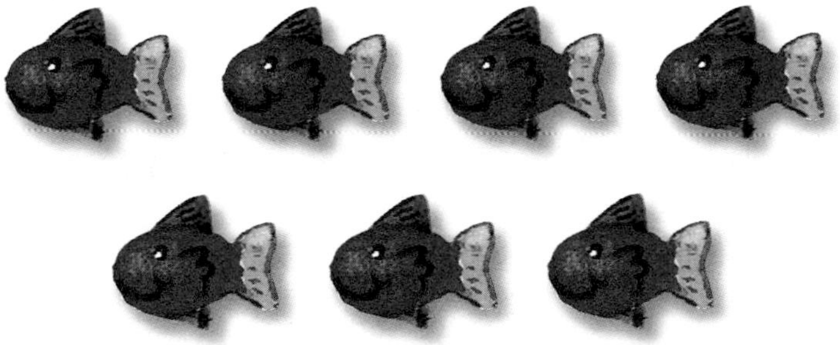

   - ⬭ 5
   - ⬭ 6
   - ⬭ 7
   - ⬭ 8

3. Which number tells how many?

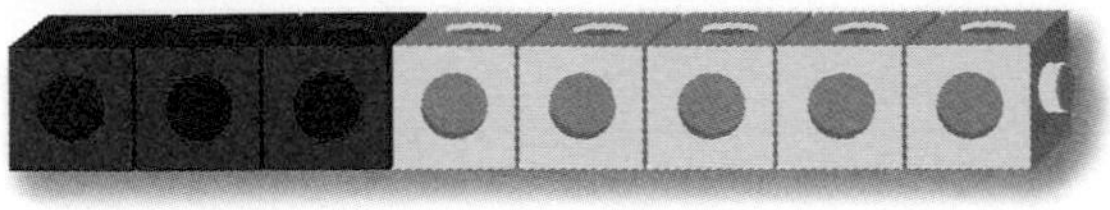

   - ⬭ 7
   - ⬭ 8
   - ⬭ 9
   - ⬭ 10

4. Which number is missing?

   6, 7, ____, 9

   - ⬭ 1
   - ⬭ 5
   - ⬭ 8
   - ⬭ 10

5. Which number is missing?

   3, ____, 5, 6

   - ⬭ 2
   - ⬭ 4
   - ⬭ 7
   - ⬭ 8

CHAPTER 12 Subtraction Facts to 12

Make up subtraction stories using the number 12.

Home Note In this chapter, your child is learning subtraction facts to 12.
ACTIVITY Have your child make up subtraction stories using the pictures on this page.

SCHOOL-HOME CONNECTION

Dear Family,

Today we started Chapter 12. We will review subtracting from 8, 9, and 10 and learn to subtract from 11 and 12. Here are the vocabulary words and an activity for us to do together at home.

Love,

____

Vocabulary

You can subtract by counting back from the greater number.

12 − 3 = 9

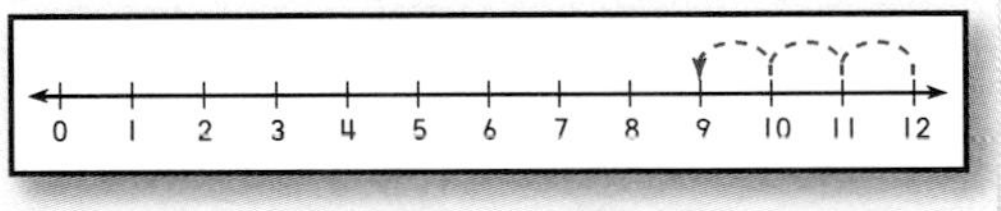

Fact families are addition and subtraction facts that use the same three numbers.

9 + 2 = 11   11 − 2 = 9

2 + 9 = 11   11 − 9 = 2

ACTIVITY

Set out twelve small snack items, such as raisins or strawberries. Have your child eat four, tell what happened, and say the subtraction sentence.

Visit our Web site for additional activities and ideas.
http://www.hbschool.com

Name ______________________________

Relating Addition and Subtraction

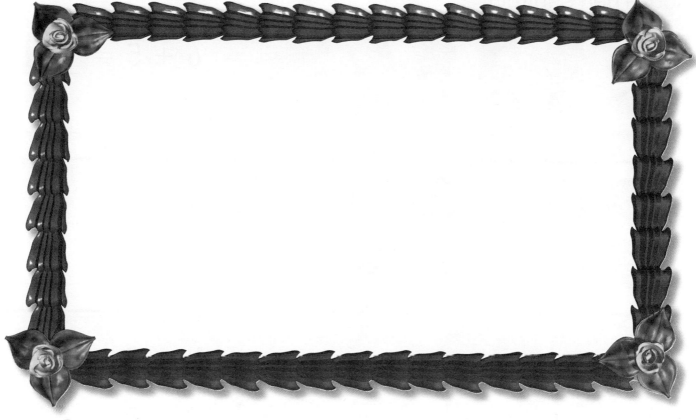

Use ● to model.
Fill in the chart.

	Put in	Put in	Write the addition sentence.	Take away	Write the subtraction sentence.
1	8	3	8 + 3 = 11	3	11 − 3 = 8
2	7	4		4	
3	9	2		2	
4	6	6		6	

Talk About It • Critical Thinking

How are 8 + 3 = 11 and 11 − 3 = 8 related?

Practice

Add. Then subtract.

1 

6 + 5 = 11

11 − 5 = 6

2 

7 + 5 = ____

12 − 5 = ____

3 

4 + 8 = ____

12 − 8 = ____

4 $\begin{array}{r} 9 \\ +\ 3 \\ \hline \end{array}$ $\begin{array}{r} 12 \\ -\ 3 \\ \hline \end{array}$

5 $\begin{array}{r} 6 \\ +\ 4 \\ \hline \end{array}$ $\begin{array}{r} 10 \\ -\ 4 \\ \hline \end{array}$

6 $\begin{array}{r} 8 \\ +\ 3 \\ \hline \end{array}$ $\begin{array}{r} 11 \\ -\ 3 \\ \hline \end{array}$

7 $\begin{array}{r} 7 \\ +\ 4 \\ \hline \end{array}$ $\begin{array}{r} 11 \\ -\ 4 \\ \hline \end{array}$

8 $\begin{array}{r} 9 \\ +\ 2 \\ \hline \end{array}$ $\begin{array}{r} 11 \\ -\ 2 \\ \hline \end{array}$

9 $\begin{array}{r} 8 \\ +\ 4 \\ \hline \end{array}$ $\begin{array}{r} 12 \\ -\ 4 \\ \hline \end{array}$

10 $\begin{array}{r} 9 \\ +\ 3 \\ \hline \end{array}$ $\begin{array}{r} 12 \\ -\ 3 \\ \hline \end{array}$

11 $\begin{array}{r} 6 \\ +\ 6 \\ \hline \end{array}$ $\begin{array}{r} 12 \\ -\ 6 \\ \hline \end{array}$

12 $\begin{array}{r} 7 \\ +\ 3 \\ \hline \end{array}$ $\begin{array}{r} 10 \\ -\ 3 \\ \hline \end{array}$

Home Note **Your child found related sums and differences.**
ACTIVITY Have your child use small objects to model the facts he or she missed.

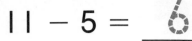

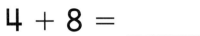

Name ______________________

Counting Back

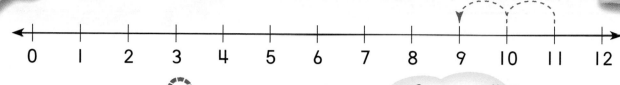

0 1 2 3 4 5 6 7 8 9 10 11 12

11 − 2 = 9

Start at 11.
Count back 2.

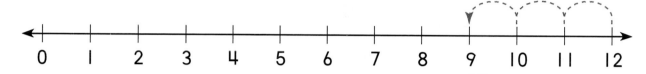

0 1 2 3 4 5 6 7 8 9 10 11 12

12 − 3 = 9

Start at 12.
Count back 3.

Use the number line to count back.

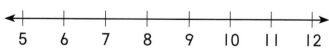

1. 5 6 7 8 9 10 11 12

   10 − 3 = ____

2. 5 6 7 8 9 10 11 12

   11 − 3 = ____

3. 5 6 7 8 9 10 11 12

   10 − 2 = ____

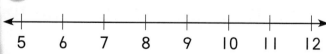

4. 5 6 7 8 9 10 11 12

   9 − 2 = ____

5. 5 6 7 8 9 10 11 12

   9 − 1 = ____

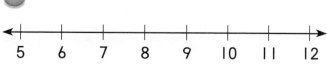

6. 5 6 7 8 9 10 11 12

   9 − 3 = ____

Practice

Count back to subtract.
Use the number line if you need it.

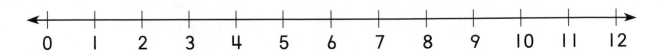

0 1 2 3 4 5 6 7 8 9 10 11 12

1

10	12	10	9	7	8
− 2	− 3	− 3	− 2	− 2	− 2
8					

2

12	11	9	8	8	7
− 3	− 2	− 3	− 1	− 3	− 1

3

10	6	9	7	6	11
− 1	− 2	− 1	− 3	− 3	− 3

Mixed Review

Color the circles to continue the pattern.

4

5

Home Note Your child used a number line to count back and subtract.
ACTIVITY Have your child use the number line on this page to practice any subtraction facts he or she missed in this lesson.

Name ____________________

Compare to Subtract

Compare. Then subtract.

1. How many more apples than oranges are there?

$$\begin{array}{r} 11 \\ -\ 9 \\ \hline 2 \end{array}$$

2. How many more plums than peaches are there?

$$\begin{array}{r} 11 \\ -\ 7 \\ \hline \end{array}$$

3. How many fewer pears than strawberries are there?

$$\begin{array}{r} 11 \\ -\ 8 \\ \hline \end{array}$$

4. How many fewer cherries than bananas are there?

$$\begin{array}{r} 12 \\ -\ 9 \\ \hline \end{array}$$

Practice

Compare. Then subtract.

1 

10 − 8 = 2

2 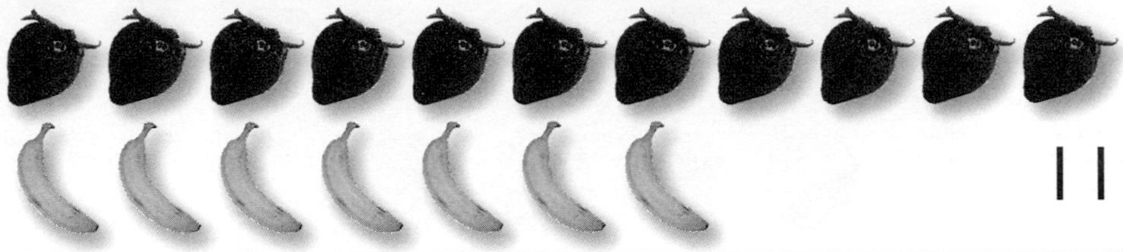

11 − 7 = ____

3 

11 − 9 = ____

4 

12 − 9 = ____

5 

12 − 8 = ____

Problem Solving

Draw a picture to show your answer.

6 Margie has 11 plums.
Jake has 9 plums.
How many more plums does Margie have?

____ more plums

Home Note **Your child compared groups to subtract.**
ACTIVITY **Set out two groups of objects, one with more objects than the other. Have your child show how to use subtraction to find out how many more one group has.**

Name ______________________________

Fact Families

Use  . Add and subtract.
Write the numbers in each fact family.

8 + 3 = 11

3 + 8 = 11

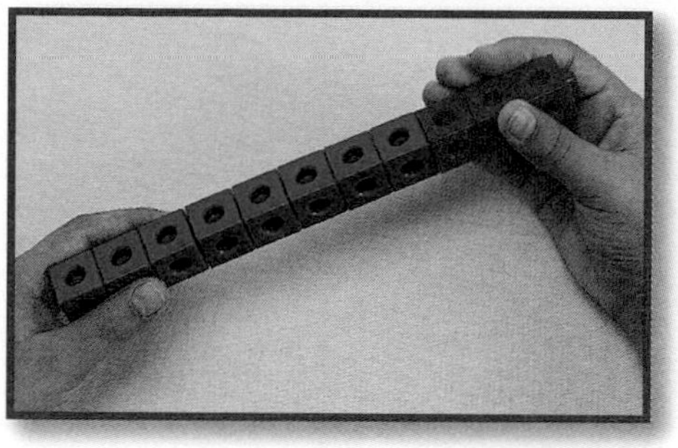

| 11 | 8 | 3 |

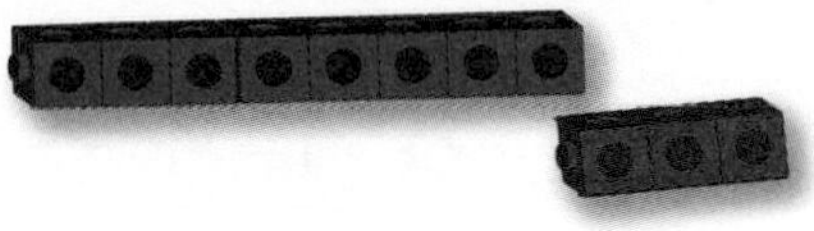

11 − 3 = 8

11 − 8 = 3

1

7 + 4 = ____
4 + 7 = ____
11 − 4 = ____
11 − 7 = ____

☐ ☐ ☐

2

9 + 2 = ____
2 + 9 = ____
11 − 2 = ____
11 − 9 = ____

☐ ☐ ☐

3

6 + 5 = ____
5 + 6 = ____
11 − 5 = ____
11 − 6 = ____

☐ ☐ ☐

4

8 + 4 = ____
4 + 8 = ____
12 − 4 = ____
12 − 8 = ____

☐ ☐ ☐

5

9 + 3 = ____
3 + 9 = ____
12 − 3 = ____
12 − 9 = ____

☐ ☐ ☐

6

7 + 5 = ____
5 + 7 = ____
12 − 5 = ____
12 − 7 = ____

☐ ☐ ☐

Talk About It • Critical Thinking

Which fact families have only 2 facts?

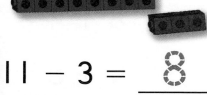

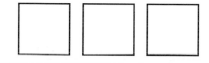

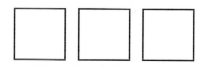

Practice

Add and subtract.
Write the numbers in each fact family.

1.
$8 + 4 = $ 12
$4 + 8 = $ 12
$12 - 4 = $ 8
$12 - 8 = $ 4
□ □ □

2.
$7 + 3 = $ ____
$3 + 7 = $ ____
$10 - 3 = $ ____
$10 - 7 = $ ____
□ □ □

3.
$6 + 4 = $ ____
$4 + 6 = $ ____
$10 - 4 = $ ____
$10 - 6 = $ ____
□ □ □

4.
$4 + 5 = $ ____
$5 + 4 = $ ____
$9 - 5 = $ ____
$9 - 4 = $ ____
□ □ □

Write About It

5. Tell a classmate a story that uses these numbers.
8 3 11
Then write all the facts for the fact family.

Home Note **Your child learned about fact families.**
ACTIVITY Have your child write some fact families.

Name ______________________

Understand • Plan • Solve • Look Back

Problem Solving
Write a Number Sentence

Write the number sentence the story shows.

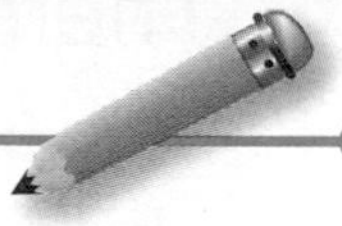

1. There were 12 apples.
The children ate 7 of them.
How many apples are left?

5 apples

12 (−) 7 = 5

2. There are 9 green apples.
There are 3 red apples. How many apples are there in all?

______ apples

___ ◯ ___ = ___

3. I had 11 pennies.
I lost 8 of them. How many pennies do I have left?

______ pennies

___ ◯ ___ = ___

4. There are 12 girls.
There are 9 boys. How many more girls than boys are there?

______ more girls

___ ◯ ___ = ___

5. I had 4 pennies.
I found 7 more pennies. How many pennies do I have now?

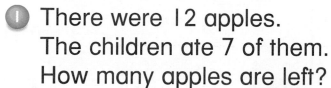

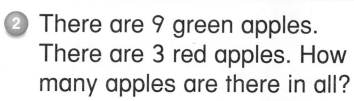

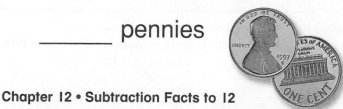

______ pennies

___ ◯ ___ = ___

Practice

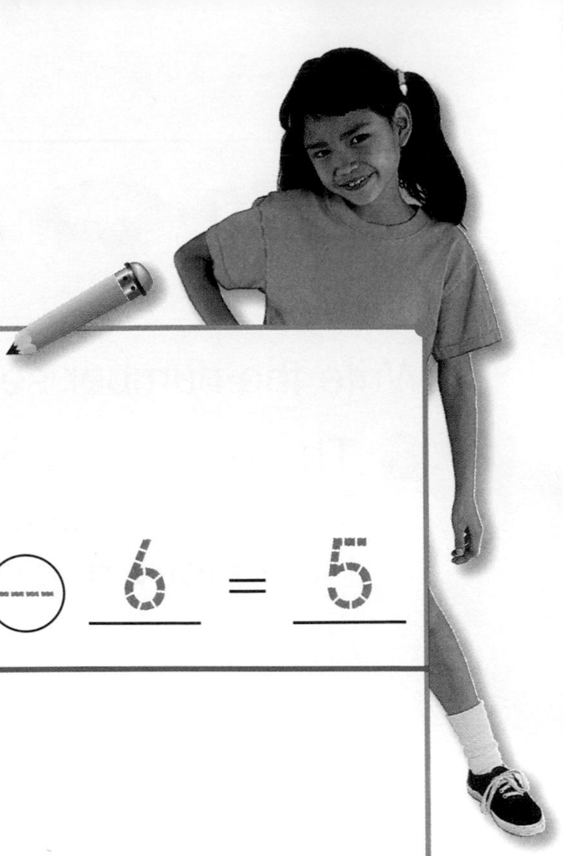

Write the number sentence the story shows.

1. There are 11 children. There are 6 balls. How many more children than balls are there?

   5 more children

   11 (−) 6 = 5

2. There are 6 girls singing. There are 5 boys singing. How many children in all are singing?

   ____ children

   ____ ◯ ____ = ____

3. I had 12 pennies. I lost 6 of them. How many pennies do I have left?

   ____ pennies

   ____ ◯ ____ = ____

4. There are 11 boys. There are 8 girls. How many more boys than girls are there?

   ____ more boys

   ____ ◯ ____ = ____

5. I had 7 pennies. I found 5 more pennies. How many pennies do I have now?

   ____ pennies

   ____ ◯ ____ = ____

Home Note Your child solved problems by writing number sentences.
ACTIVITY Have your child use 12 pennies to model some subtraction story problems. Then have him or her write the number sentence the story problem shows.

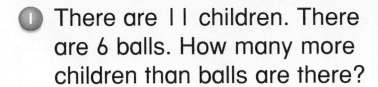

Name ____________________

Concepts and Skills

Add. Then subtract.

1 

$6 + 5 =$ ____

$11 - 5 =$ ____

Count back to subtract.

2 

$12 - 3 =$ ____

Compare. Then subtract.

3 

$11 - 7 =$ ____

Add or subtract.
Write the numbers for each fact family.

4
$8 + 4 =$ ____ ☐
$4 + 8 =$ ____ ☐
$12 - 4 =$ ____ ☐
$12 - 8 =$ ____ ☐

5
$3 + 9 =$ ____ ☐
$9 + 3 =$ ____ ☐
$12 - 9 =$ ____ ☐
$12 - 3 =$ ____ ☐

Problem Solving

Write the number sentence the story shows.

6 There were 11 oranges.
The children ate 7 of them.
How many oranges are left?

____ oranges

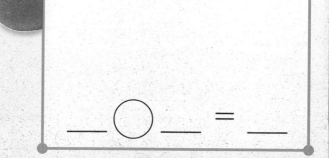

___ ◯ ___ = ___

Name ______________________________

TAAS Prep

Mark the best answer.

Cumulative Chapters 1–12

1. Look at the picture.
Which subtraction sentence tells how many are left?

- ⬭ 5 + 3 = 8
- ⬭ 5 − 3 = 2
- ⬭ 8 − 5 = 3
- ⬭ 8 − 3 = 5

2. Which number sentence tells how many are left?
Erica has 6 pennies.
She gives 4 pennies to Gil.

- ⬭ 4 + 6 = 10
- ⬭ 6 + 4 = 10
- ⬭ 10 − 4 = 6
- ⬭ 6 − 4 = 2

3. Which number is missing?

7, 8, ____, 10

- ⬭ 6
- ⬭ 7
- ⬭ 8
- ⬭ 9

4. How many sides?

- ⬭ 4
- ⬭ 3
- ⬭ 2
- ⬭ 8

5. Which shape does not belong?

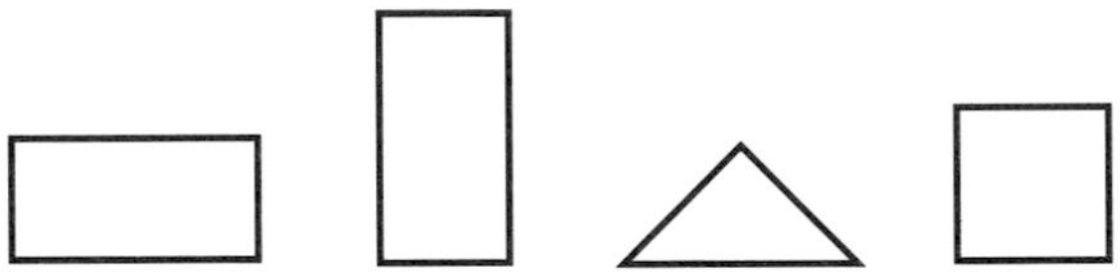

- ⬭ 
- ⬭ 
- ⬭ 
- ⬭ 

6. Which shape comes next?

- ⬭ 
- ⬭ 
- ⬭ 
- ⬭ 

Name ______________________________

MATH FUN

Use: dot cube, counter, game piece

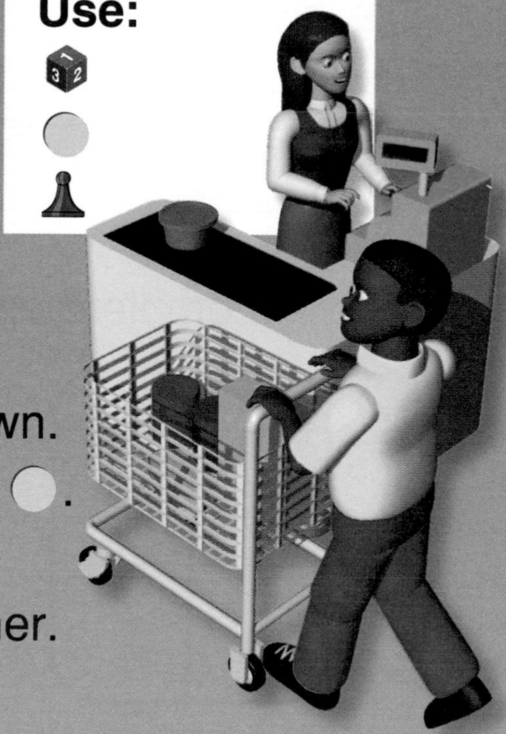

Shopping for Sums and Differences

1. Roll the dot cube.
2. Move your game piece the number of spaces shown.
3. Add or subtract. Check your answer with a counter.
4. Take turns with your partner.
5. The first person to reach END is the winner.

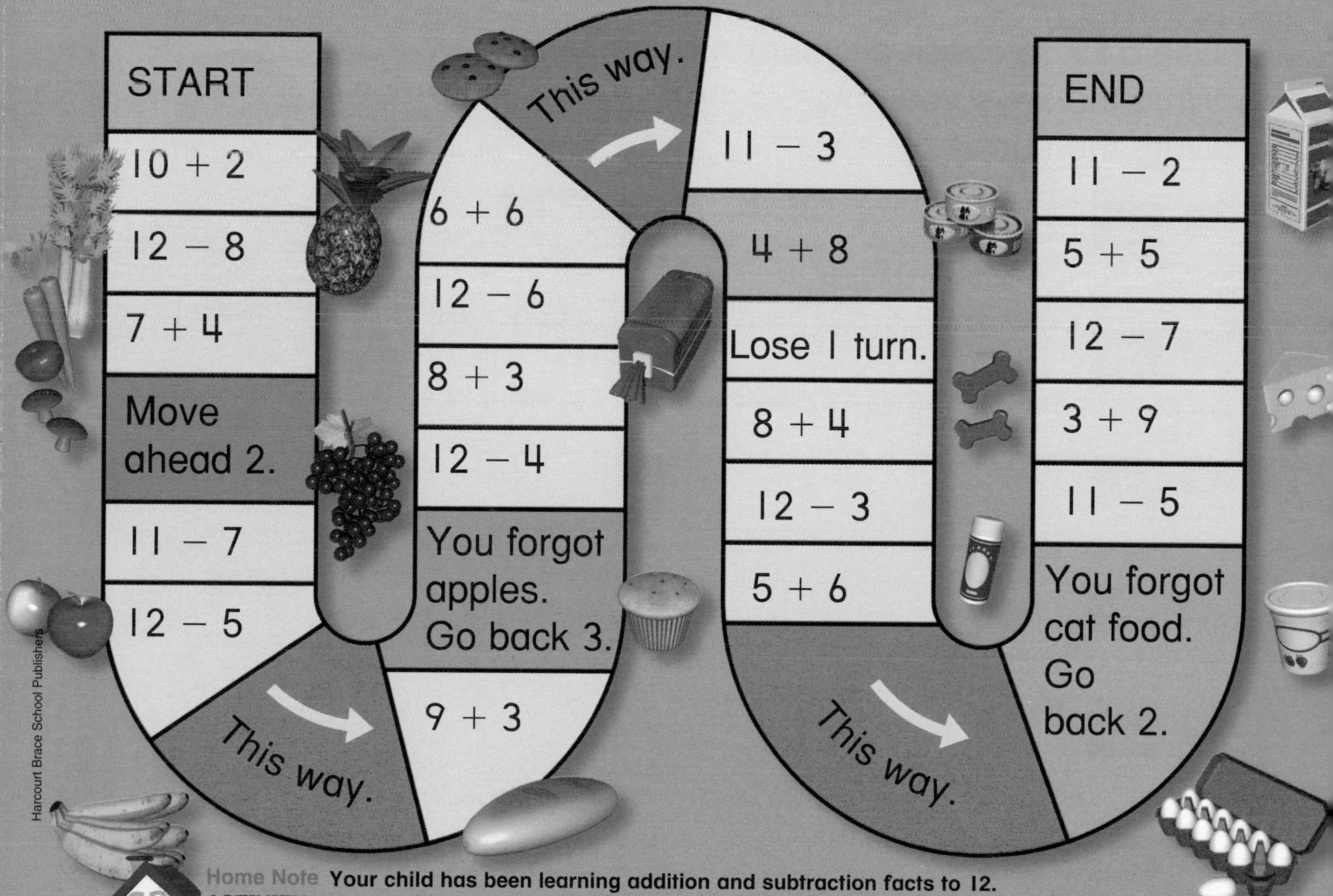

Home Note Your child has been learning addition and subtraction facts to 12.
ACTIVITY Play this game at home and have your child count back, count on, and use doubles and fact families to add or subtract.

Name ______________________________

Calculator | Computer

Add to complete each table.

1

Add 3	
6	9
8	

2

Add 4	
5	
8	

3

Add 5	
6	
7	

Use a calculator to check all your answers.
Write the keys you press.
Write the sum.

4	ON/C	6	+	3	=	9
5	ON/C				=	
6	ON/C				=	
7	ON/C				=	
8	ON/C				=	
9	ON/C				=	

The Animals' Picnic

BY DAVID McPHAIL

This book will help me review doubles.

This book belongs to ________________.

A

All the animals were having a picnic.

One mouse invited one mouse.

They went by bike.

B

Two sheep invited two sheep.

They went by car.

Three rabbits invited three rabbits.

They went by wheelbarrow.

D

Four chickens invited four chickens.

They went by plane.

Five ducks invited five ducks.

They went by canoe.

Six ants invited six ants.

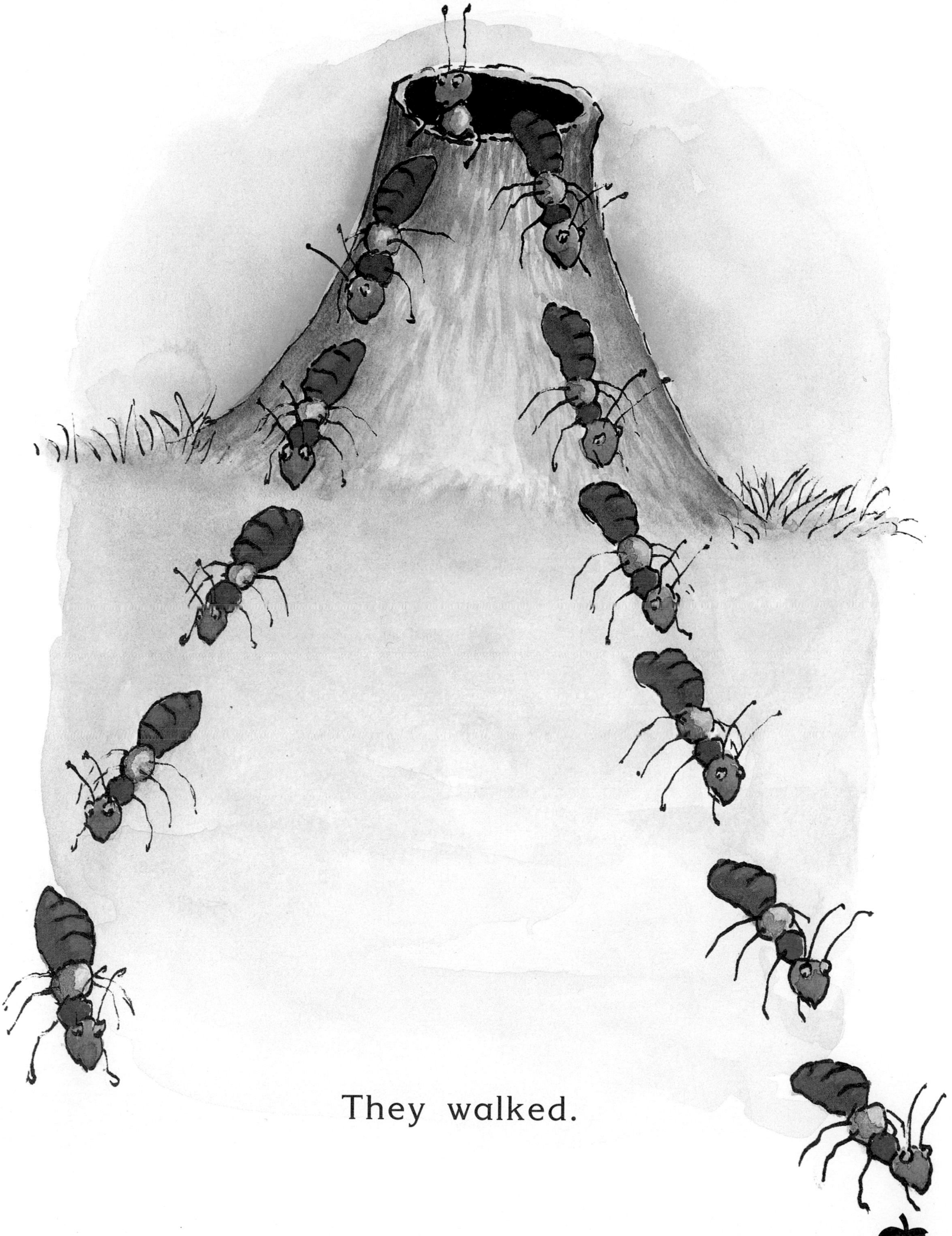

They walked.

The ants got there first.

Name ______________________________

Concepts and Skills

Circle the greater number.
Count on to add.

1. 9 + 1 = ____
2. 2 + 7 = ____
3. 3 + 8 = ____
4. 6 + 3 = ____
5. 5 + 2 = ____
6. 3 + 9 = ____

Write the sums.
Circle the doubles.

7. 5 + 5 = ____
8. 6 + 1 = ____
9. 3 + 3 = ____
10. 9 + 2 = ____

Write the sum.

11. 3 + 3 + 1 = ____
12. 2 + 4 + 4 = ____
13. 5 + 5 + 2 = ____
14. 3 + 2 + 1 = ____
15. 4 + 3 + 2 = ____

Add.
Complete each table.

16.

Add 2.	
6	
8	
9	

17.

Add 3.	
4	
9	
7	

18.

Add 4.	
5	
6	
8	

Add. Then subtract.

19 

8 + 4 = _____

12 − 4 = _____

Subtract.

20 Count back.

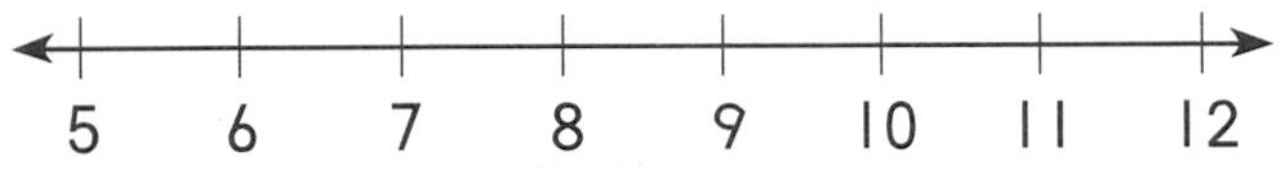

11 − 3 = _____

21 Compare.

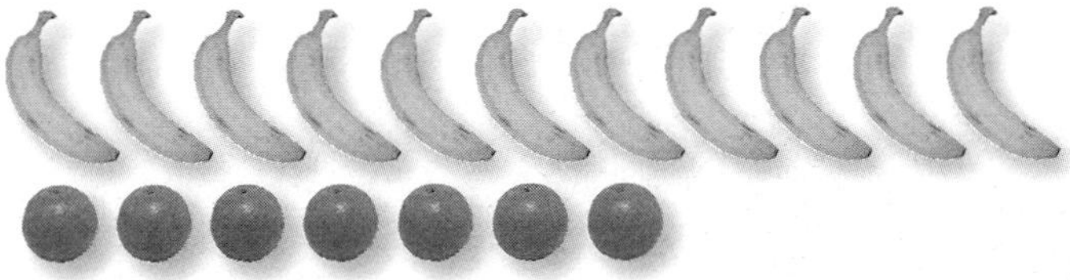

11 − 7 = _____

Add or subtract.

22 6 + 5 = _____ 11 − 5 = _____

5 + 6 = _____ 11 − 6 = _____

Problem Solving

Write the number sentence.

23 I had 12 pennies.
I lost 5 of them.
How many pennies
do I have left?

_____ − _____ = _____

_____ pennies

24 There are 11 boys.
There are 9 girls.
How many more boys
than girls are there?

_____ − _____ = _____

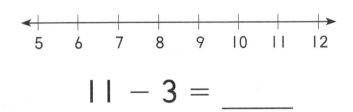

_____ more boys

Name ____________________

Performance Assessment

Use Workmat 2 and 11 ●.

1. Put ● on one side.
Put ○ on the other side.
Draw and color the counters.

Write an addition fact.

___ + ___ = 11

Write a related subtraction fact.

11 − ___ = ___

Now use 12 counters to show related facts with 12.

2. ___ + ___ = 12   12 − ___ = ___

3. ___ + ___ = 12   12 − ___ = ___

Write About It

4. Tell a classmate an addition story and a subtraction story for the related facts.
Draw pictures of the stories.

4 + 7 = 11	11 − 7 = 4

Name ________________________

Chapters 1-12 Cumulative Review

Fill in the ⬭ for the correct answer.

1 Find the sum.

$2 + 3 =$ ____

⬭ 3
⬭ 4
⬭ 5
⬭ 6

2 Add or subtract.
Use ●.

4 birds are in a tree.
2 fly away.
How many are left?

⬭ 2
⬭ 3
⬭ 4
⬭ 6

3

$5 + 3 =$ ____

⬭ 2
⬭ 7
⬭ 8
⬭ 9

4

$$\begin{array}{r} 5 \\ +3 \\ \hline \end{array}$$

⬭ 2
⬭ 7
⬭ 9
⬭ 8

5

$$\begin{array}{r} 7 \\ +4 \\ \hline \end{array}$$

⬭ 9
⬭ 10
⬭ 11
⬭ 12

6

$$\begin{array}{r} 6 \\ +6 \\ \hline \end{array}$$

⬭ 6
⬭ 10
⬭ 11
⬭ 12

7

$$\begin{array}{r} 12 \\ -3 \\ \hline \end{array}$$

⬭ 7
⬭ 8
⬭ 9
⬭ 10

8

$$\begin{array}{r} 11 \\ -8 \\ \hline \end{array}$$

⬭ 2
⬭ 3
⬭ 4
⬭ 5

9 Which is the same shape as this box?

⬭ 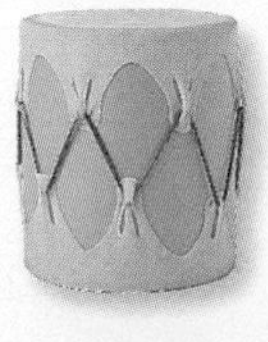

⬭ 

⬭ 

⬭ 

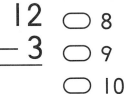

10 Which figure rolls?

⬭ 

⬭  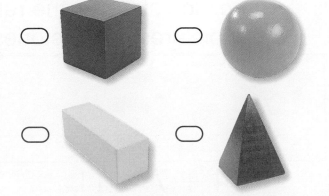

⬭ 

⬭ 

CHAPTER 13

Building Numbers to 100

How many beans are in each group?
How can you find out how many beans in all?

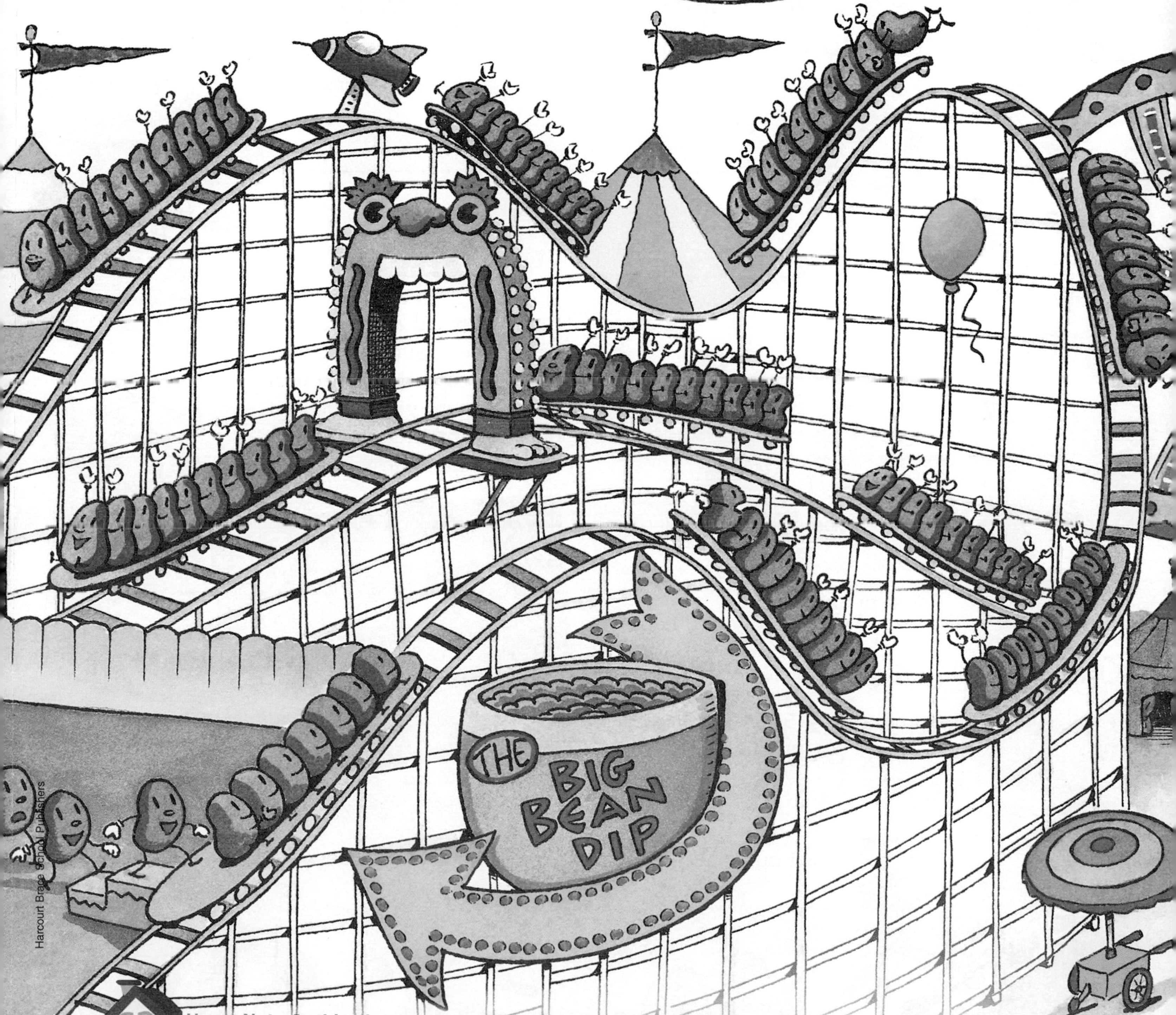

Home Note In this chapter, your child will learn about numbers to 100.
ACTIVITY Have your child start at any number and count by ones or tens to 100.

SCHOOL-HOME CONNECTION

Dear Family,

Today we started chapter 13. We will count, read, and write numbers to 100. Here are the new vocabulary words and an activity for us to do together at home.

Love,

____

Vocabulary

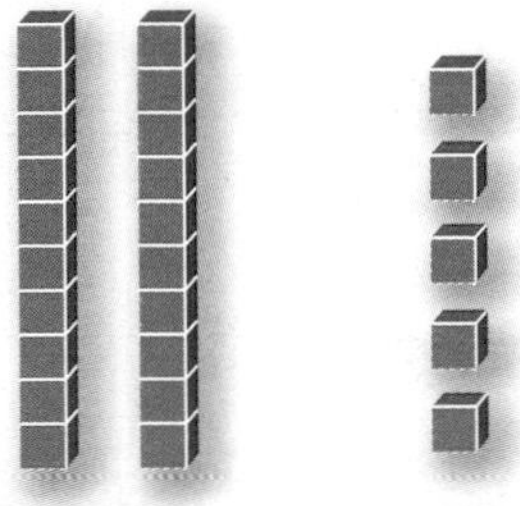

2 **tens** 5 **ones**

Count by **tens** to 60.

10, 20, 30, 40, 50, 60

Count by **ones** to 10.

1, 2, 3, 4, 5, 6, 7, 8, 9, 10

ACTIVITY

Give your child a handful of small objects such as pennies or paper clips. Have him or her form as many groups of 10 as possible and then name and write the number.

Visit our Web site for additional ideas and activities.
http://www.hbschool.com

Name ______________________

Tens

10

1 ten

10 objects in a group make 1 ten.

Use .
Put in one to show each ten. Draw.
Write how many tens.
Write the number.

1 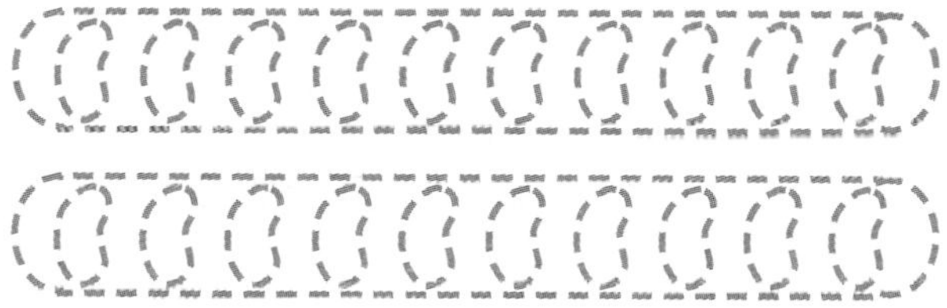

2 tens = 20
twenty

2

____ tens = ____
thirty

3

____ tens = ____
forty

Talk About It • Critical Thinking

How many ones are there in 50?
How do you know?

Practice

Write how many tens.
Write the number.

1 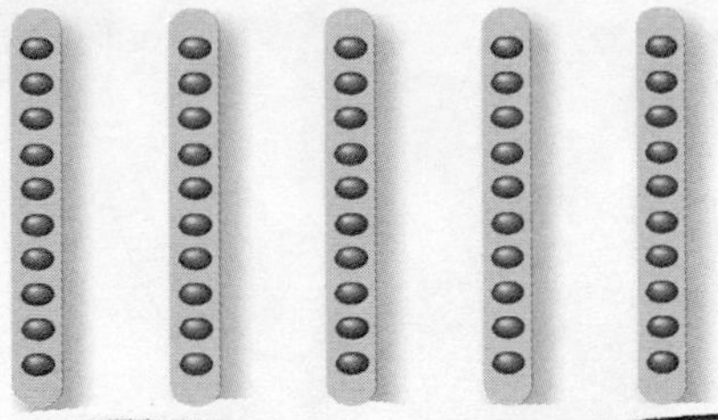

5 tens = 50
fifty

2 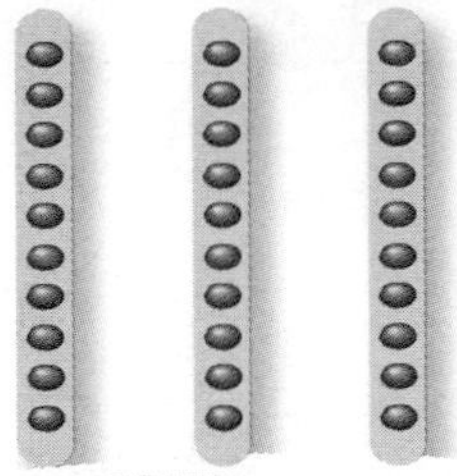

_____ tens = _____
thirty

3 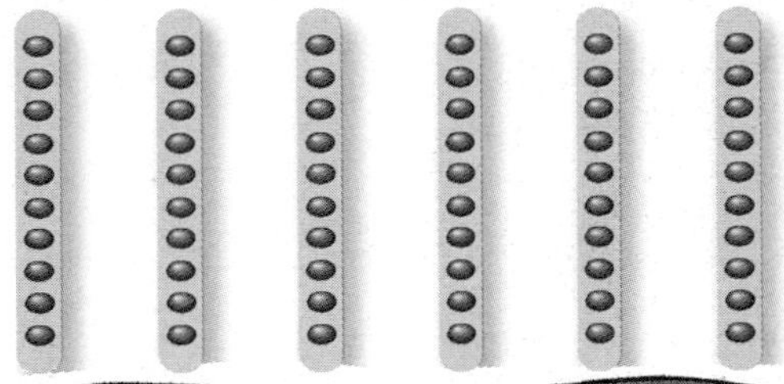

_____ tens = _____
sixty

4 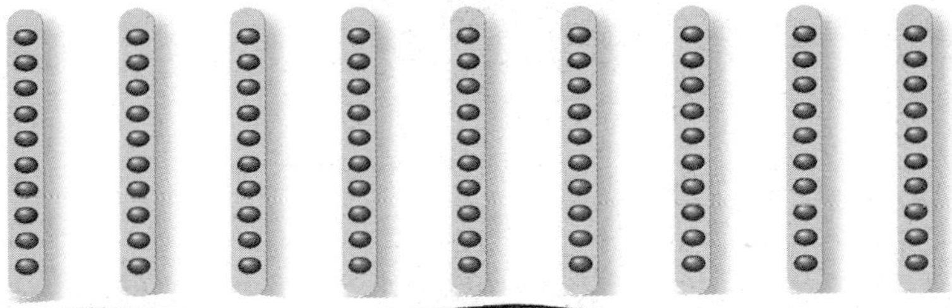

_____ tens = _____
ninety

5 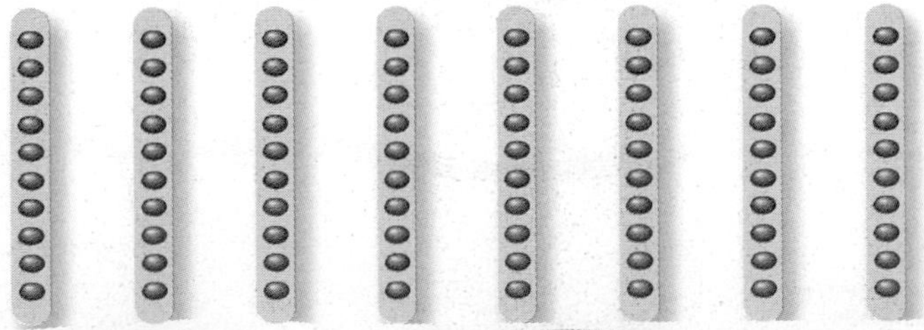

_____ tens = _____
eighty

6 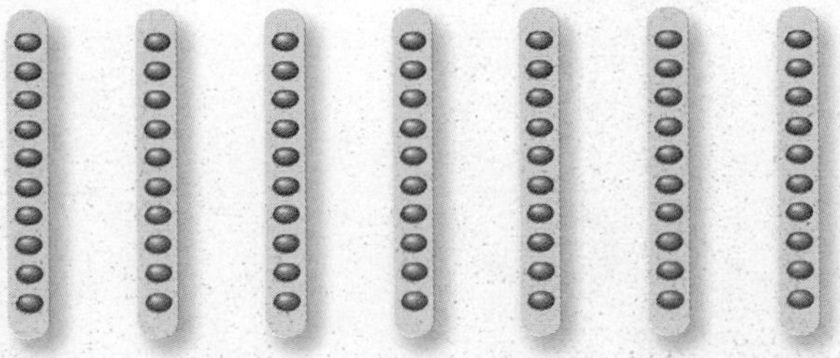

_____ tens = _____
seventy

Home Note Your child modeled groups of tens and counted by tens to 90.
ACTIVITY Have your child group objects in tens and tell how many in all.

Name ______________________________

Tens and Ones to 20

1 ten 0 ones = 10

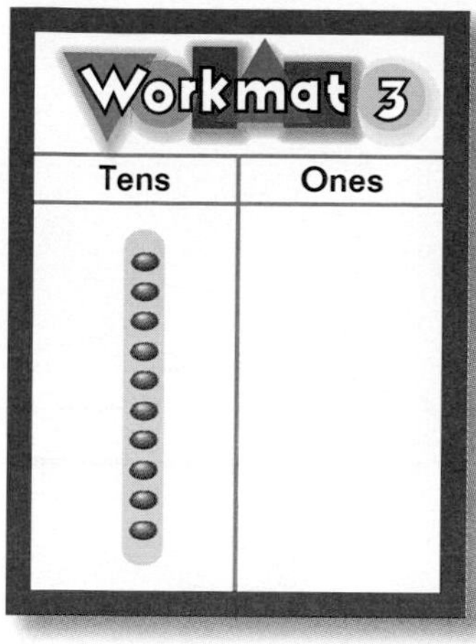

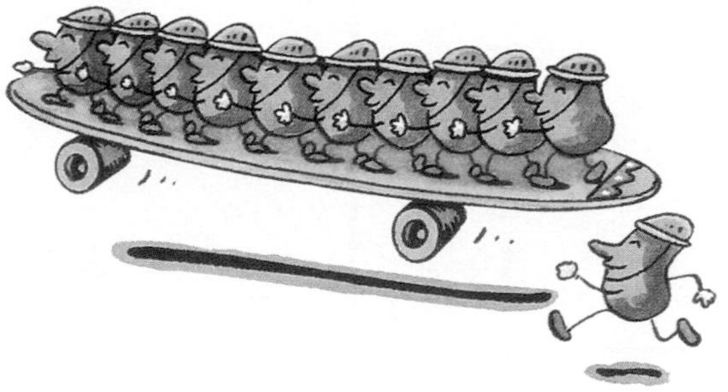

1 ten 1 one = 11

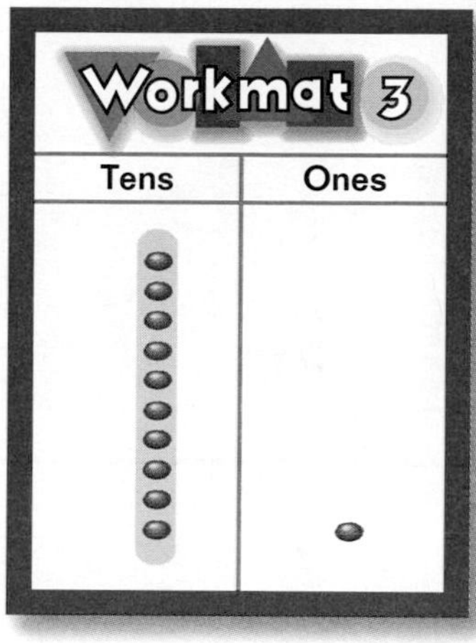

Use Workmat 3 and . Show each model.
Write how many tens and ones. Write the number.

1. 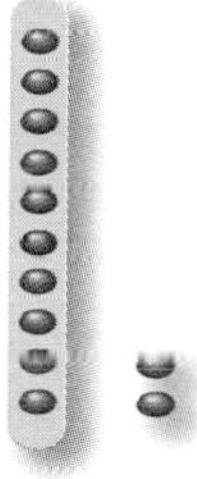

1 ten 2 ones = 12

2. 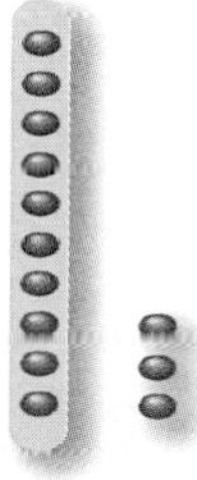

____ ten ____ ones = ____

3. 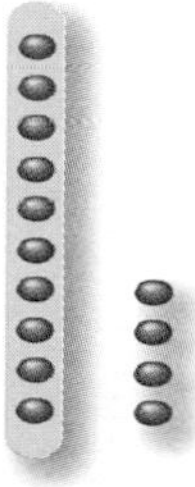

____ ten ____ ones = ____

4. 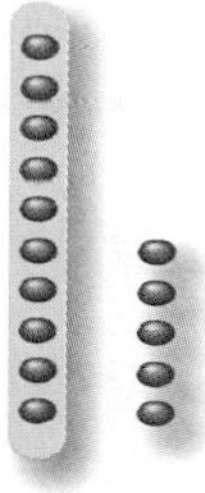

____ ten ____ ones = ____

5. 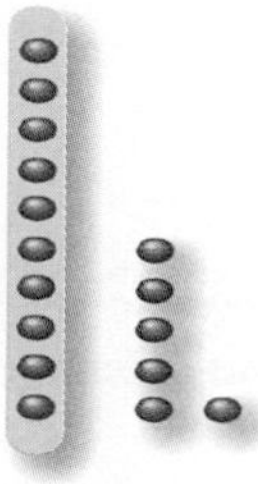

____ ten ____ ones = ____

6. 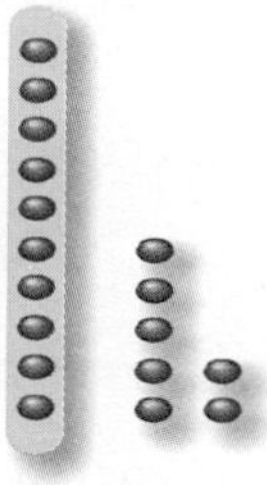

____ ten ____ ones = ____

Practice

Use Workmat 3 and . Show each model.
Write how many tens and ones. Write the number.

1. 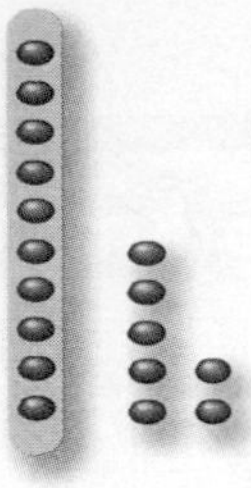

1 ten 7 ones = 17

2. 

____ ten ____ ones = ____

3. 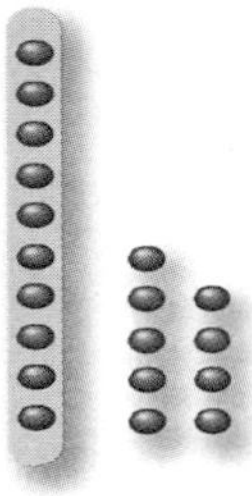

____ ten ____ ones = ____

4. 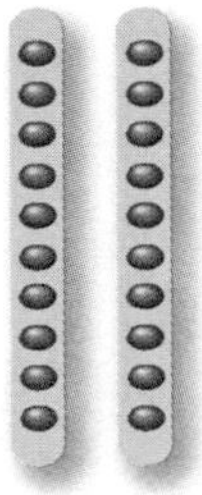

____ tens ____ ones = ____

5. 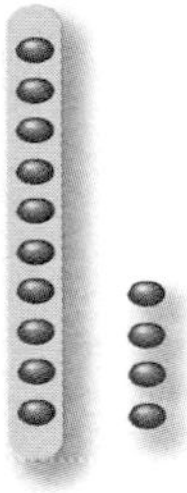

____ ten ____ ones = ____

6. 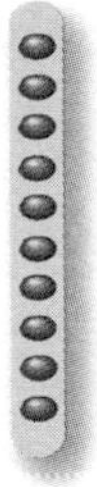

____ ten ____ ones = ____

Problem Solving • **Visual Thinking**

Solve. Then draw a picture.

7. Jane has 10 beans in a bag.
She puts in 4 more beans.
How many beans are in the bag?

____ beans

Home Note Your child modeled and wrote numbers to 20.
ACTIVITY Have your child use small objects to show numbers between 10 and 20. Ask him or her how many tens and ones are in each group.

Name ____________________

Tens and Ones to 50

2 tens 4 ones = 24

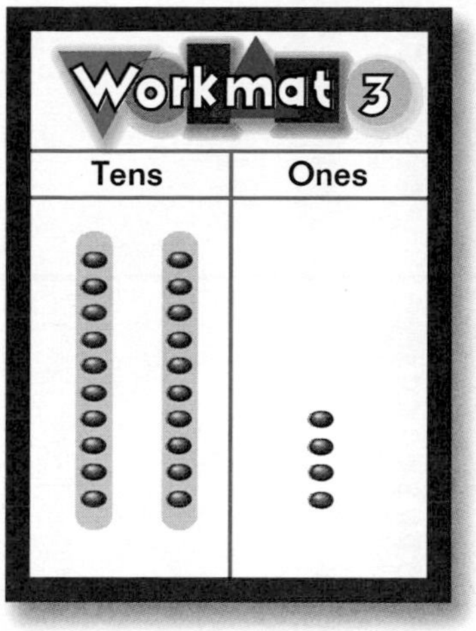

4 tens 2 ones = 42

Use Workmat 3 and to show each model.
Write how many tens and ones. Write the number.

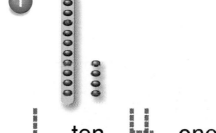

1. 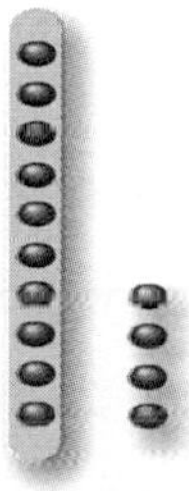

1 ten 4 ones = 14

2. 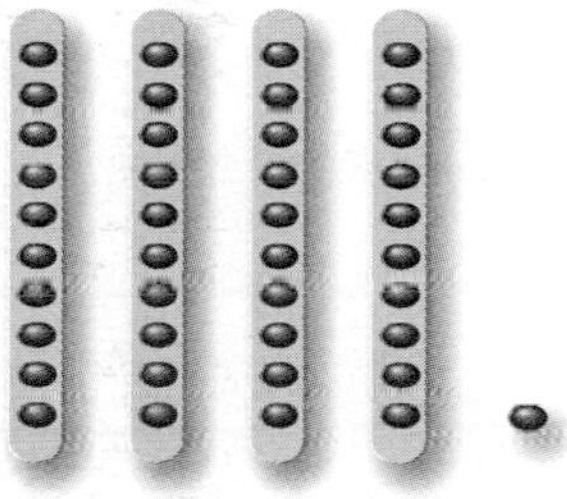

____ tens ____ one = ____

3. 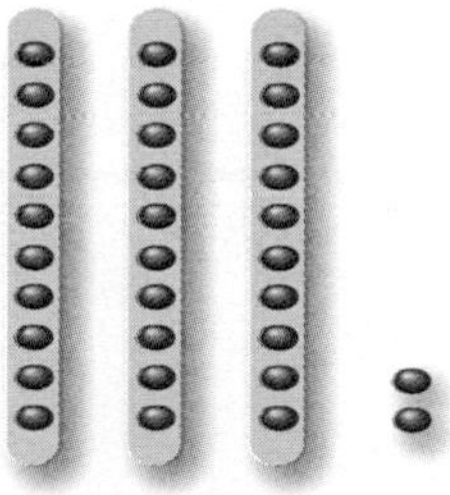

____ tens ____ ones = ____

4. 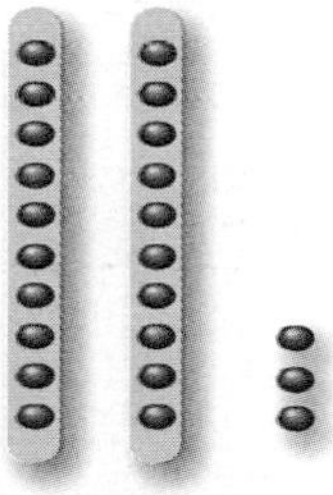

____ tens ____ ones = ____

5. 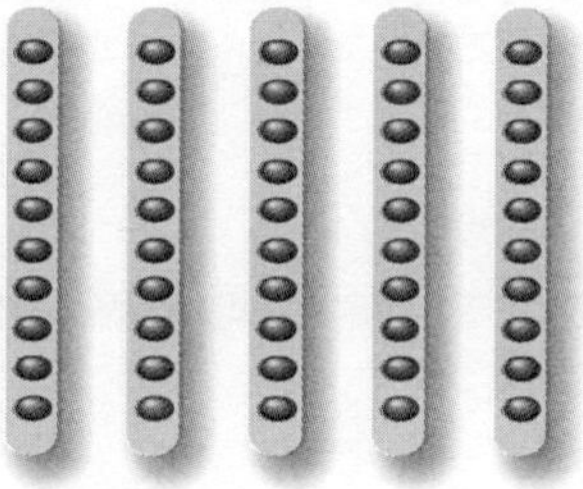

____ tens ____ ones = ____

6. 

____ tens ____ ones = ____

Practice

Write the number.

1. 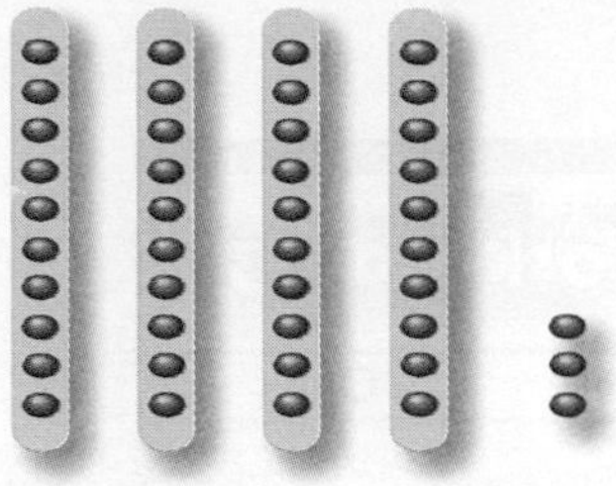 43

2. 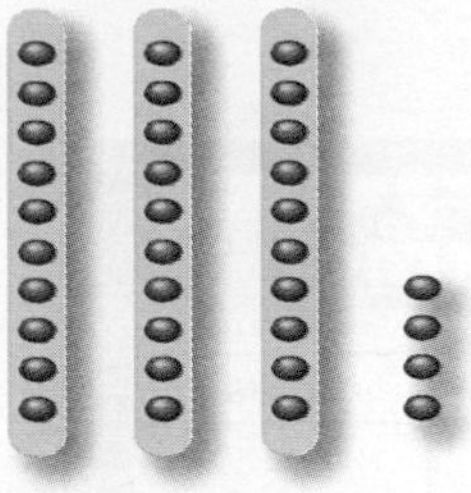 ____

3. 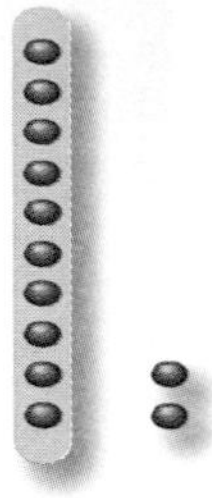 ____

4. 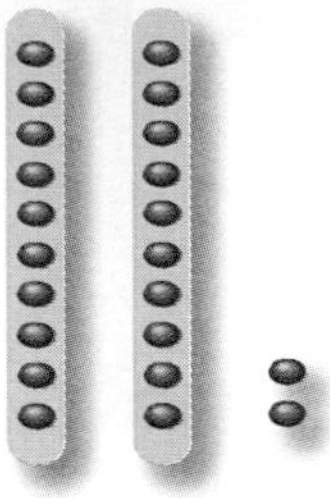 ____

5. 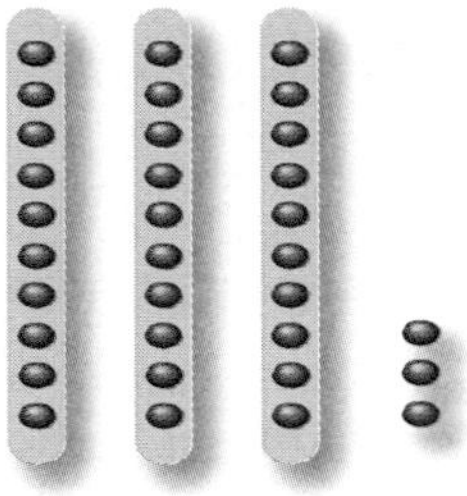 ____

6. 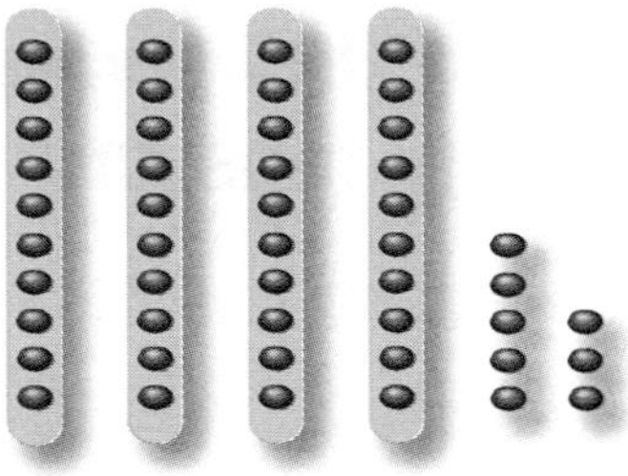 ____

7.  ____

8. 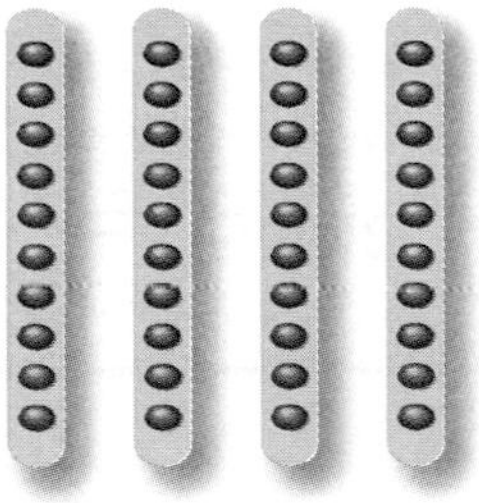 ____

Mixed Review

Add or subtract.

9.

2	7	5	8	6	4
+ 6	− 2	+ 5	− 6	+ 3	− 0

Home Note Your child modeled and wrote numbers to 50.
ACTIVITY Set out 50 pennies. Have your child take some, put them in stacks of ten plus extras, and tell how many he or she took.

Name ______________________________

Tens and Ones to 80

Use Workmat 3 and ■ to show each model.
Write how many tens and ones.
Write the number.

1. 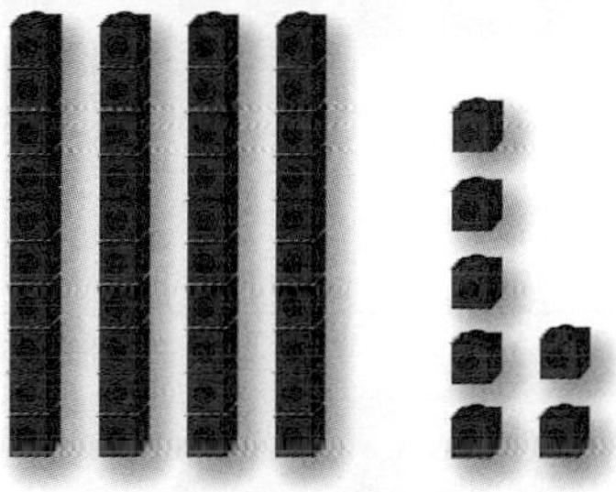

4 tens 7 ones = 47

2. 

____ tens ____ ones = ____

3. 

____ tens ____ ones = ____

4. 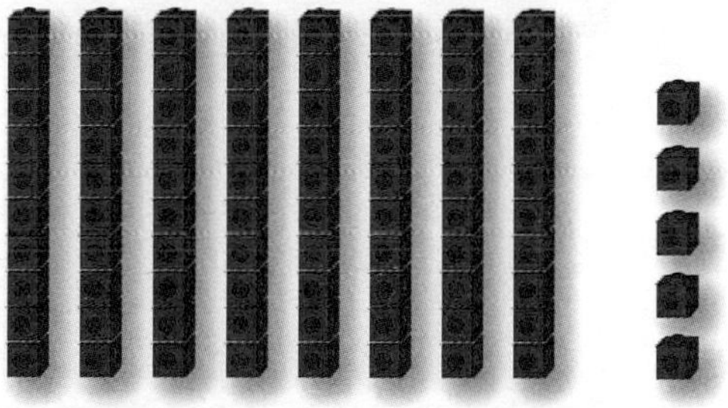

____ tens ____ ones = ____

5. 

____ tens ____ ones = ____

6. 

____ tens ____ ones = ____

Practice

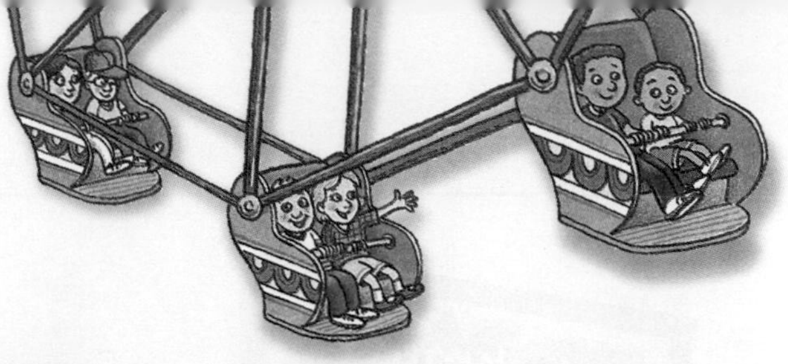

Write the number.

1. 

62

2. 

____

3. 

____

4. 

____

5. 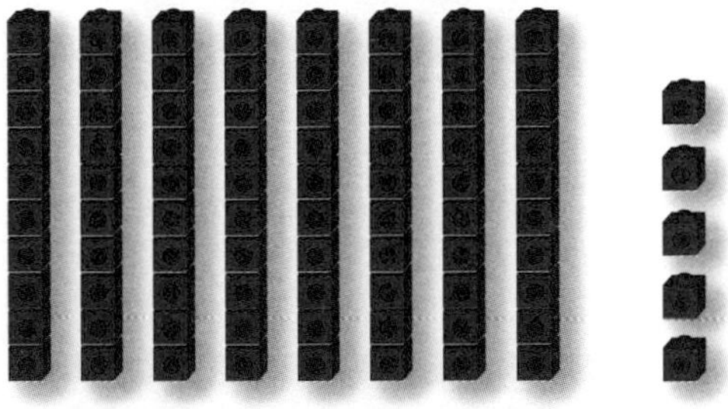

____

6. 

____

Write About It

7. Draw pictures showing 63 and 36. Use the words **tens** and **ones** to write about each number.

Home Note Your child modeled and wrote numbers to 80.
ACTIVITY Say some numbers between 10 and 80. Have your child tell how many tens and ones there are in each number.

Name ______________________________

56

Use Workmat 3 and .
Show each model.
Write the number.

1. 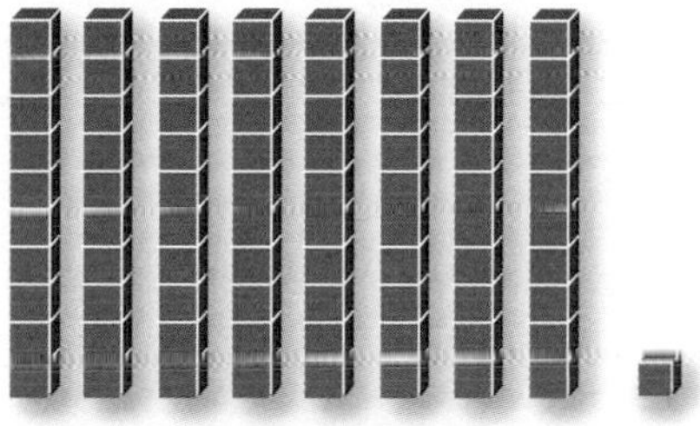 81

2.  ______

3. 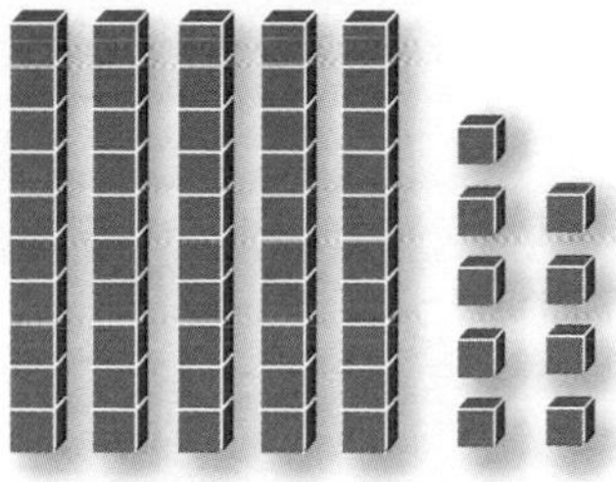 ______

4. 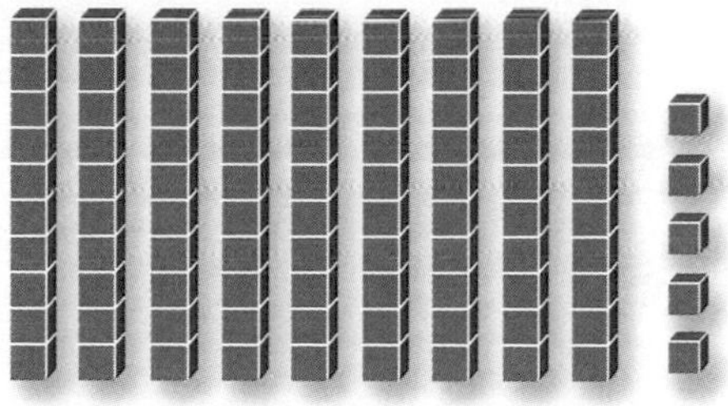 ______

5. 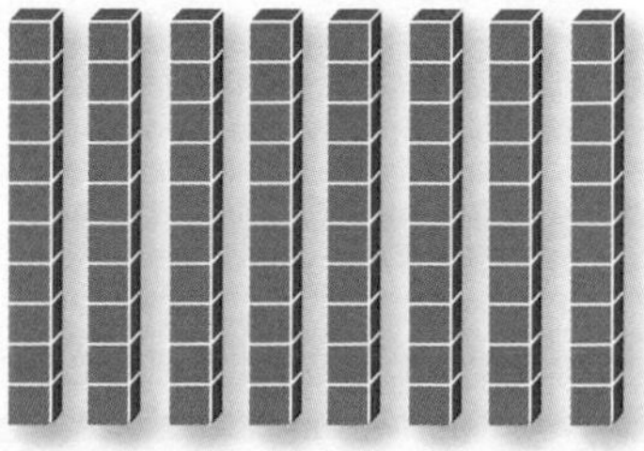 ______

6.  ______

Talk About It • Critical Thinking

How is 57 different from 75?

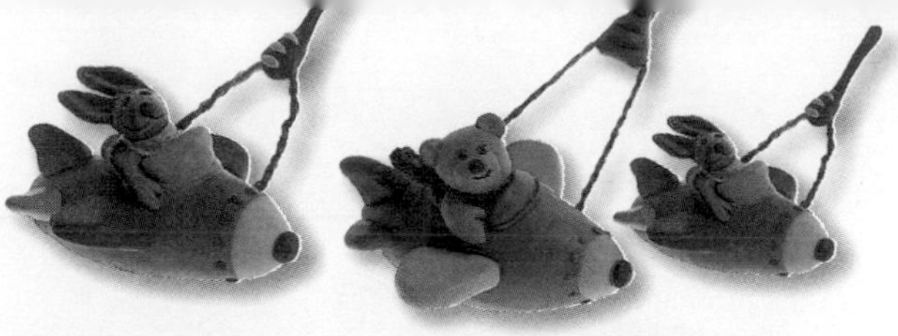

Practice

Write the number.

1. 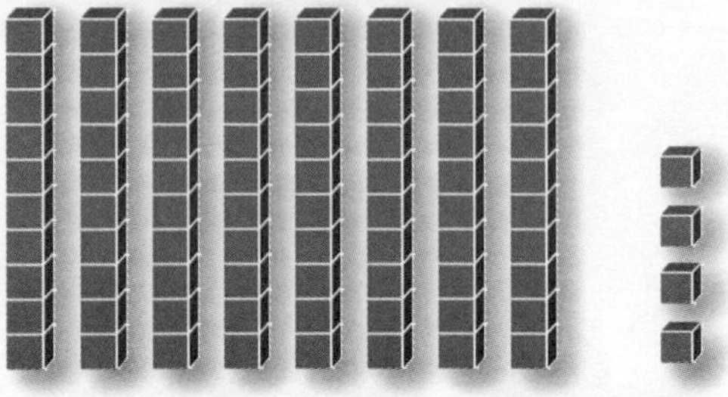
84 ______

2. 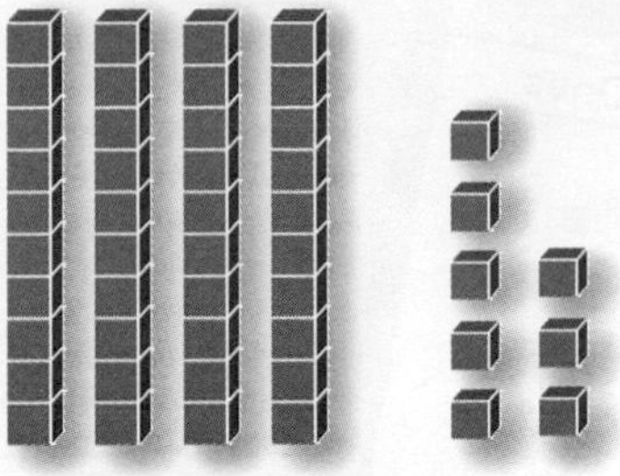
______

3. 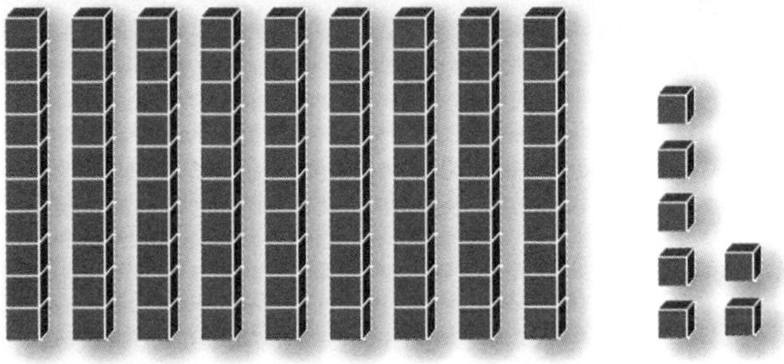
______

4. 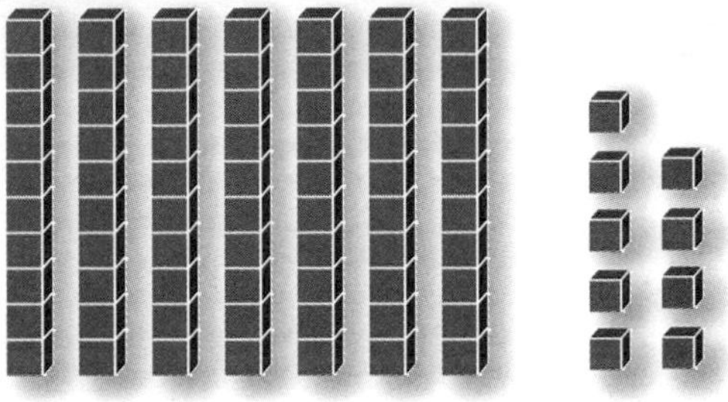
______

5. 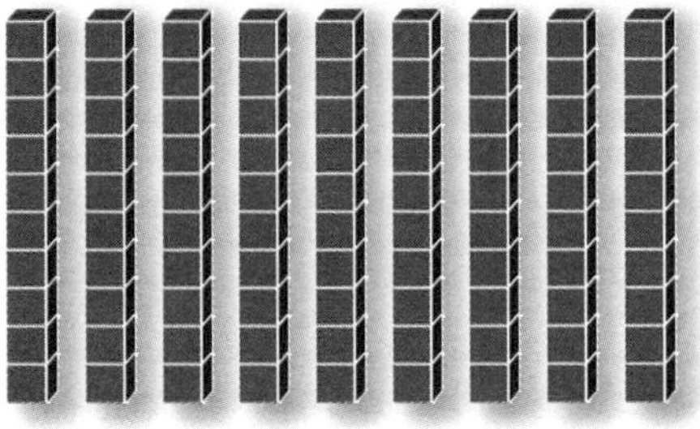
______

6. 
______

Problem Solving • Visual Thinking

Solve.
Draw a picture to show your answer.

7. Molly has 4 .
Ted has 3 .
How many do they have in all?

______

Home Note Your child modeled and wrote numbers to 100.
ACTIVITY Have your child count aloud from 1 to 100.

Name ______________________

Understand • Plan • Solve • Look Back

Problem Solving
Use Estimation

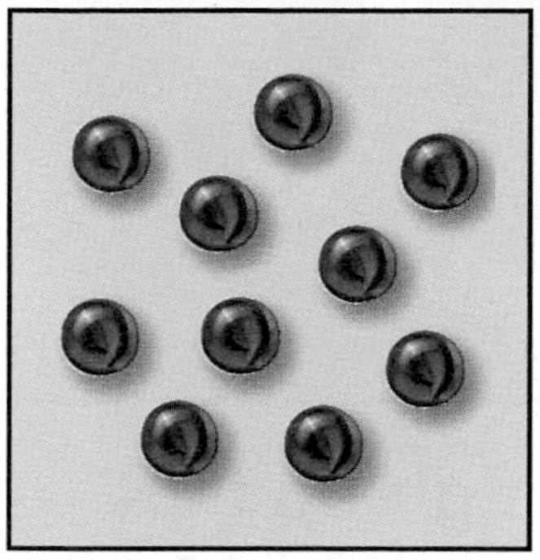

This group has 10 objects.

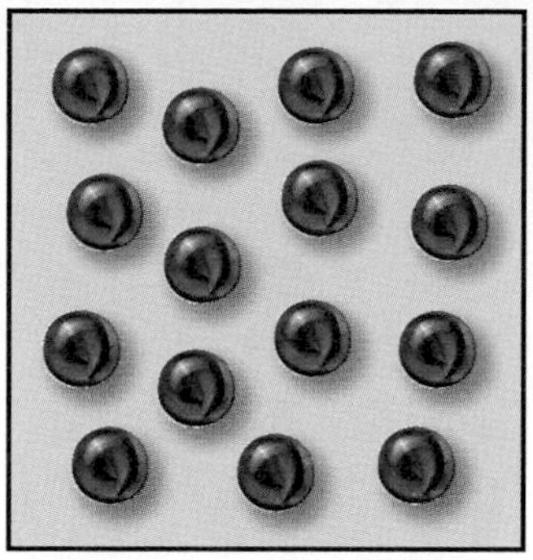

This group has more than 10.

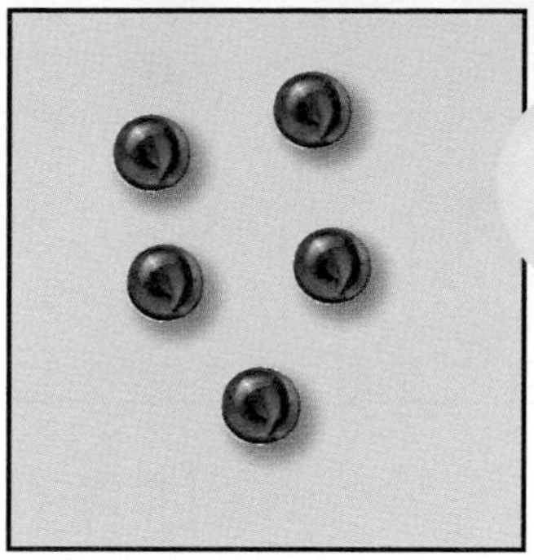

This group has fewer than 10.

Use these pictures to help you estimate.

Which is the better estimate?
Circle it.

1

more than 10 fewer than 10

2

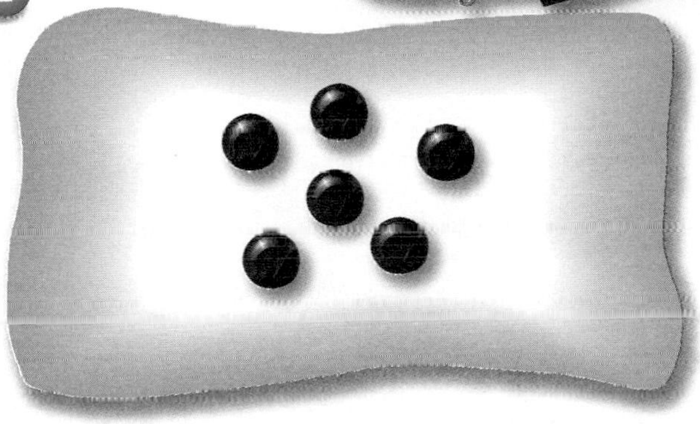

more than 10 fewer than 10

3

more than 10 fewer than 10

4

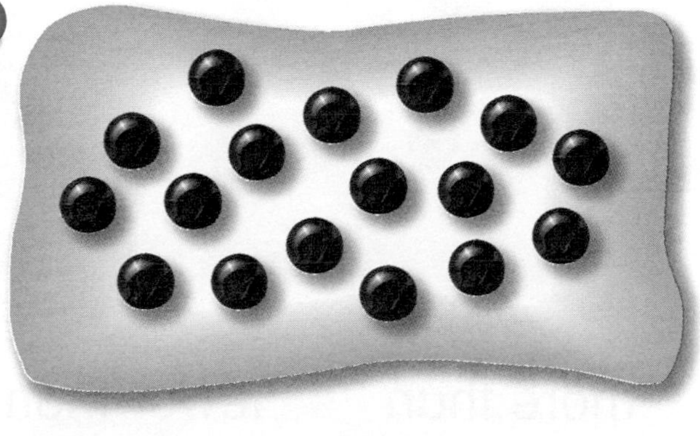

more than 10 fewer than 10

Practice

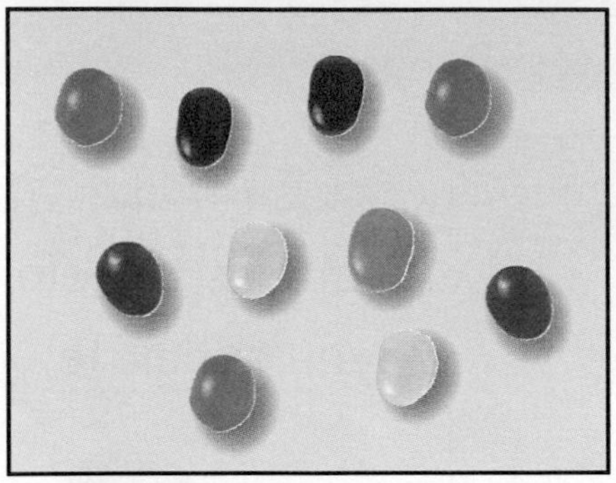

This group has 10 objects.

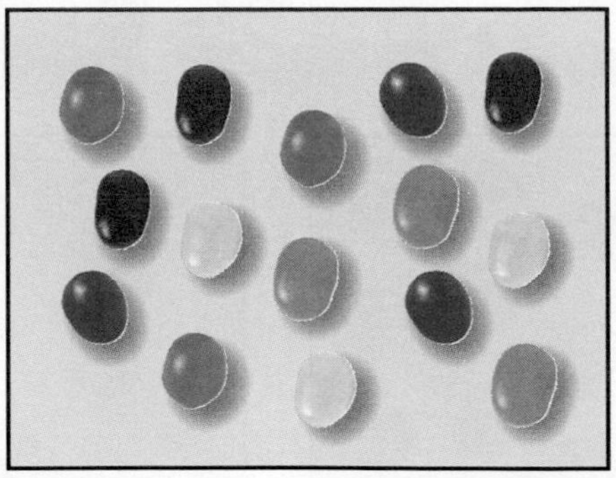

This group has more than 10.

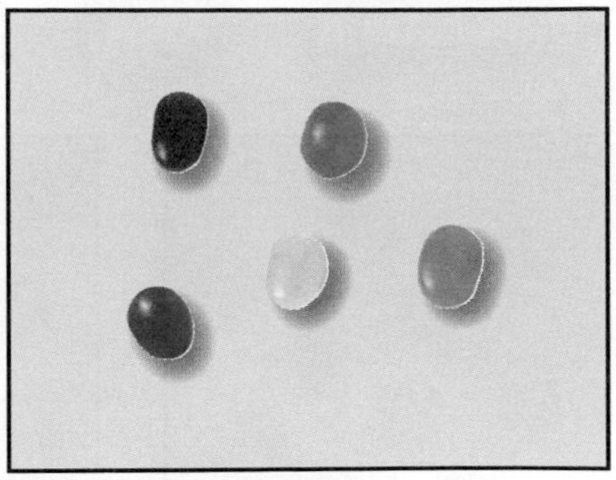

This group has fewer than 10.

Which is the better estimate? Circle it.

1 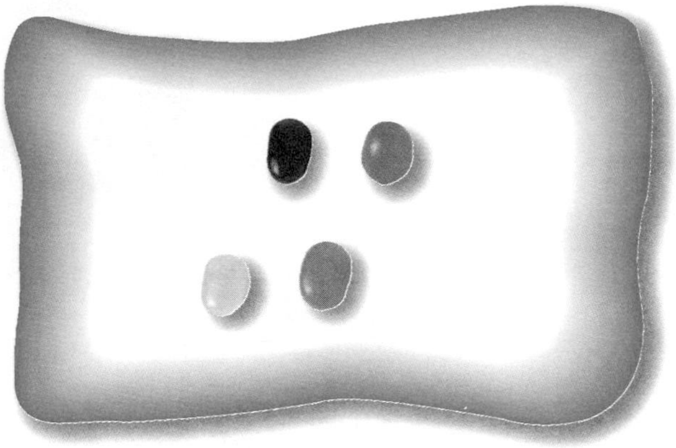

more than 10 | fewer than 10

2 

more than 10 | fewer than 10

3 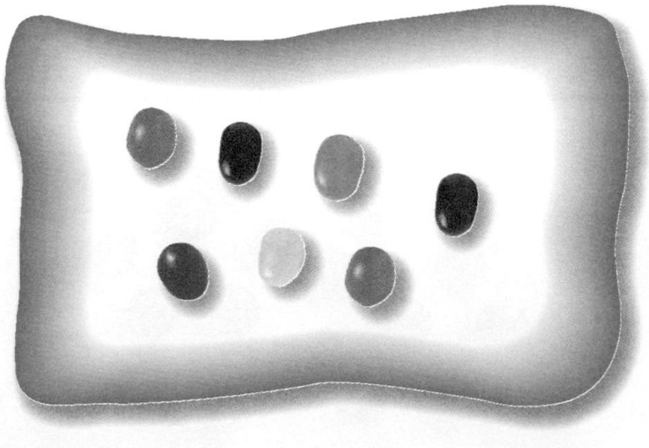

more than 10 | fewer than 10

4 

more than 10 | fewer than 10

Home Note **Your child estimated whether a group had more or fewer than 10 objects.**
ACTIVITY **Set out a group of 5 objects, and have your child estimate, without counting, whether there are more or fewer than 10. Repeat, setting out groups of 7, 11, and 15.**

Name ____________________

Chapter 13

Review/Test

Concepts and Skills

Write how many tens and ones.

1

____ ten ____ ones

2
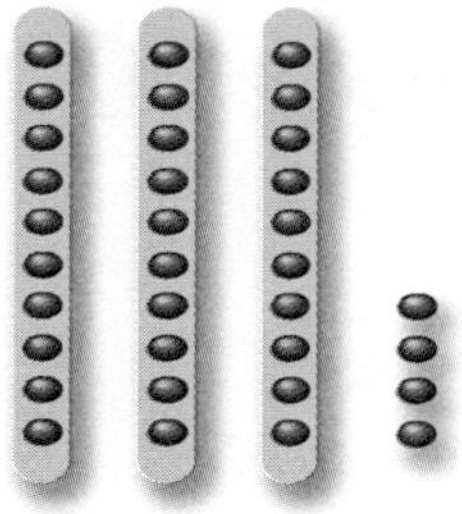

____ tens ____ ones

3
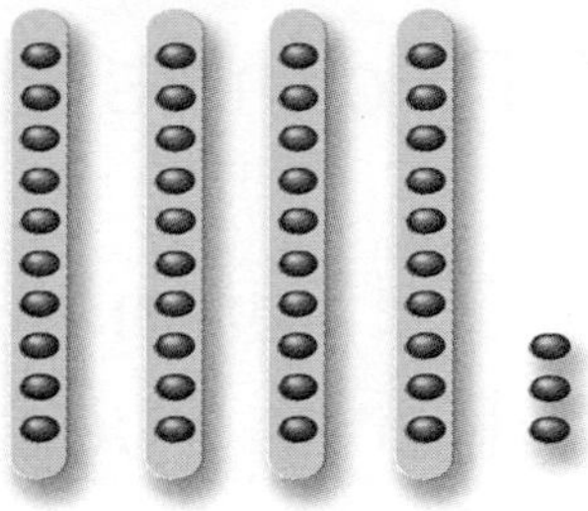

____ tens ____ ones

Write the number.

4
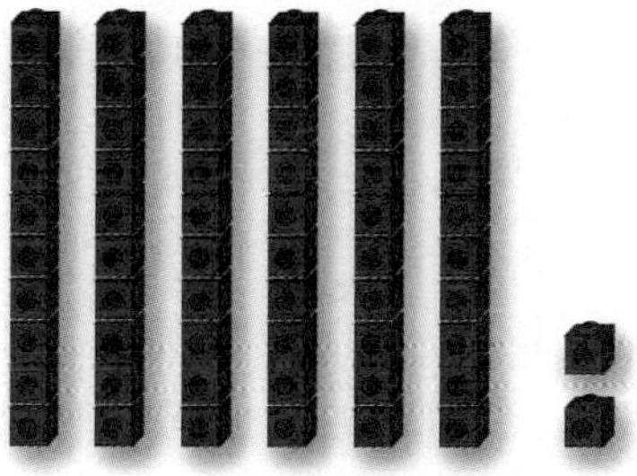

______

5

______

6
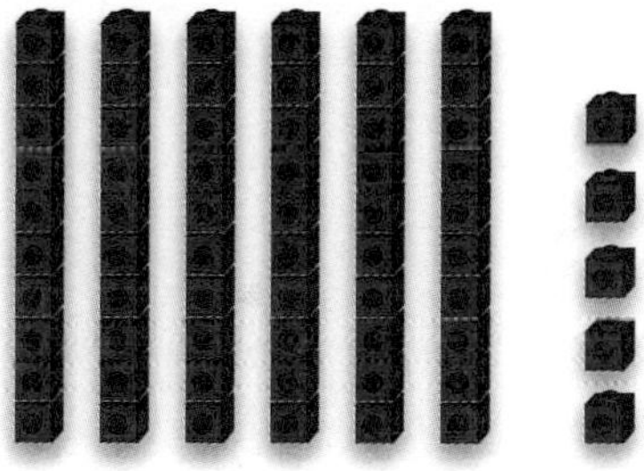

______

7
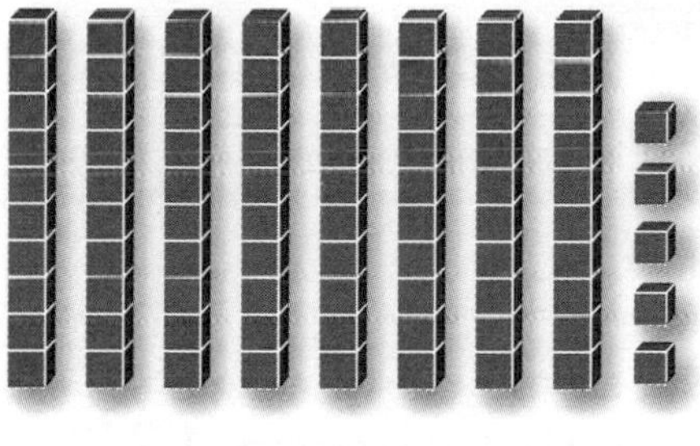

______

8
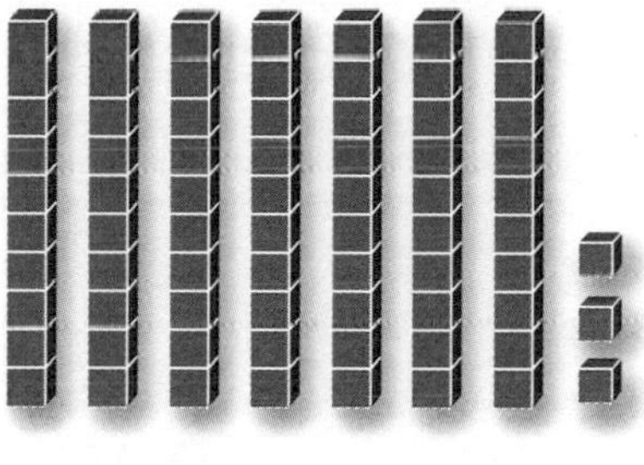

______

9
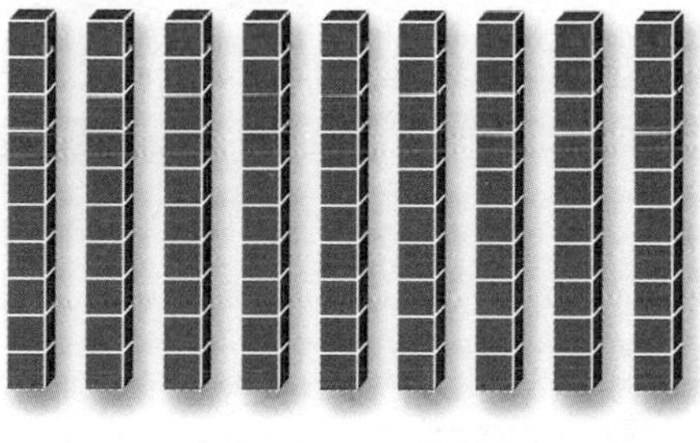

______

Problem Solving

Which is the better estimate?
Circle it.

10 more than 10

fewer than 10

Name ________________________________

TAAS Prep

Read each question.

1. Find the sum or difference.

   5 children play.
   1 child goes home.
   How many are left?

   - ⬭ 3
   - ⬭ 4
   - ⬭ 5
   - ⬭ 6

2. Which solid figure has 6 faces?

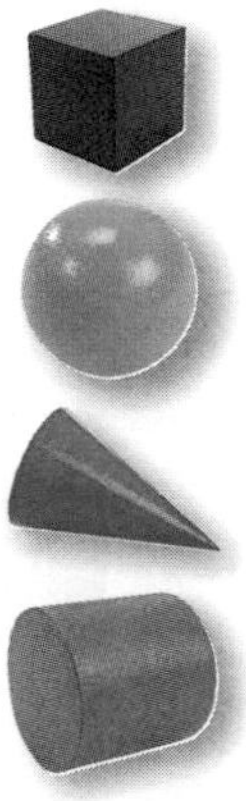

3. Which model shows 32?

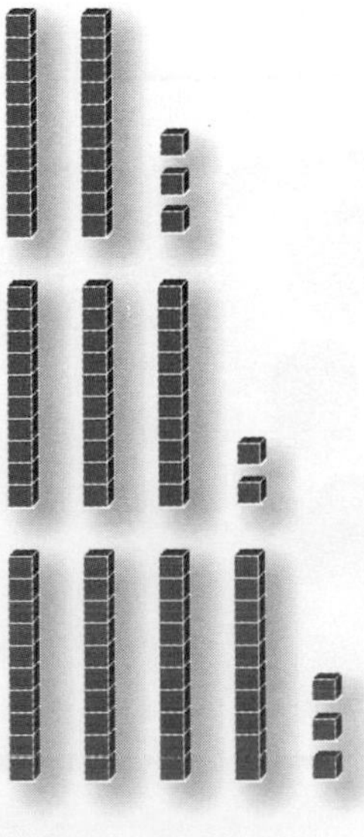

4. Find the sum or difference.

   There are 4 blue birds.
   There are 4 black birds.
   How many birds in all?

   - ⬭ 4
   - ⬭ 6
   - ⬭ 8
   - ⬭ 10

5. Which has the same shape?

6. Which model shows 17?

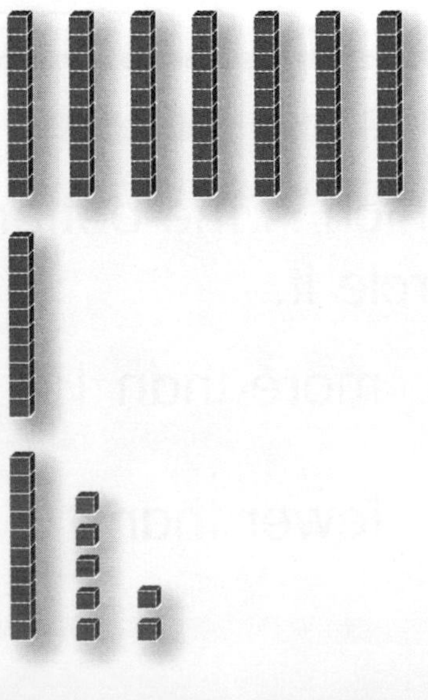

CHAPTER 14 Comparing and Ordering Numbers

Home Note In this chapter, your child will learn to compare and order numbers to 100.
ACTIVITY Say a number from 2 to 99, and have your child say the number that comes before and the number that comes after your number.

SCHOOL-HOME CONNECTION

Dear Family,

Today we started Chapter 14. We will learn to compare and order numbers to 100. Here are the new vocabulary words and an activity for us to do together at home.

Love,

____

Vocabulary

The numbers are in order from **least** to **greatest.**

24 is **less than** 46.

46 is **greater than** 24.

24 is **before** 46.

68 is **after** 46.

46 is **between** 24 and 68.

ACTIVITY

Write the numbers from 50 to 60 on separate slips of paper. Mix up the slips. Have your child put the numbers in order and read them.

Visit our Web site for additional ideas and activities.
http://www.hbschool.com

Name ______________________

Ordinals

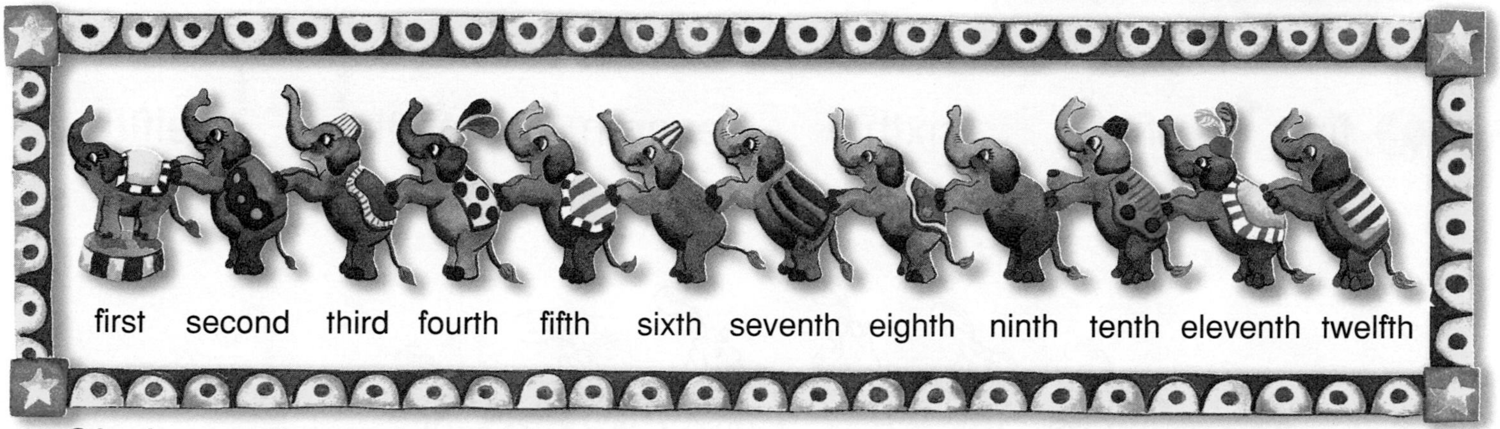

Circle to show the order.

1

second fifth seventh eleventh

2

first fourth eighth tenth

3

third sixth ninth twelfth

4

second fourth seventh ninth

Talk About It • Critical Thinking

If you are eighth in line, how many people are in front of you?

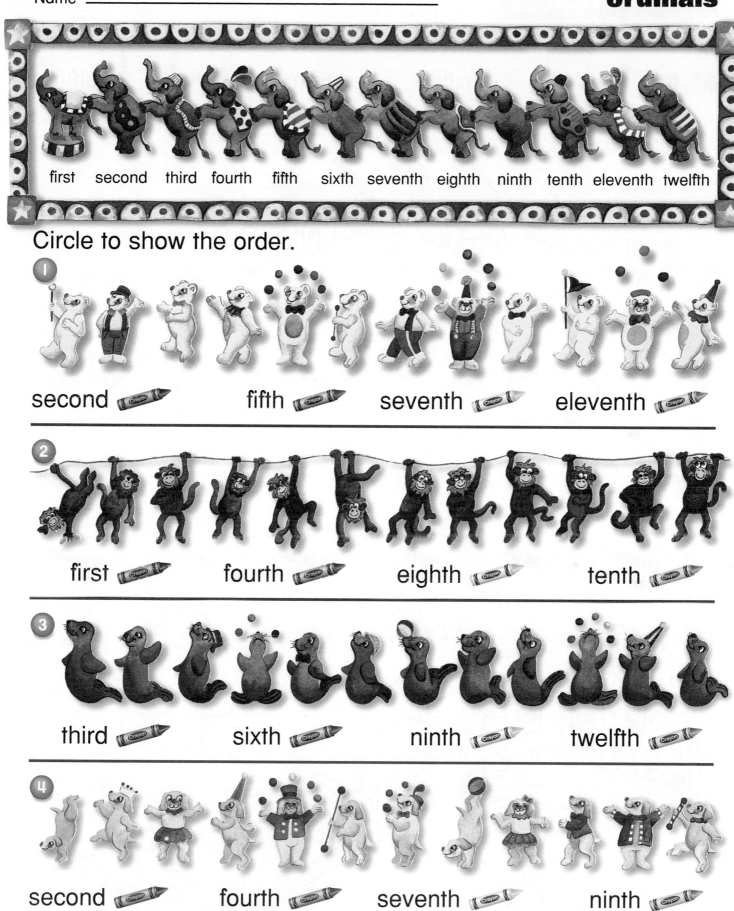

Practice

Color.

1. first fifth sixth seventh ninth twelfth

2. second third fourth eighth tenth eleventh

Home Note Your child colored to show ordinal numbers to twelfth.
ACTIVITY Have your child place twelve objects in a row and name the position of each object.

Name ______________________________

Greater Than

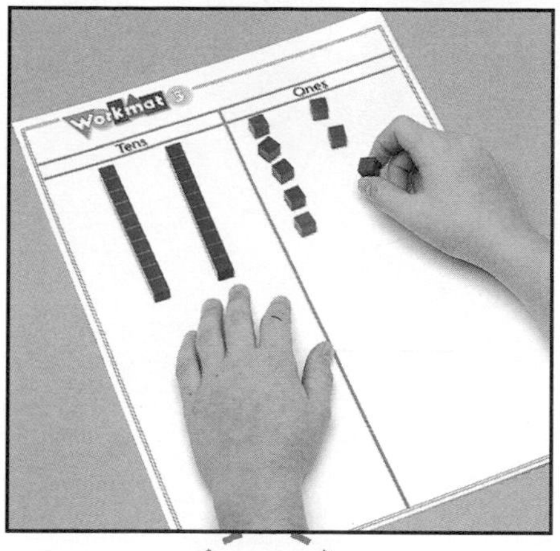

28

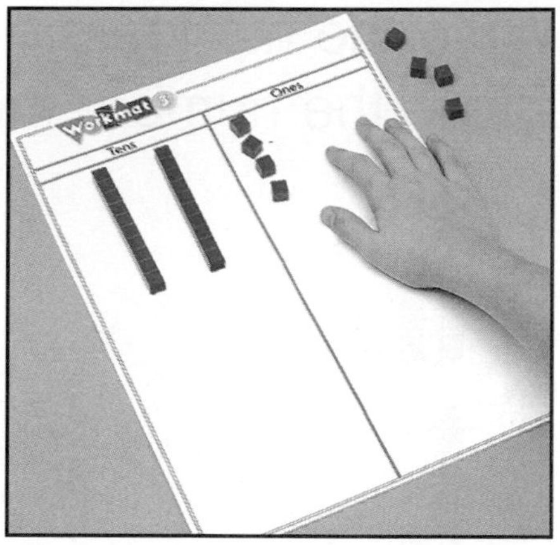

24

Use Workmat 3 and tens and ones blocks.
Show each pair of tens and ones.
Write the numbers.
Circle the number that is greater.

	tens	ones	number
1	4	7	____
	3	7	____
2	1	3	____
	3	1	____
3	5	3	____
	3	5	____

Talk About It • Critical Thinking

How do you know when a number is greater than another number?

Practice

Write the numbers.
Circle the number that is greater.

1 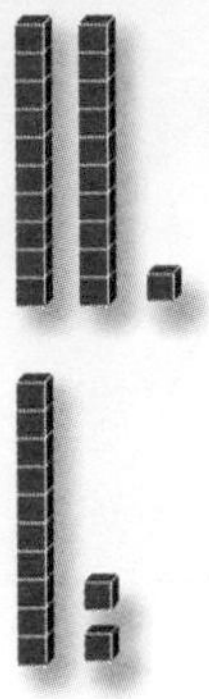

21

12

2 

______

______

3 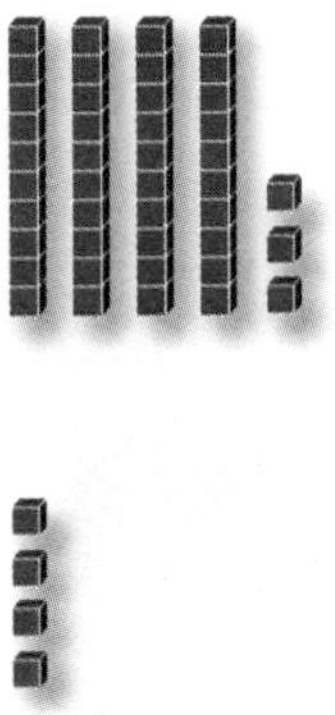

______

______

4 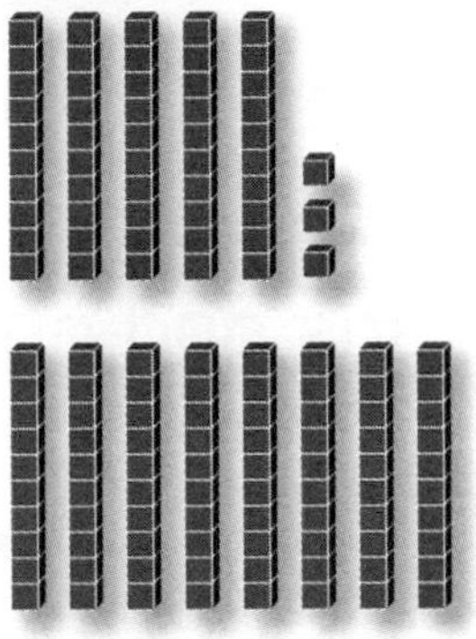

______

______

5 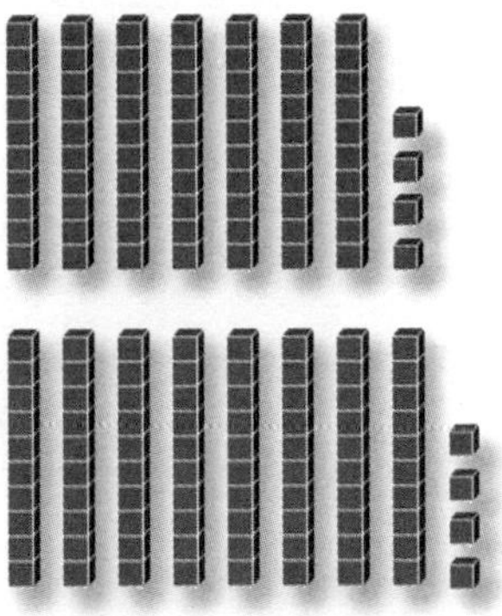

______

______

6 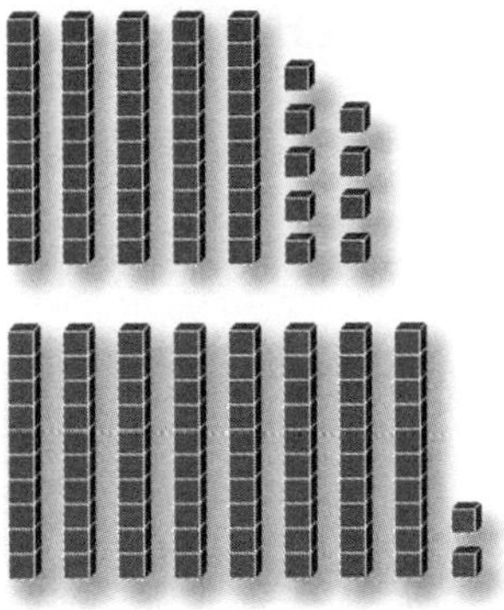

______

______

Problem Solving • Reasoning

Circle the numbers that are greater than 50.

7

14	83	94	44	62	70
35	53	72	19	28	59

Home Note Your child modeled and compared two-digit numbers to determine which was greater. ACTIVITY Name a two-digit number such as 16. Have your child name three numbers that are greater than your number.

Name ______________________

Less Than

35

37

Use Workmat 3 and .
Show each pair of tens and ones.
Write the numbers.
Circle the number that is less.

	tens	ones	number
1	2	6	____
	3	6	____
2	4	5	____
	4	3	____
3	3	0	____
	2	5	____

Practice

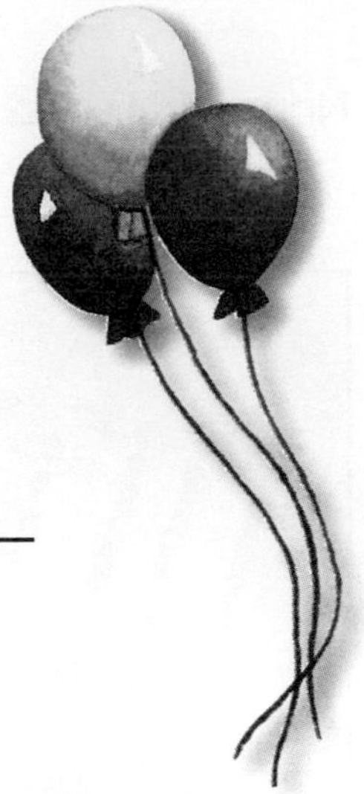

Write the numbers. Circle the number that is less.

1. 
   32
   (23)

2. 
   ____
   ____

3. 
   ____
   ____

4. 
   ____
   ____

5. 
   ____
   ____

6. 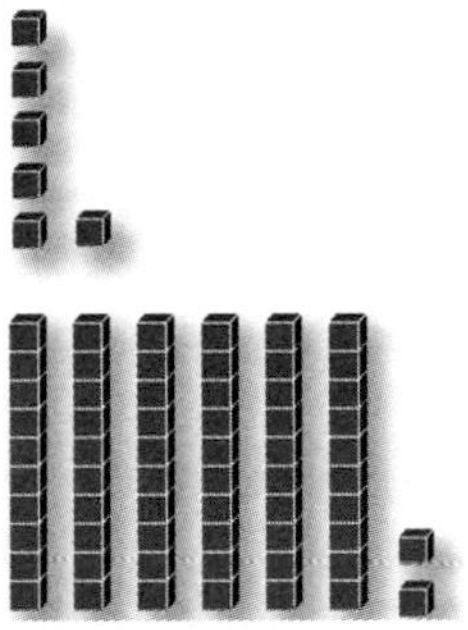
   ____
   ____

Mixed Review

Solve.

7. Beth found 4 pennies on the floor and 3 pennies on the table. How many pennies in all did she find?

   ____ pennies

Home Note Your child modeled and compared two-digit numbers to determine which was less.
ACTIVITY Name a two-digit number. Have your child name three numbers that are less than your number.

Name ______________________________

Before, After, Between

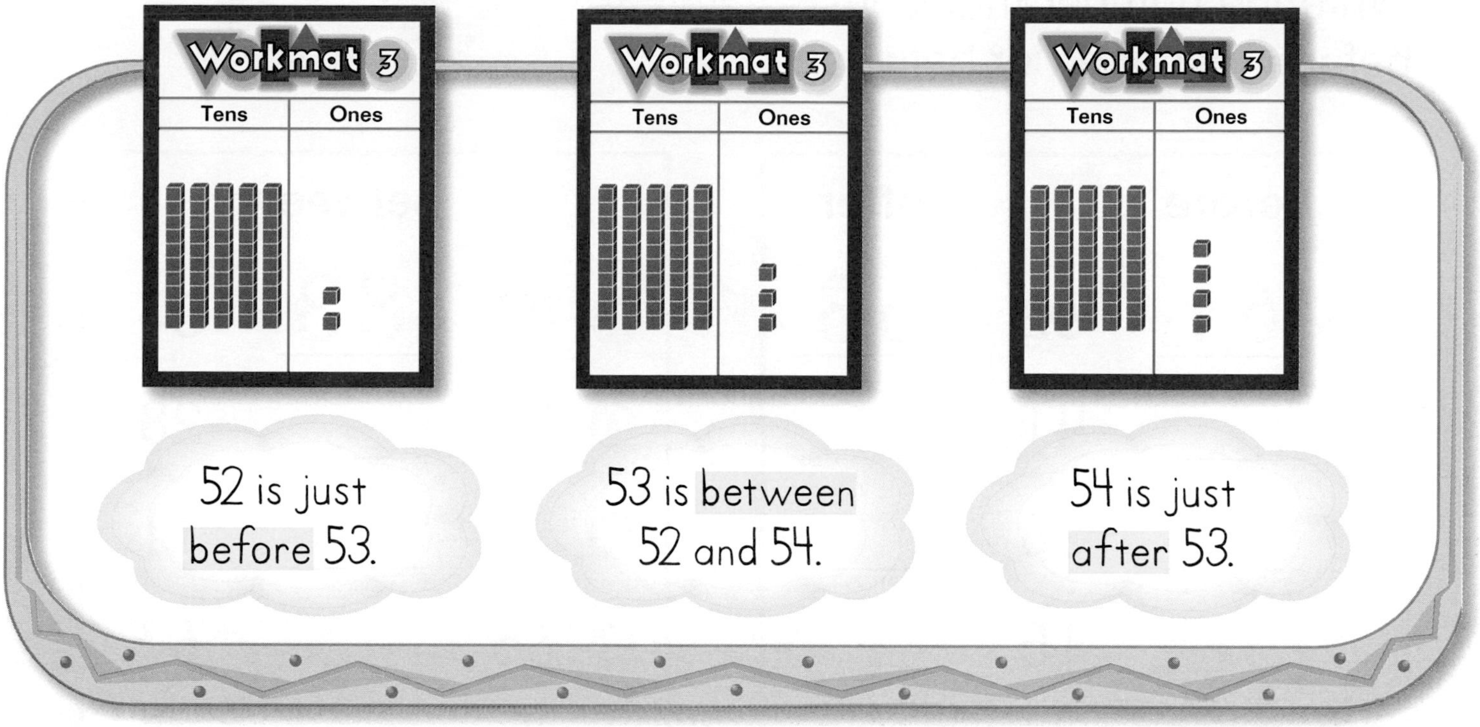

Use Workmat 3 and [tens and ones blocks].
Show the numbers.
Write the number that is just before, just after, or between.

	before		after
1	46	47	48
2	____	98	____
3	____	32	____
4	____	20	____

		between	
5	25	26	27
6	60	____	62
7	83	____	85
8	89	____	91

Practice

Write the number that is just before, just after, or between.

	before		after
1	61	62	63
2	____	41	____
3	____	36	____
4	____	15	____
5	____	88	____
6	____	23	____
7	____	94	____
8	____	50	____
9	____	79	____

		between	
10	82	83	84
11	26	____	28
12	9	____	11
13	65	____	67
14	81	____	83
15	39	____	41
16	50	____	52
17	77	____	79
18	98	____	100

Write About It

19 Write three numbers between 10 and 99. Use **greater than, less than, before, after,** and **between** to tell about the numbers.

Home Note Your child modeled two-digit numbers and then wrote the numbers before, after, or between them.
ACTIVITY Name a two-digit number. Have your child name the numbers that come just before and just after your number.

Name ______________________

Order to 100

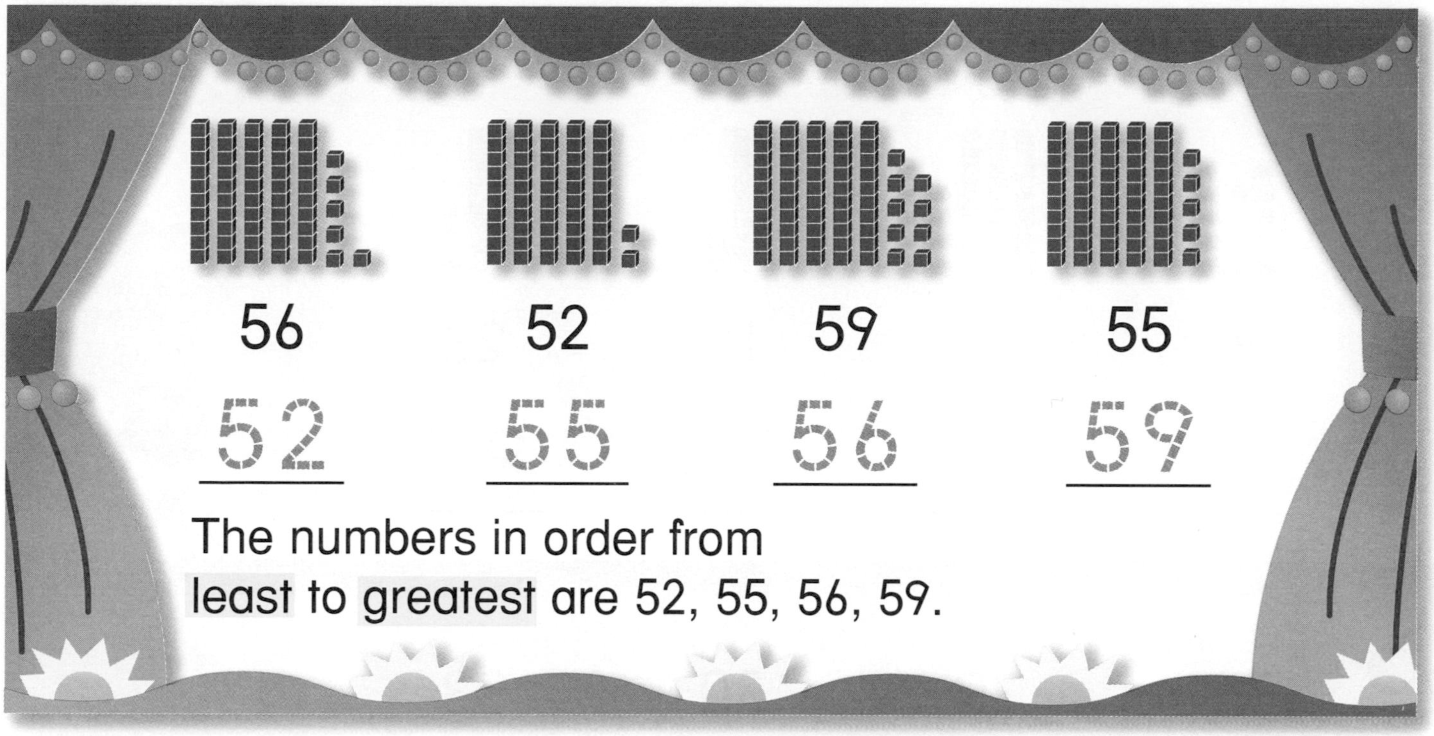

56 52 59 55
52 55 56 59
The numbers in order from least to greatest are 52, 55, 56, 59.

Write the numbers in order from least to greatest.

1. 

51 31 61 41

____ ____ ____ ____

2. 

25 40 39 72

____ ____ ____ ____

3. 

21 12 30 13

____ ____ ____ ____

Practice

Write the numbers in order from least to greatest.

1.

9	32	12	23
9	12	23	32

2. 36 39 30 33

____ ____ ____ ____

3. 78 79 72 75

____ ____ ____ ____

4. 52 32 22 92

____ ____ ____ ____

5. 28 88 98 58

____ ____ ____ ____

6. 62 26 16 61

____ ____ ____ ____

Home Note Your child wrote numbers in order from least to greatest.
ACTIVITY Write five two-digit numbers in any order on a sheet of paper. Have your child write them in order from least to greatest.

Name ______________________________

Chapter 14
Review/Test

Concepts and Skills

1. Circle to show the order.

first third fourth eighth tenth eleventh

Write the numbers. Circle the number that is greater.

2. 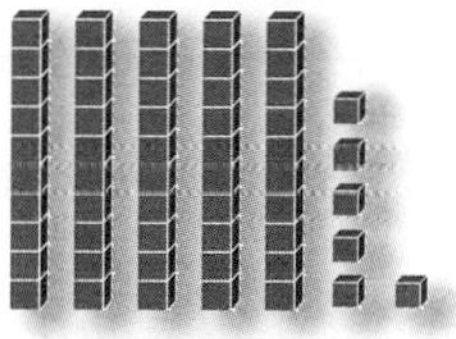 ______

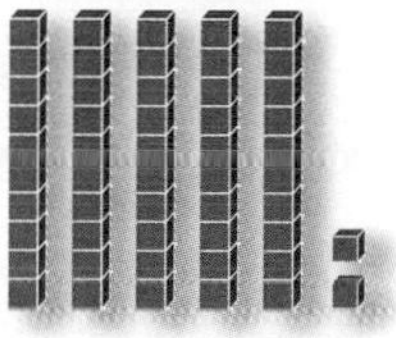 ______

Write the numbers. Circle the number that is less.

3. 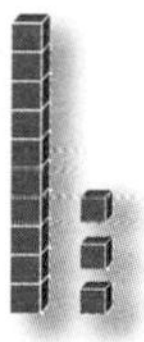 ______

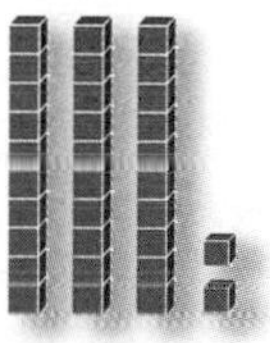 ______

Write the number that is just before, just after, or between.

	before		after
4.	______	32	______
5.	______	20	______

		between	
6.	83	______	85
7.	60	______	62

Write the numbers in order from least to greatest.

8. 62 41 59 37

______ ______ ______ ______

Name ______________________________

Cumulative Chapters 1–14

TAAS Prep

Mark the best answer.

1 Which shows the numbers in order from least to greatest?

44   25   51

- ◯ 44, 51, 25
- ◯ 25, 51, 44
- ◯ 25, 44, 51
- ◯ 51, 44, 25

2 Greg has 20 crayons. Sam has 14 crayons. Who has more?

- ◯ Greg
- ◯ Sam

3 How many are there?

- ◯ 4
- ◯ 3
- ◯ 43
- ◯ 7

4 Which shows the numbers in order from least to greatest?

- ◯ 
- ◯ 
- ◯ 
- ◯ 

5 Which comes next?

- ◯ 
- ◯ 
- ◯ 
- ◯ 

6 Which addition sentence tells how many in all?

- ◯ $2 + 6 = 8$
- ◯ $3 + 5 = 8$
- ◯ $5 + 3 = 8$
- ◯ $4 + 4 = 8$

CHAPTER 15

Patterns on a Hundred Chart

How many in each group?
Count different ways to find out.

Home Note In this chapter, your child is learning different ways to count to 100.
ACTIVITY Have your child count to 50 by tens, by fives, and then by twos.

SCHOOL-HOME CONNECTION

Dear Family,

Today we started Chapter 15. We will learn to count by tens, fives, and twos. We will also find out about even and odd numbers. Here are the new vocabulary words and an activity for us to do together at home.

Love,

____

Vocabulary

6 is an even number.

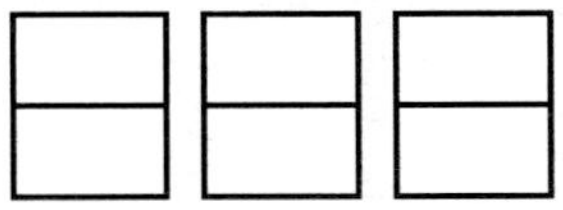

6 can be shown as three pairs of objects.

5 is an odd number.

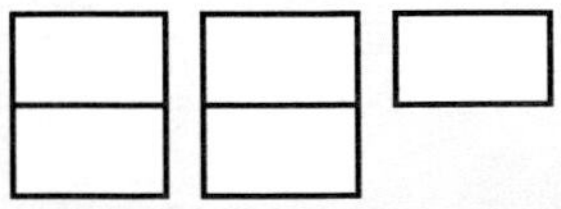

5 can be shown as two pairs of objects with one left over.

ACTIVITY

Have your child place 100 pennies in stacks of 10 pennies each and count the pennies by tens. Then have him or her repeat the activity, stacking and counting pennies by fives and then by twos.

Visit our Web site for additional ideas and activities.
http://www.hbschool.com

Name ______________________

Counting by Tens

Count by tens.
Write how many.

1

10 20 30 40 50 crayons

2

10 ____ ____ ____ ____

60 ____ ____ crayons

3

____ ____ ____ ____ ____

____ ____ ____ ____ crayons

Practice

Write the missing numbers.
Count by tens. Color the tens.

1

1	2	3	4	5	6	7	8	9	10
11	12	13	14	15	16	17	18	19	
21	22	23	24	25	26	27	28	29	
31	32	33	34	35	36	37	38	39	
41	42	43	44	45	46	47	48	49	
51	52	53	54	55	56	57	58	59	
61	62	63	64	65	66	67	68	69	
71	72	73	74	75	76	77	78	79	
81	82	83	84	85	86	87	88	89	
91	92	93	94	95	96	97	98	99	

Problem Solving

Fill in the table.

2 Rosa saves 10 pennies every day.
How many pennies does
she have on Tuesday?

______ pennies

Sunday	Monday	Tuesday
10	20	

Home Note Your child counted by tens.
ACTIVITY Have your child stack pennies in groups of 10, count the groups by tens, and tell how many pennies in all.

Name ______________________________

Counting by Fives

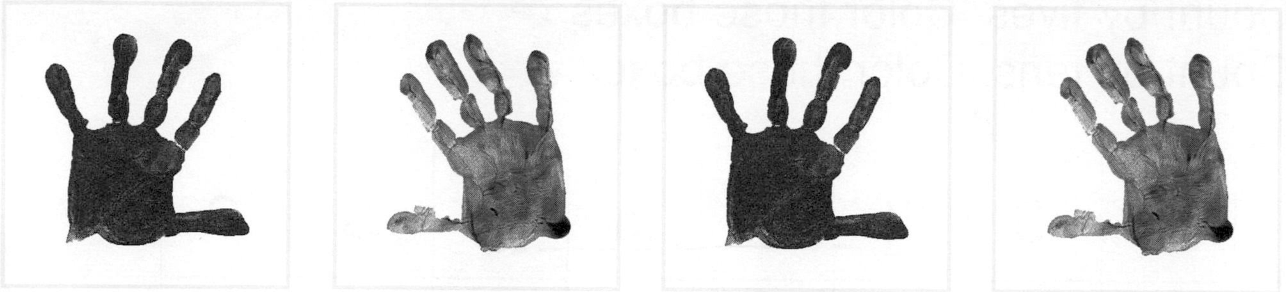

5 10 15 20 fingers

Count by fives. Write how many.

1

____ ____ ____ ____ ____

____ ____ ____ ____ fingers

2

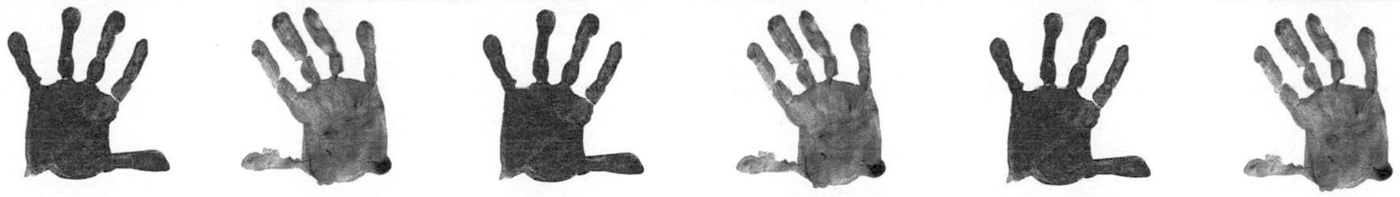

____ ____ ____ ____ ____ ____

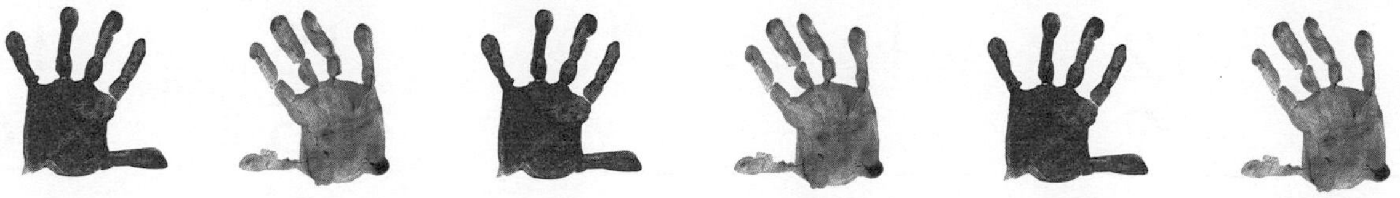

____ ____ ____ ____ ____ ____ fingers

Talk About It • Critical Thinking

How do you know what number comes next when you count by fives?

Practice

1. Write the missing numbers.
Count by fives. Color those boxes (crayon).
Count by tens. Color those boxes (crayon).

1	2	3	4	5	6	7	8	9	
11	12	13	14		16	17	18	19	
21	22	23	24		26	27	28	29	
31	32	33	34		36	37	38	39	
41	42	43	44		46	47	48	49	

Mixed Review

Add.

2. $7 + 3 = \square$ $\quad 7 + 2 = \square$ $\quad 8 + 1 = \square$ $\quad 9 + 2 = \square$ $\quad 5 + 5 = \square$ $\quad 6 + 6 = \square$

Subtract.

3. $11 - 3 = \square$ $\quad 12 - 9 = \square$ $\quad 11 - 5 = \square$ $\quad 10 - 3 = \square$ $\quad 10 - 5 = \square$ $\quad 12 - 6 = \square$

Home Note **Your child counted by fives and tens.**
ACTIVITY **Have your child stack pennies in groups of 5, count the groups by fives, and tell how many pennies he or she has in all.**

Name ____________________

Counting by Twos

Count by twos. Write how many.

1 

2 4 6 8 10

12 14 16 18 20 red crayons

2 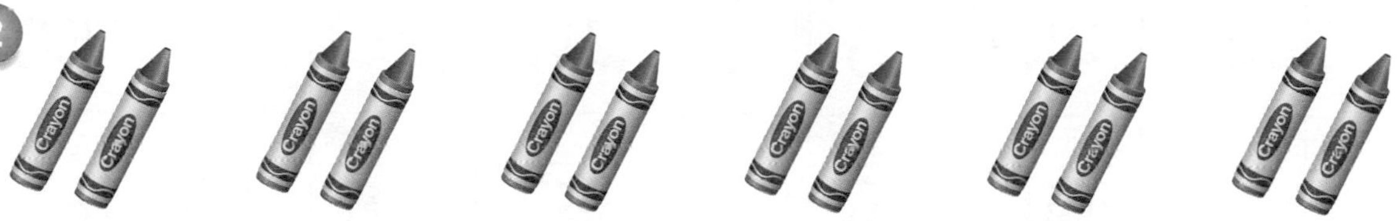

____ ____ ____ ____ ____ ____

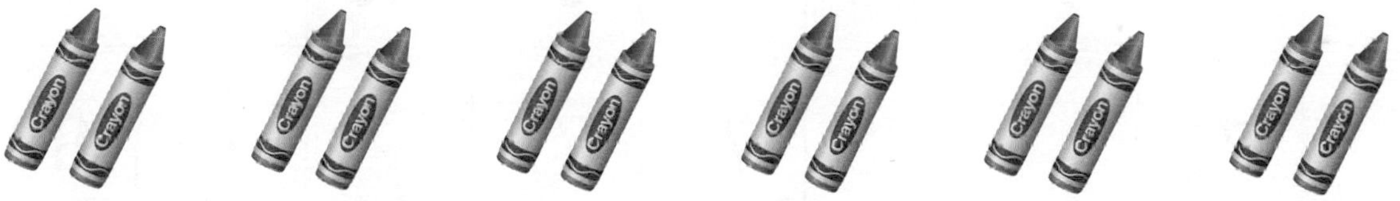

____ ____ ____ ____ ____ ____ blue crayons

3 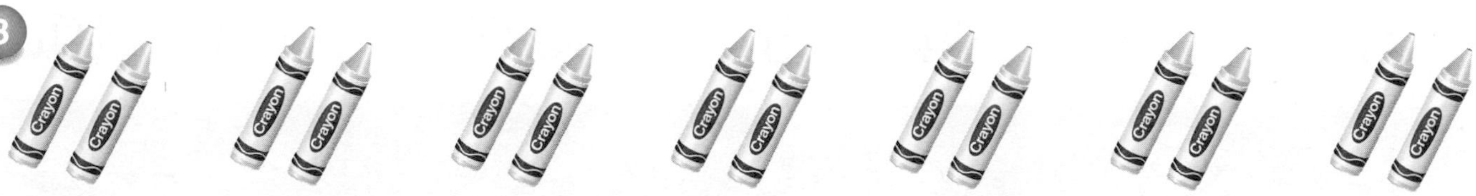

____ ____ ____ ____ ____ ____ ____

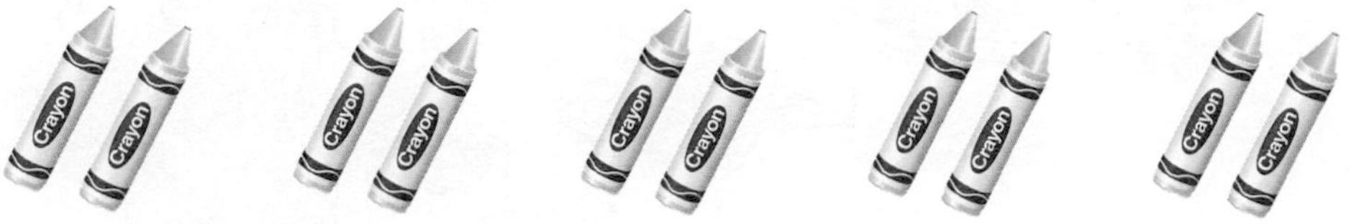

____ ____ ____ ____ ____ yellow crayons

Practice

Count by twos. Write the missing numbers.

1	2	3	4	5		7		9	
11	12	13		15		17		19	20
21		23	24	25		27		29	30
31	32	33		35		37	38	39	40
41		43	44	45	46	47		49	50
51	52	53	54	55		57	58	59	60
61		63	64	65		67	68	69	
71		73		75	76	77		79	80
81	82	83		85		87	88	89	
91		93	94	95	96	97		99	100

Home Note Your child counted by twos.
ACTIVITY Have your child stack pennies in groups of 2, count the groups by twos, and tell how many pennies he or she has in all.

Name ______________________________

Even and Odd Numbers

Use [cube].

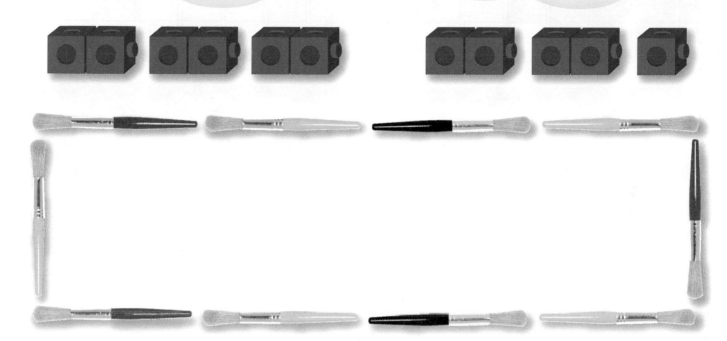

Snap 6 cubes together in pairs. There are none left over. 6 is an even number.

Snap 5 cubes together in pairs. There is one left over. 5 is an odd number.

Use [cube] to show each number.
Circle even or odd.

1	7	even	(odd)	2	8	even	odd
3	10	even	odd	4	13	even	odd
5	3	even	odd	6	6	even	odd
7	4	even	odd	8	2	even	odd

Practice

Color the squares to show each number.
Circle even or odd.

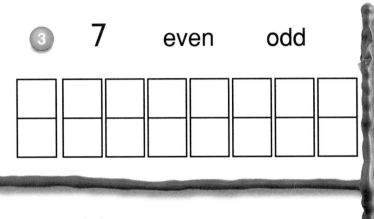

1. 9 even odd

2. 12 even odd

3. 7 even odd

4. 14 even odd

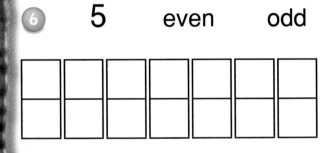

5. 11 even odd

6. 5 even odd

Write About It

7. Draw pictures showing 13 and 16. Use the words even and odd to write about each picture.

Home Note Your child identified numbers as even or odd.
ACTIVITY Give your child a group of about 20 small objects. Have him or her count the objects and tell how many in all. Then have your child pair the objects and tell if the number is even or odd.

Name ______________________

Chapter 15

Review/Test

Concepts and Skills

Count by tens. Write how many.

1

____ ____ ____ ____ ____ crayons

Count by fives. Write how many.

2

____ ____ ____ ____ ____ fingers

Count by twos. Write how many.

3
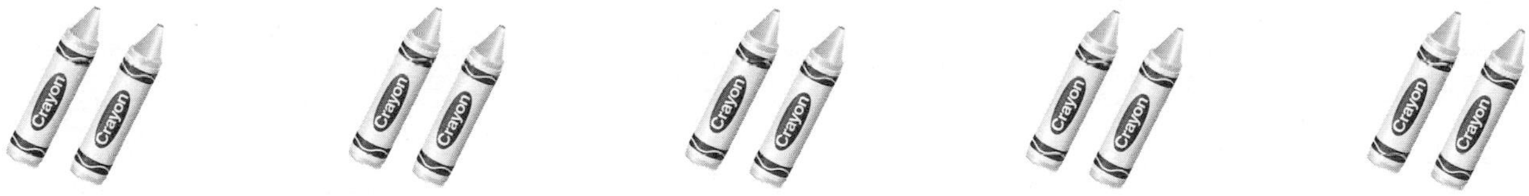

____ ____ ____ ____ ____ crayons

Color the squares to show each number.
Circle even or odd.

4 6 even odd

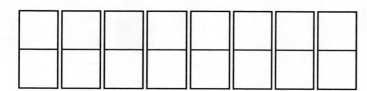

5 11 even odd

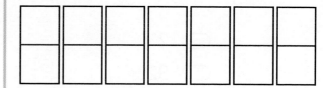

Name ______________________

TAAS Prep

Mark the best answer.

1. Count by twos. Which numbers come next?

4, 6, 8, 10, ___, ___

- ⬭ 9, 8
- ⬭ 11, 12
- ⬭ 12, 13
- ⬭ 12, 14

2. Count by tens. How many pennies in all?

- ⬭ 60
- ⬭ 70
- ⬭ 80
- ⬭ 90

3. Which number is greater?

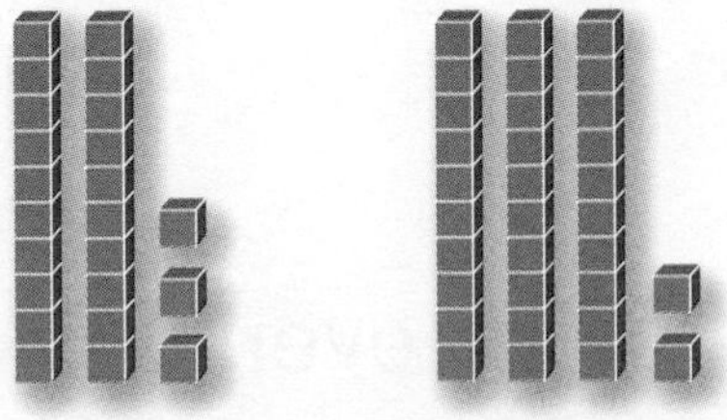

23　　32

- ⬭ 23
- ⬭ 32

4. Which number is less?

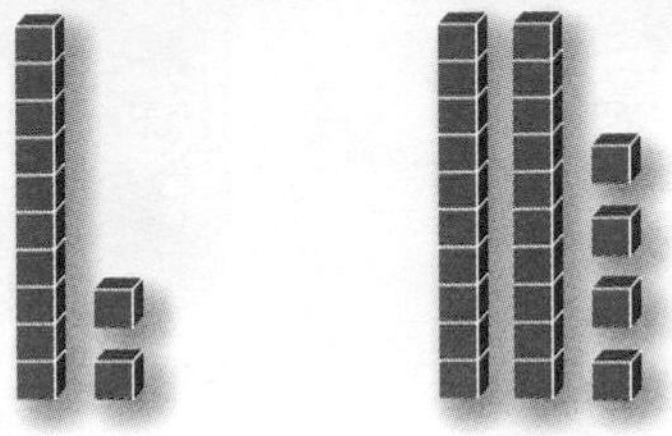

12　　24

- ⬭ 12
- ⬭ 24

5. How many?

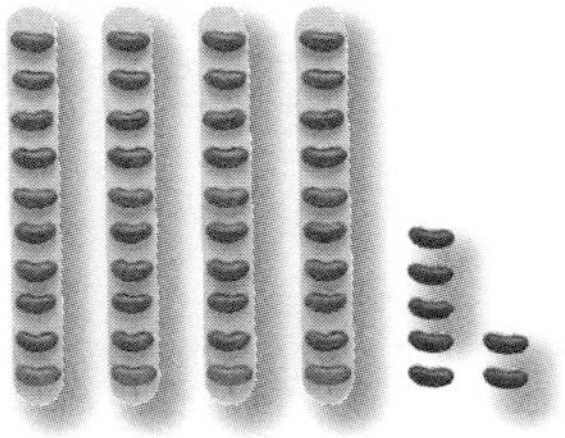

- ⬭ 37
- ⬭ 47
- ⬭ 52
- ⬭ 67

6. Which figure has the same shape?

- ⬭ 
- ⬭ 
- ⬭ 
- ⬭ 

Name ____________________________________

MATH FUN

Use: , 2 0-5, 2 4-9,

Roll and Cover

Play with a partner.

1. Roll the , .
2. Use 1 or 2 of the numbers to make a number on the chart.
3. Color that number on the chart.
4. Play until all numbers are covered.

1	2	3	4	5	6	7	8	9	10
11	12	13	14	15	16	17	18	19	20
21	22	23	24	25	26	27	28	29	30
31	32	33	34	35	36	37	38	39	40
41	42	43	44	45	46	47	48	49	50
51	52	53	54	55	56	57	58	59	60
61	62	63	64	65	66	67	68	69	70
71	72	73	74	75	76	77	78	79	80
81	82	83	84	85	86	87	88	89	90
91	92	93	94	95	96	97	98	99	100

Home Note Your child has been learning about numbers to 100.
ACTIVITY Play this game with your child. Then have your child count by twos, fives, and tens to 100.

Technology

Name ____________________

Calculator **Computer**

Use a .
Press the keys.
Write what you see.

1. Write a number between 1 and 9: 4

AC + [4] = = = [12]

2. Write another number between 1 and 9: ______

AC + [ ] = = = [ ]

3. Write another number between 1 and 9: ______

AC + [ ] = = = [ ]

4. What 3 answers did you get?

______ ______ ______

5. Write the 3 answers in order from least to greatest.

______ ______ ______

written by Linda Cave

Bugs!

illustrated by Anna Rich

This book will help me review number patterns.

This book belongs to ______________________.

A

Harcourt Brace School Publishers

How many wings do you see?

B

Count by twos and then tell me.

____ wings

How many black spots do you see?

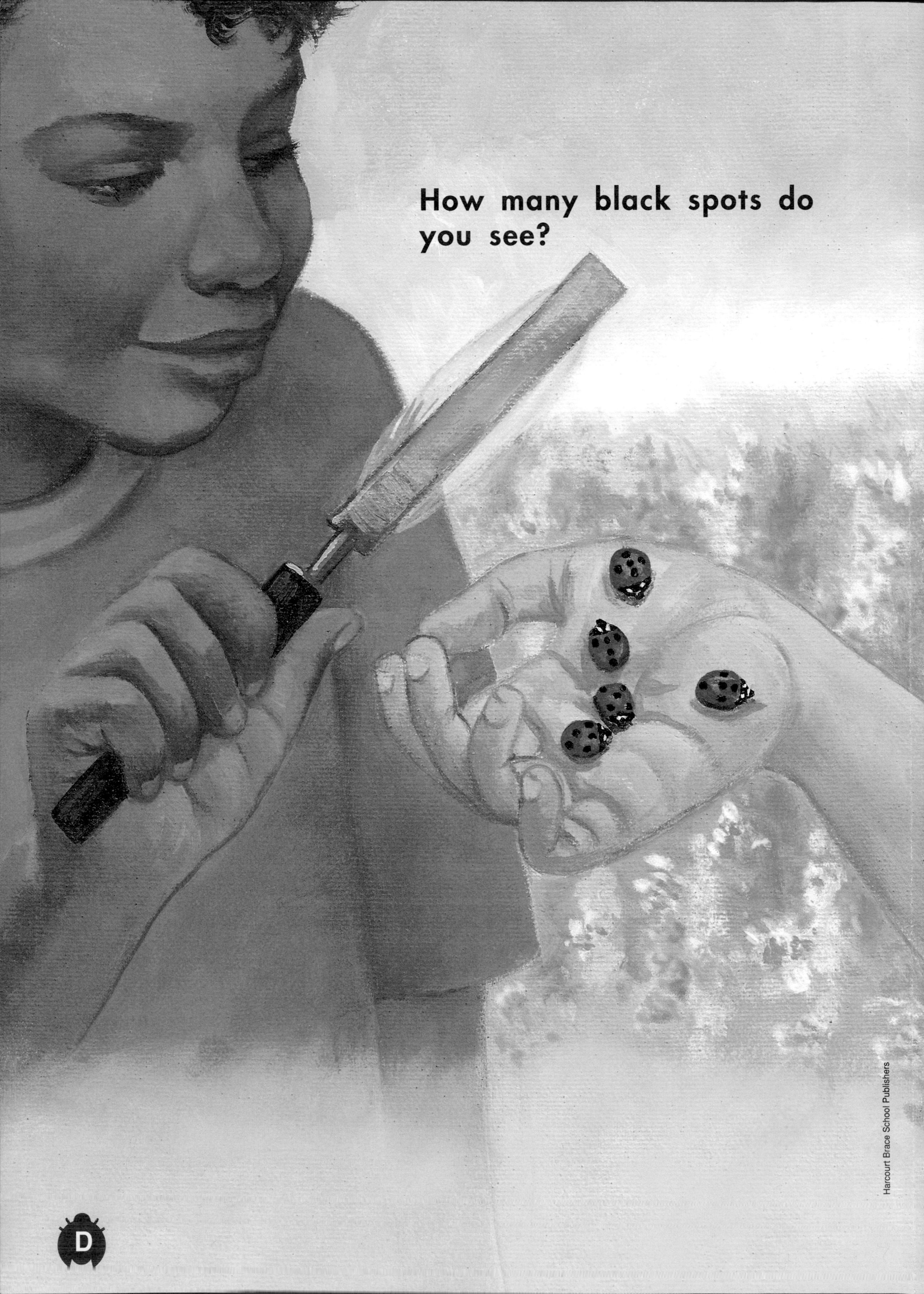

Count by fives and then tell me.

_____ black spots

How many bugs do you see?

F

Count by tens and then tell me.

______ **bugs**

Take the bugs and let them go.
They will help the garden grow!

Name ______________________

Concepts and Skills

Write how many.

1 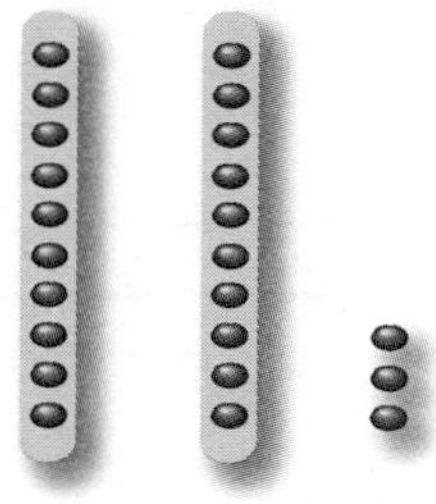

_____ tens _____ ones

2 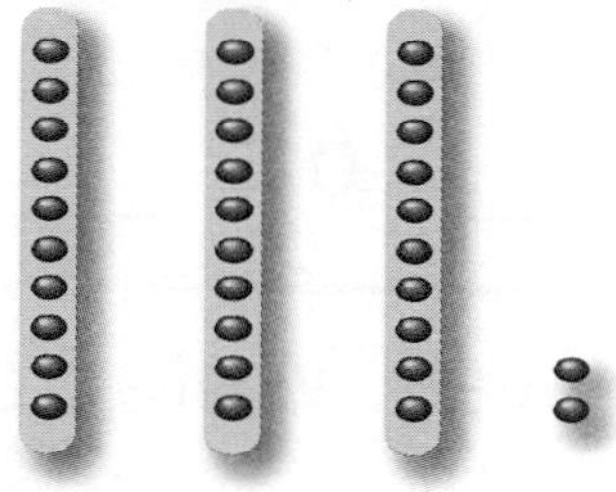

_____ tens _____ ones

3 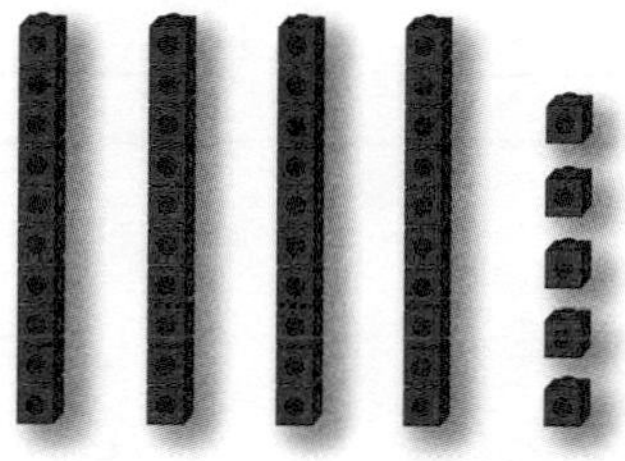 _____

4 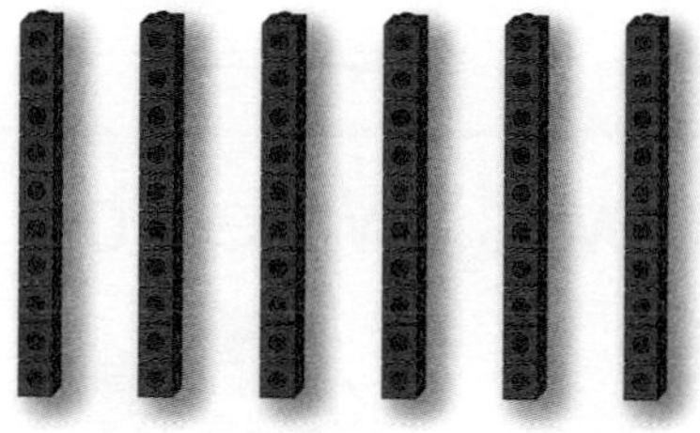 _____

5 Write the numbers.
Circle the greater number.

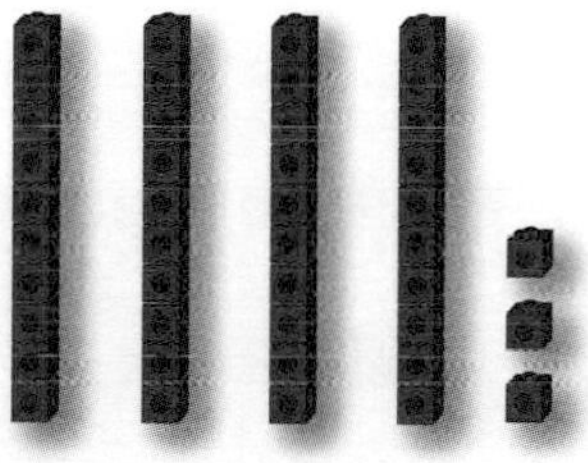 _____

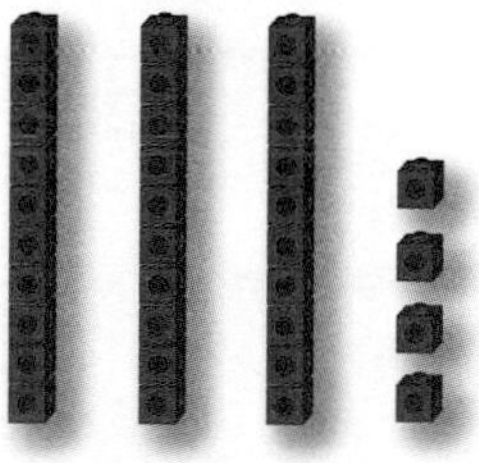 _____

6 Write the numbers. Circle the number that is less.

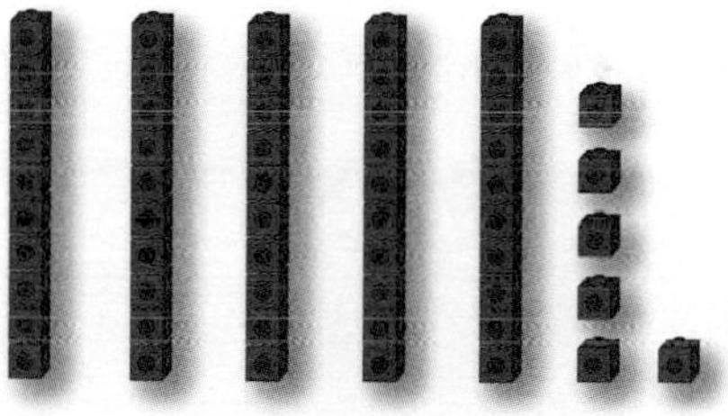 _____

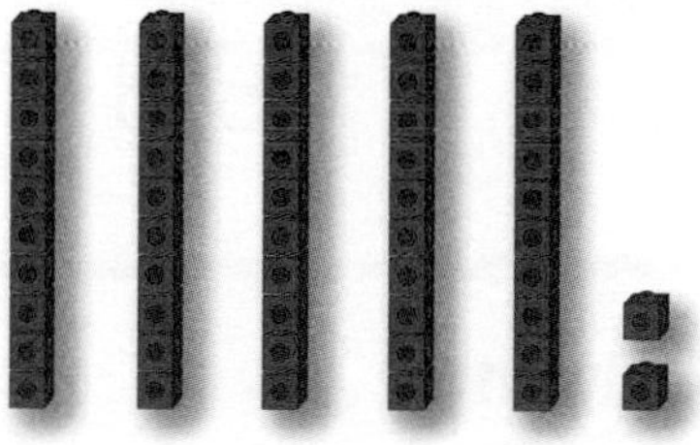 _____

Circle to show the order.

7 second  fifth  seventh  tenth 

Write the number that is just before, just after, or between.

before after	between
8 ____, 63, ____	9 39, ____, 41

Count by fives. Write how many.

10

5 ____ ____ ____ ____

Count by twos. Write how many.

11

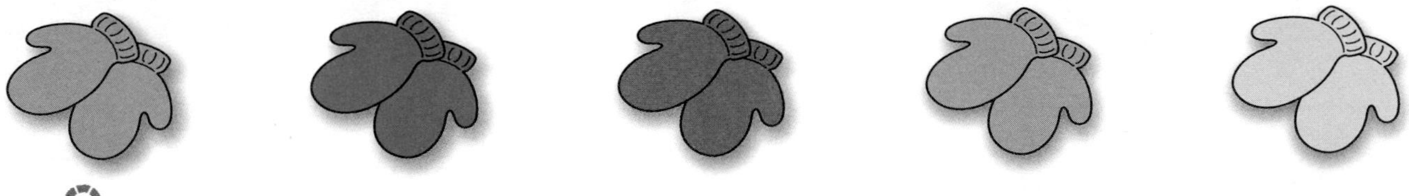

2 ____ ____ ____ ____

Circle even or odd.

10

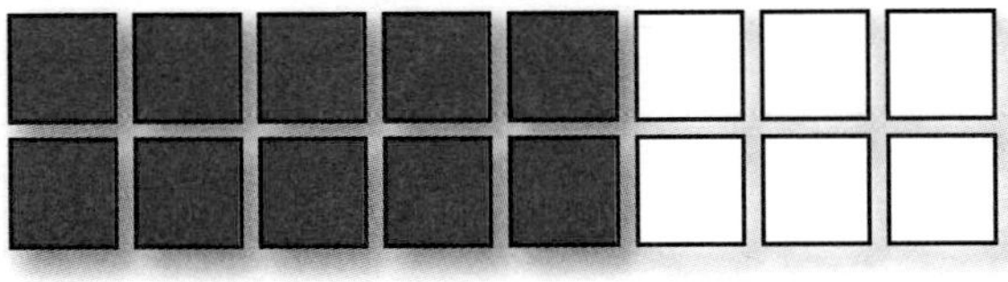

12 even odd

15

13 even odd

Problem Solving

Circle the better estimate.

14 more than 10

fewer than 10

Name ______________________________

Performance Assessment

Use Workmat 3 and .
Build four numbers between 11 and 99.
Draw a picture to show each number.
Write how many tens and ones.
Write the number.

1. ____ tens ____ ones = ____

2. ____ tens ____ ones = ____

3. ____ tens ____ ones = ____

4. ____ tens ____ ones = ____

5. Write the numbers in order from least to greatest.

____ ____ ____ ____

Write About It

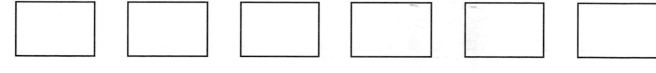

6. Draw pictures of  to show counting by twos.
Write how many.

____ ____ ____ ____ ____ ____

Name ________________________

Chapters 1–15 Cumulative Review

Fill in the ⬭ for the correct answer.

1

$3 + 2 =$ ____

- ⬭ 1
- ⬭ 4
- ⬭ 5
- ⬭ 6

2

$4 - 1 =$ ____

- ⬭ 2
- ⬭ 3
- ⬭ 4
- ⬭ 6

3 How many?

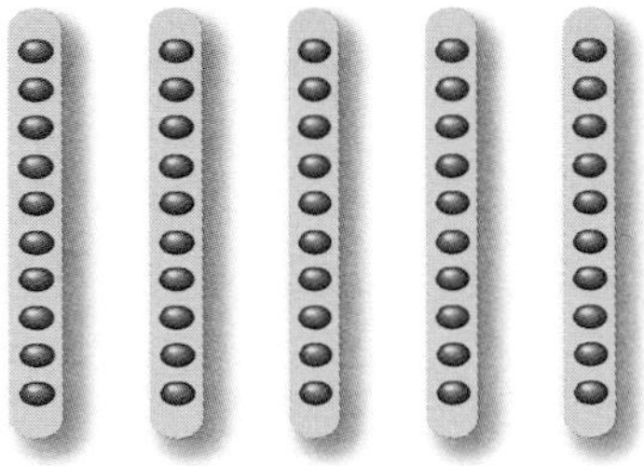

- ⬭ 5
- ⬭ 50
- ⬭ 60
- ⬭ 65

4 Which is the closed figure?

⬭  ⬭ 

⬭  ⬭ 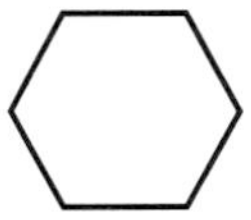

Which number sentence tells the story?

5 There are 7 big dogs.
There are 2 little dogs.
How many dogs in all?

- ⬭ $7 + 2 = 9$
- ⬭ $9 - 2 = 7$
- ⬭ $5 + 2 = 7$
- ⬭ not here

6 I had 9 nuts.
I ate 2 of them.
How many nuts are left?

- ⬭ $7 + 2 = 9$
- ⬭ $9 - 2 = 7$
- ⬭ $7 - 2 = 5$
- ⬭ not here

7 How many more  ?

$7 - 3 =$ ____

- ⬭ 3
- ⬭ 5
- ⬭ 4
- ⬭ 10

8 Which number is less?

⬭ 35 ⬭ 33

CHAPTER 16

Counting Pennies, Nickels, and Dimes

10¢
1¢
10¢
1¢
10¢
5¢
You have 16¢.
What could you buy?

Home Note In this chapter, your child will learn to count pennies, nickels and dimes.
ACTIVITY Have your child use the picture to tell what he or she could buy for 16 cents.

SCHOOL-HOME CONNECTION

Dear Family,

Today we started Chapter 16. In this chapter we will learn to count pennies, nickels, and dimes. Here are the new vocabulary words and an activity for us to do together at home.

Love,

____

Vocabulary

penny
one cent
1¢

nickel
five cents
5¢

dime
ten cents
10¢

ACTIVITY

Cut paper into squares. On each square, write an amount from 1¢ to 50¢. Have your child take a square and use coins to show the amount. Continue until all the squares have been used.

Visit our Web site for additional ideas and activities.
http://www.hbschool.com

Name ______________________

Pennies and Nickels

or

penny

1 cent

1¢

or

nickel

5 cents

5¢

Use penny and nickel. Draw them.
Count by ones or fives. Write the amount.

1. 2 pennies

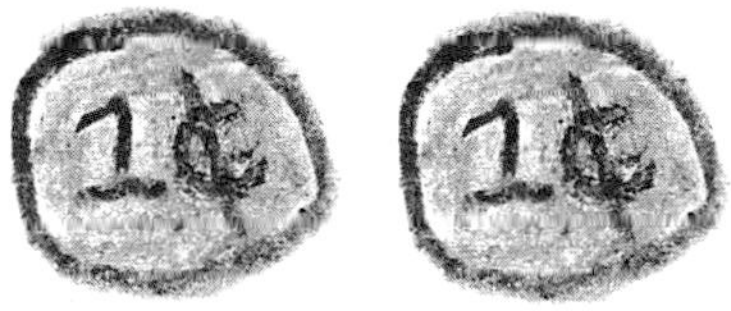

2 ¢

2. 3 pennies ☐ ¢

3. 4 pennies ☐ ¢

4. 2 nickels ☐ ¢

5. 3 nickels ☐ ¢

Talk About It • Critical Thinking

Which coin is worth more money?

Practice

Count by tens.
Write the amount.

1 

10¢, 20¢, 30¢ — 30¢

2 

____¢, ____¢ — ☐¢

3 

____¢, ____¢, ____¢, ____¢ — ☐¢

4 

____¢, ____¢, ____¢, ____¢, ____¢, ____¢ — ☐¢

Mixed Review

5. $4 + 2 =$ ☐
6. $3 + 3 =$ ☐
7. $2 + 6 =$ ☐
8. $5 + 1 + 3 =$ ☐
9. $6 + 1 + 2 =$ ☐
10. $3 + 1 + 4 =$ ☐

Home Note Your child counted collections of dimes to 90¢.
ACTIVITY Have your child count groups of pennies by ones and groups of dimes by tens to find their value.

Name ______________________________

Counting Collections of Nickels and Pennies

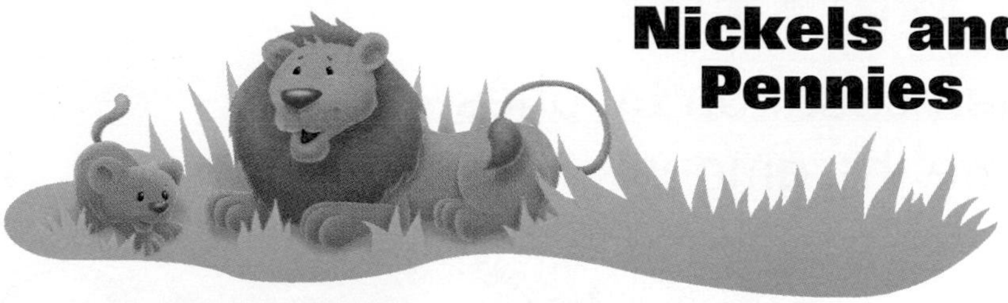

Count by fives.
Then count on by ones.
Write the amount.

1 

5 ¢, 10 ¢, 11 ¢, 12 ¢, 13 ¢, 14 ¢, 15 ¢ [15] ¢

2 

____¢, ____¢, ____¢, ____¢, ____¢, ____¢, ____¢ [ ] ¢

3 

____¢, ____¢, ____¢, ____¢, ____¢ [ ] ¢

4 

____¢, ____¢, ____¢, ____¢, ____¢, ____¢ [ ] ¢

5 

____¢, ____¢, ____¢, ____¢, ____¢, ____¢ [ ] ¢

Practice

Count by fives.
Then count on by ones.
Write the amount.

1  17 ¢

2  ___ ¢

3  ___ ¢

4  ___ ¢

5  ___ ¢

Problem Solving

Solve.
Draw a picture to show your answer.

6 Jack has 4 nickels.
He spends 1 nickel.
How much money
does he have left?

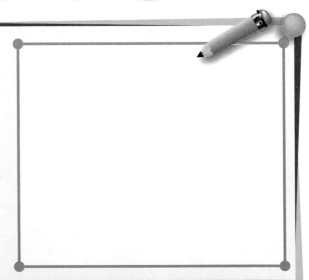

_______ ¢

Home Note **Your child counted mixed collections of pennies and nickels.**
ACTIVITY **Set out mixed groups of nickels and pennies in different ways, and have your child tell each amount.**

Name ______________________________

Counting Collections of Dimes and Pennies

Count by tens.
Then count on by ones.
Write the amount.

1 

10¢, 20¢, 30¢, 31¢, 32¢, 33¢, 34¢ [34]¢

2 

____¢, ____¢, ____¢, ____¢, ____¢, ____¢, ____¢ [ ]¢

3 

____¢, ____¢, ____¢, ____¢, ____¢, ____¢, ____¢ [ ]¢

4 

____¢, ____¢, ____¢, ____¢, ____¢, ____¢ [ ]¢

5 

____¢, ____¢, ____¢, ____¢, ____¢ [ ]¢

Practice

Count by tens.
Then count on by ones.
Write the amount.

1  41 ¢

2  ¢

3  ¢

4  ¢

5  ¢

Write About It

6 Bob buys a toy car for 23¢.
Use penny, dime, and dime.
Draw coins to show two ways
to make 23¢.

Home Note **Your child counted mixed collections of dimes and pennies.**
ACTIVITY Have your child count to find the value of mixed groups of dimes and pennies.

Name ______________________________

Understand • Plan • Solve • Look Back

Problem Solving

Choose the Model

Which two groups in each row add up to the amount on the tag? Color them.

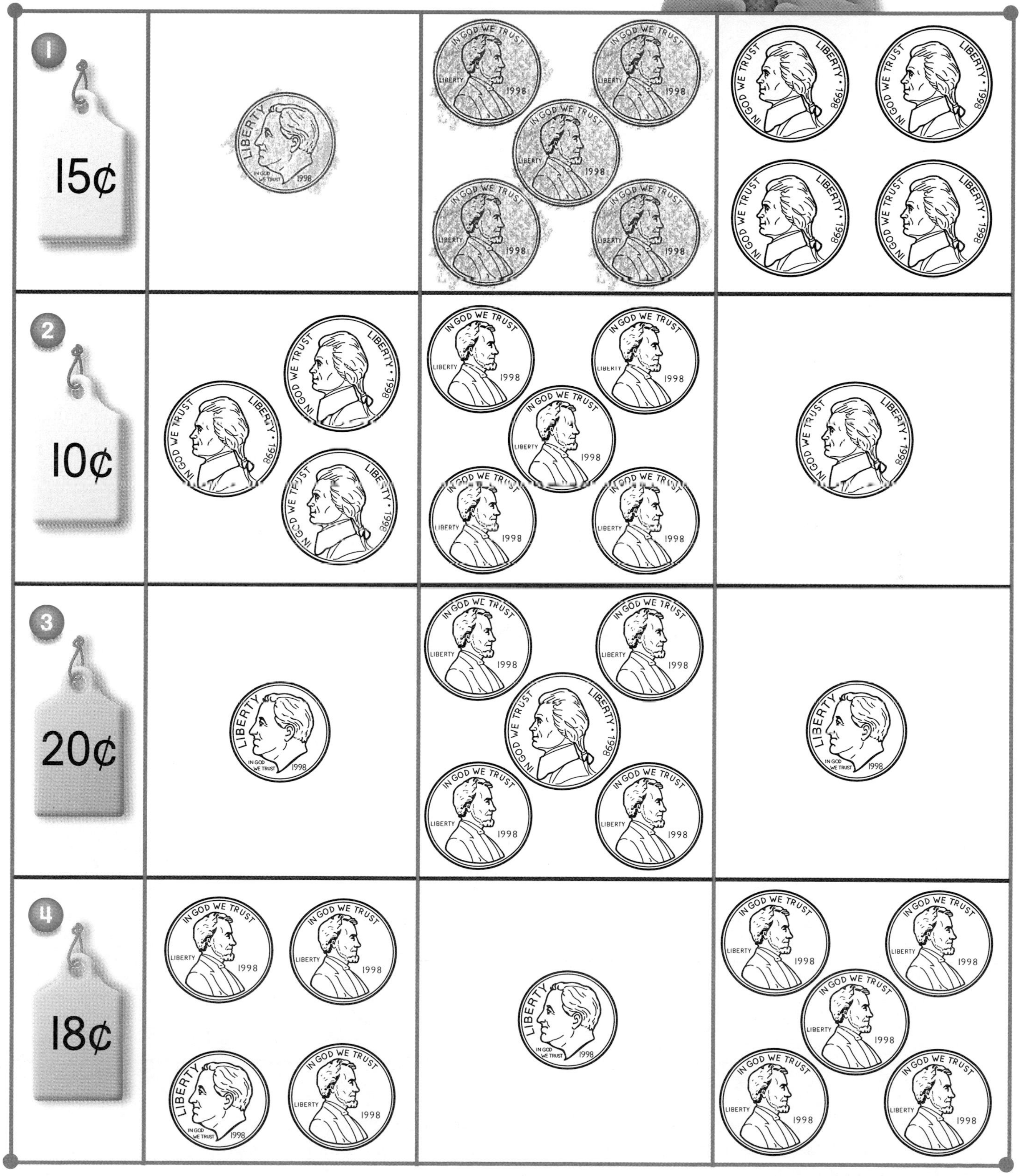

Practice

Which two groups in each row add up to the amount on the tag? Color them.

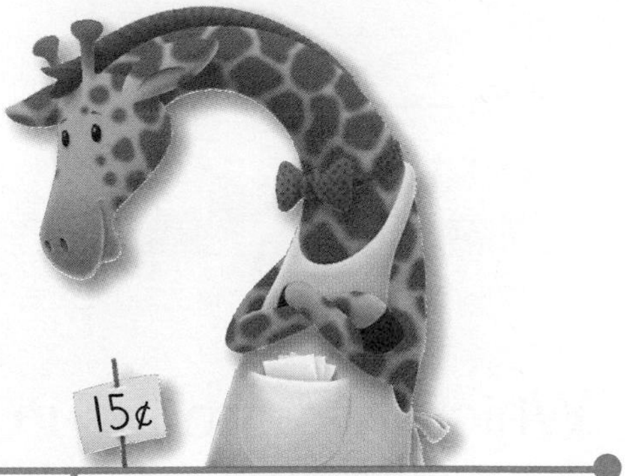

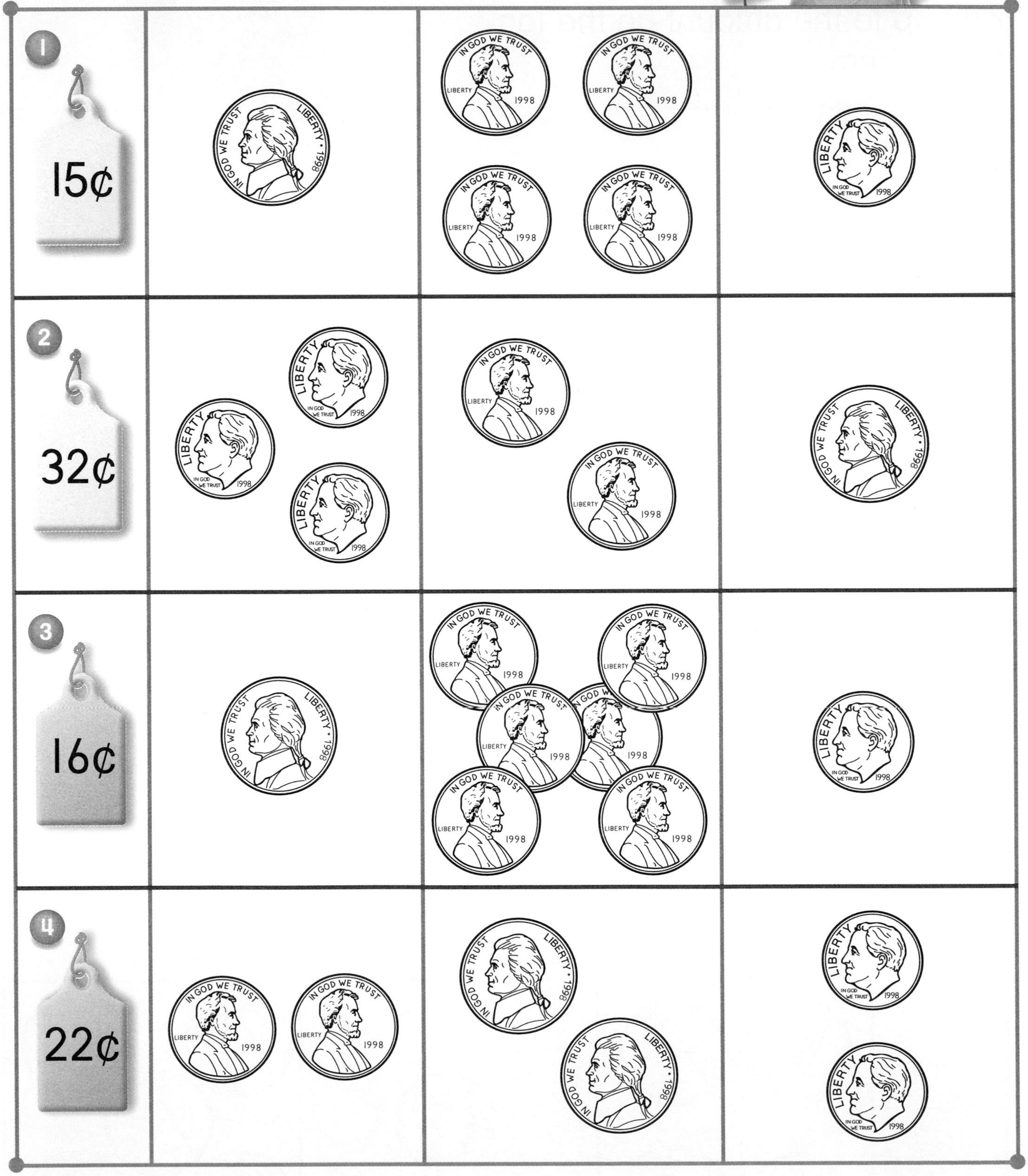

Home Note **Your child chose groups of coins to show amounts of money.**
ACTIVITY **Have your child use coins to show different ways to make amounts of money you name.**

Name ______________________________

Chapter 16

Review/Test

Concepts and Skills

Use  and . Draw them.
Count by ones or fives.
Write the amount.

1. 3 pennies

☐¢

2. 4 nickels

☐¢

Count. Write the amount.

3. 

____¢, ____¢, ____¢, ____¢, ____¢

☐¢

4. 

____¢, ____¢, ____¢, ____¢, ____¢, ____¢

☐¢

Problem Solving

Which two groups add up to the amount on the tag? Color them.

5. 

Name ______________________

Cumulative Chapters 1–16

TAAS Prep

Mark the best answer.

1. Which shape is a mistake in the pattern?

- ⬭ 
- ⬭ 
- ⬭ 

2. Which is a way to make 8?

- ⬭ 5 + 2
- ⬭ 6 + 3
- ⬭ 3 + 0
- ⬭ 7 + 1

3. Which number sentence tells how many are left?

- ⬭ 5 − 5 = 0
- ⬭ 5 − 2 = 3
- ⬭ 5 − 3 = 2
- ⬭ 5 − 0 = 5

4. Which coins add up to the amount on the tag?

- ⬭ 
- ⬭ 
- ⬭ 

5. Which number is greater?

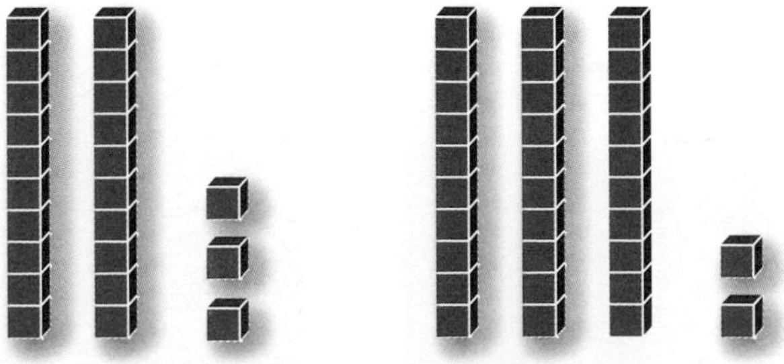

- ⬭ 23
- ⬭ 32

CHAPTER 17

Using Pennies, Nickels, and Dimes

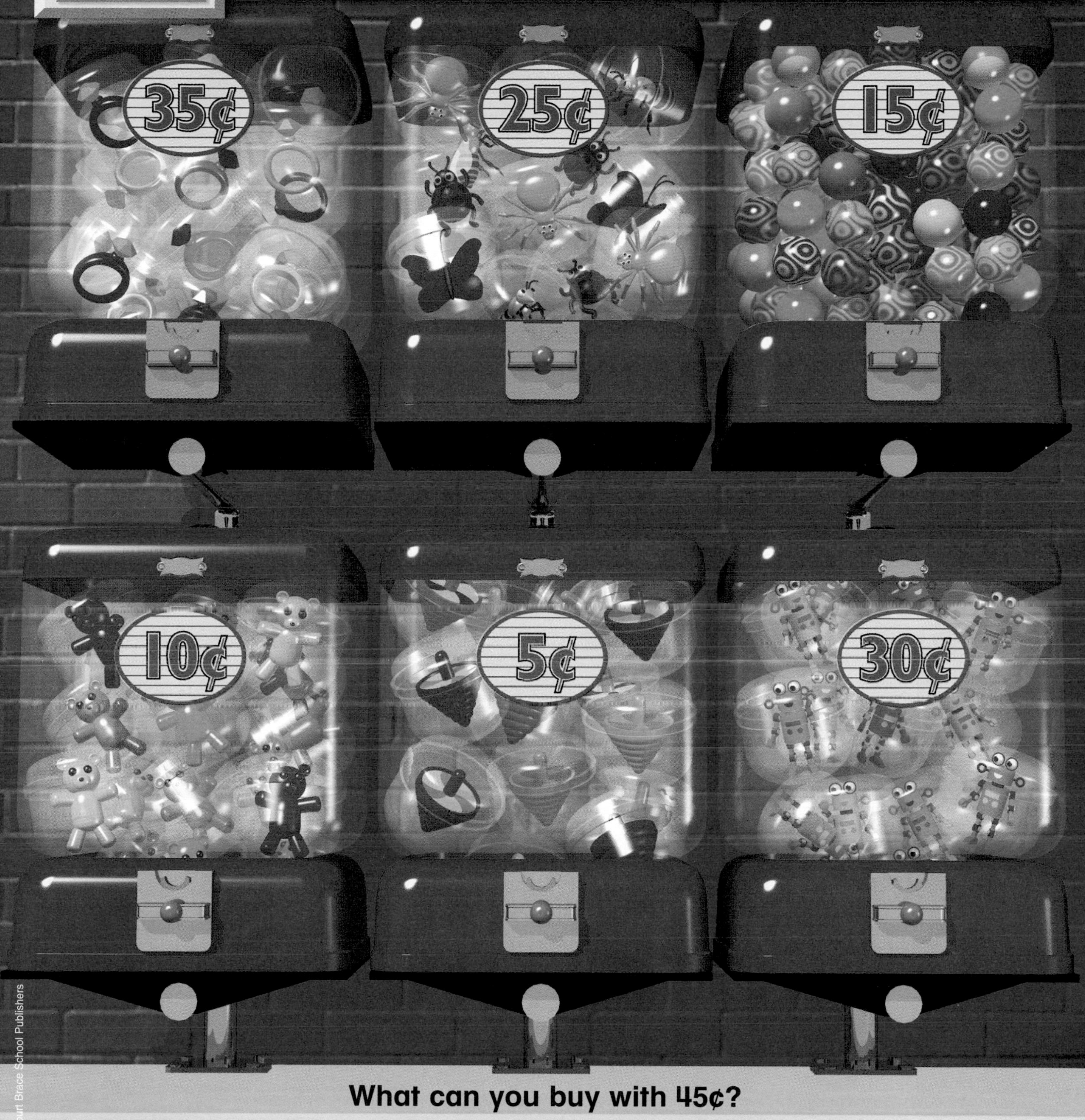

What can you buy with 45¢?
How could you use the fewest coins?

Home Note In this chapter, your child will learn to trade coins for equal amounts and to recognize the value of a quarter.
ACTIVITY Have your child name different ways to use pennies, nickels, dimes, and quarters to buy items in the picture.

SCHOOL-HOME CONNECTION

Dear Family,

Today we started Chapter 17. We will trade coins and learn the value of a quarter. Here are the new vocabulary words and an activity for us to do together.

Love,

____

Vocabulary

You can **trade** 5 pennies for 1 nickel.

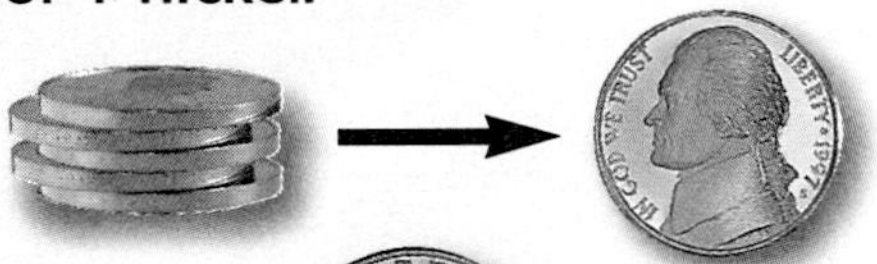

quarter  25 cents 25¢

2 dimes and 1 nickel are the **fewest** coins you can trade for a quarter.

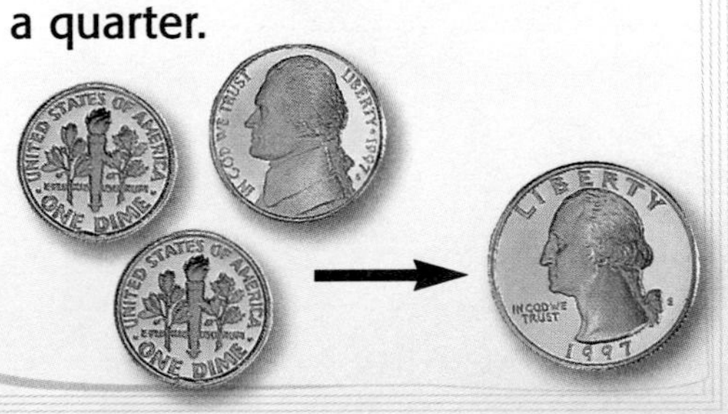

ACTIVITY

Name amounts of money up to 49¢. Have your child use coins to show each amount in several ways. Ask your child to point out the way that uses the fewest coins.

Visit our Web site for additional activities and ideas.
http://www.hbschool.com

Name ______________________________

Trading Pennies, Nickels, and Dimes

You can trade pennies for nickels and dimes.

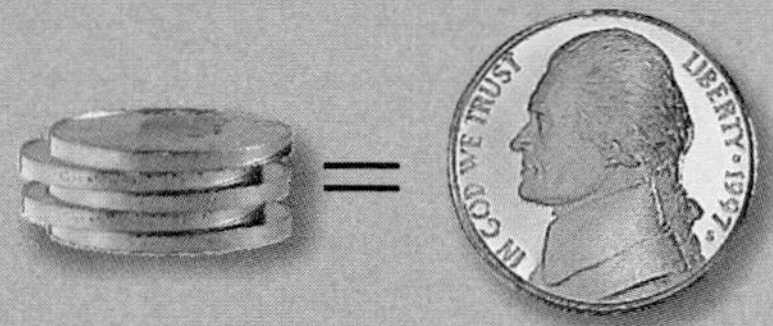

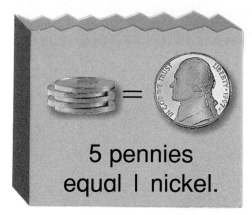

5 pennies equal 1 nickel.

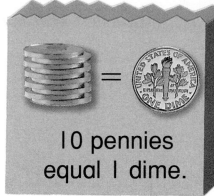

10 pennies equal 1 dime.

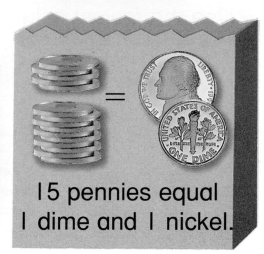

15 pennies equal 1 dime and 1 nickel.

Use coins. Trade pennies for nickels and dimes. Draw the coins.

1 

5¢

2 

3 

4 

Use coins. Trade pennies for nickels and dimes. Use the fewest coins. Draw the coins.

5 

6 

Talk About It • **Critical Thinking**

If you wanted the fewest coins in your pocket, what would you trade 20 pennies for?

Practice

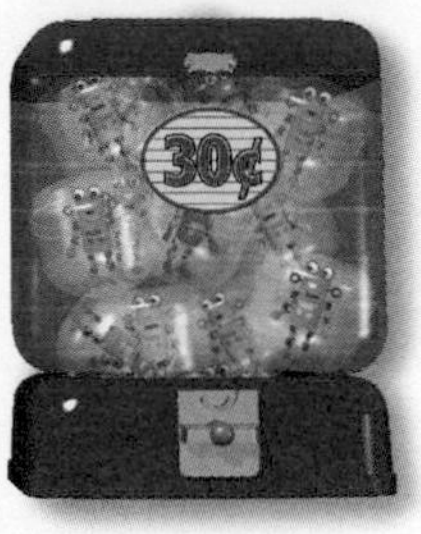

Use coins. Trade for nickels and dimes.
Use the fewest coins. Draw the coins.

1 

2 

3 

4 

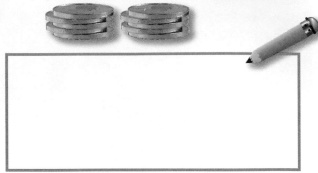

Problem Solving • **Reasoning**

Solve.
Draw a picture to show your answer.

5 Jenny bought a toy dog.
She used 5 nickels.

Did she use the fewest coins that equal 25¢? ______

Show the price using the fewest coins.

Home Note **Your child traded groups of coins for other groups that equaled the same amount of money.**
ACTIVITY **Have your child show you different groups of coins that equal the same amount of money as a nickel and a dime.**

Name ______________________________

Equal Amounts

Use coins.
Show the amount in two ways.
Draw the coins.
Circle the way that uses fewer coins.

1

15¢

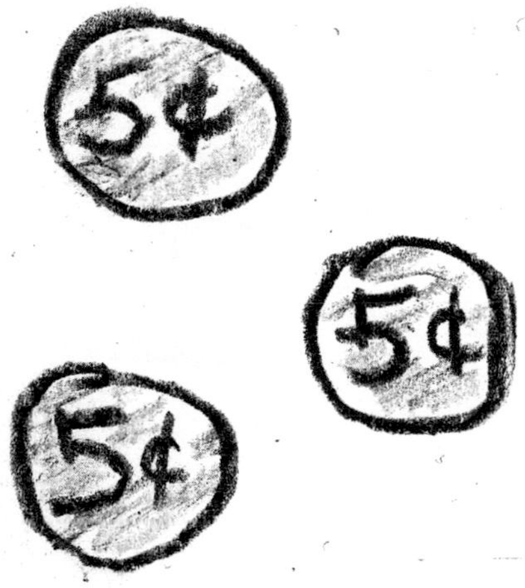

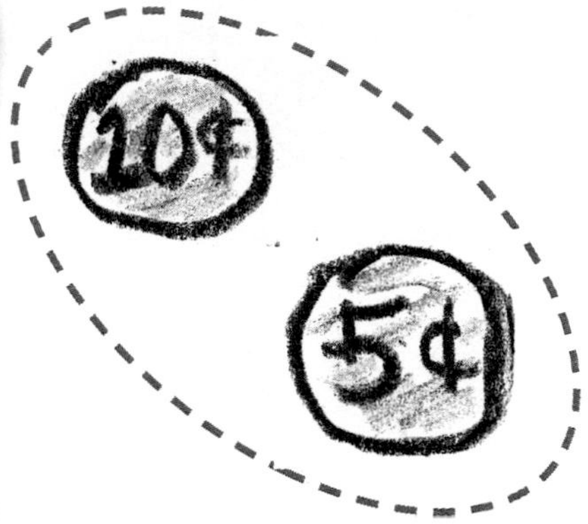

2

20¢

3

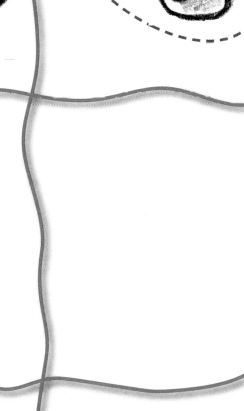

25¢

Practice

Use coins.
Show the amount in two ways.
Draw the coins.
Circle the way that uses fewer coins.

1

10¢

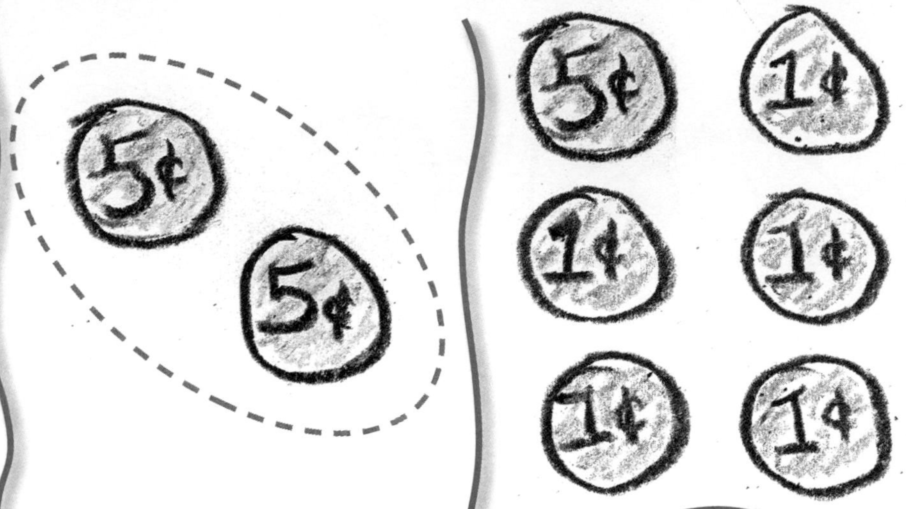

2

30¢

3

35¢

Harcourt Brace School Publishers

Home Note **Your child showed an amount of money in two ways and identified the way that used fewer coins.**
ACTIVITY **Name an amount of money. Have your child show you the fewest coins that will equal that amount.**

Name ______________________________

How Much Is Needed?

Circle the coins you need.
Use the fewest coins.

1

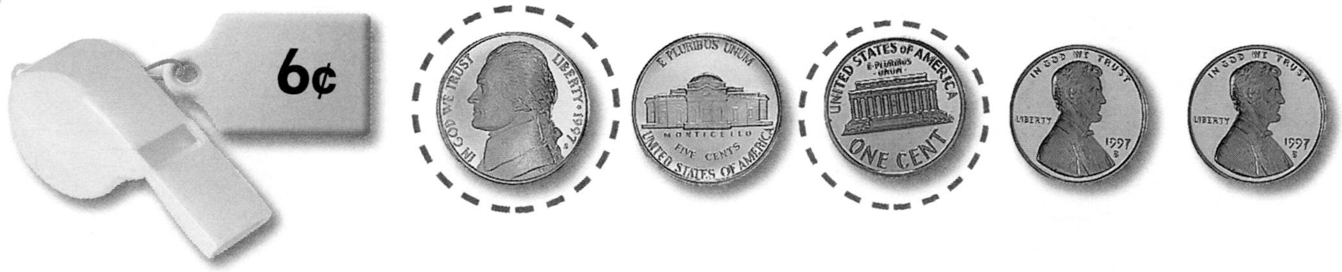

2

3

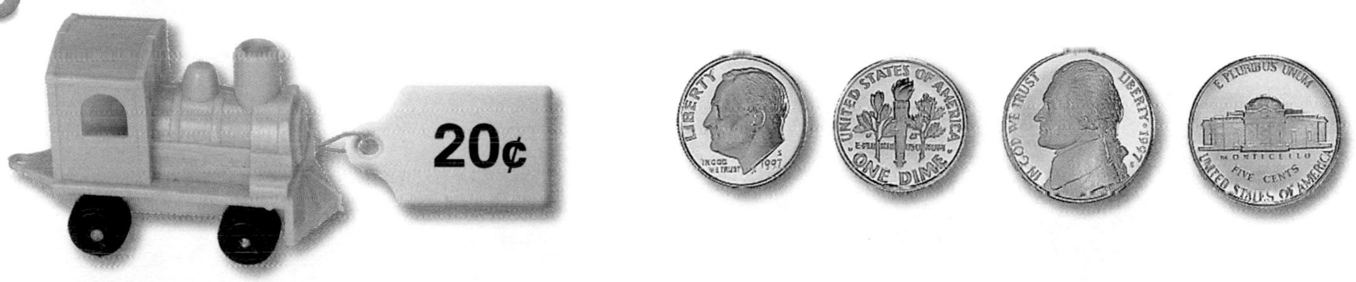

4

5

Practice

Circle the coins you need.
Use the fewest coins.

1

2

3

4

Mixed Review

Write the numbers that come just before, just after, and between.

5 ____, 98, ____ 6 ____, 72, ____ 7 ____, 30, ____

8 12, ____, 14 9 23, ____, 25 10 ____, 53, ____

Home Note Your child chose the fewest coins needed to make up an amount of money.
ACTIVITY As you shop, have your child tell you the fewest coins you could use to pay for items.

Name ______________________________

Quarter

 or 

1 quarter = 25¢

Use Workmat 4 and coins.
Show ways to make 25¢.
Write how many of each coin you need.

1	0	2	1	0
2				
3				
4				
5				

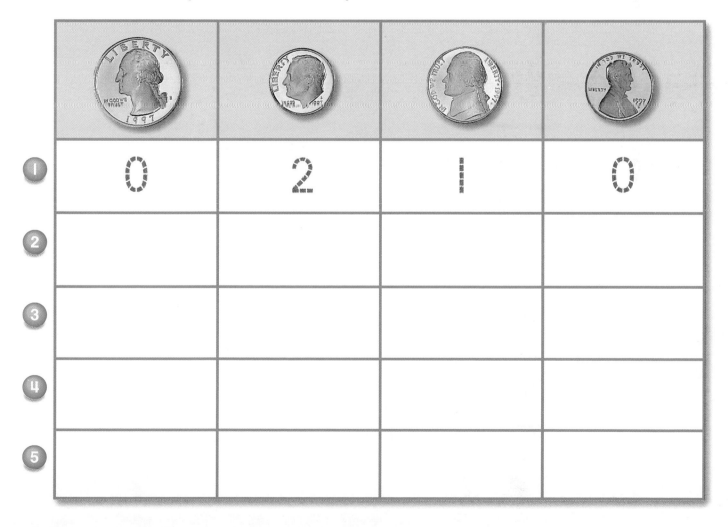

Talk About It • Critical Thinking

Which way uses the fewest coins?

Practice

Write each amount.
Circle the coins that equal a  .

1 

25 ¢

2 

_____ ¢

3 

_____ ¢

4 

_____ ¢

5 

_____ ¢

6 

_____ ¢

Write About It

7 Winnie buys a toy car for 20¢.
Draw 4 different groups of coins
she could use.

Home Note Your child learned the value of a quarter.
ACTIVITY Have your child show you four different groups of coins that equal a quarter.

Name ____________________

Understand • Plan • Solve • Look Back

Problem Solving
Act It Out

Take turns with a partner.
One person can be the shopper.
The other can be the store clerk.
Use the fewest coins to buy things.

Draw what you bought.	Draw the coins you used.
1	
2	

Practice

Play store again.
Use the fewest coins
to buy things.

Draw what you bought.	Draw the coins you used.
1	
2	

Home Note Your child acted out buying items in a store.
ACTIVITY Cut out items from magazines and price them from 5¢ to 49¢. Have your child "buy" the items with pennies, nickels, dimes, and quarters.

Name ______________________________

Concepts and Skills

Use coins.
Trade pennies for nickels and dimes.
Draw the coins.

1 

2 

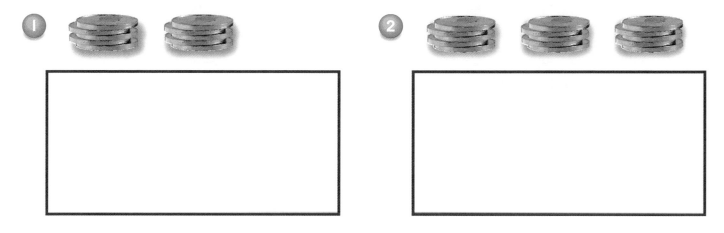

Show the amount in two ways.
Draw the coins.
Circle the way that uses fewer coins.

3 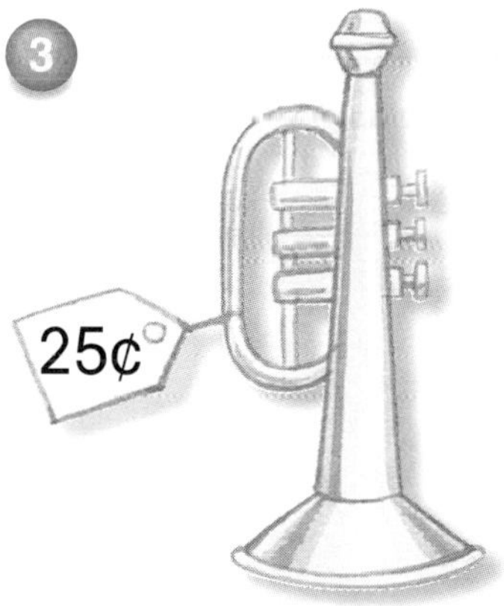

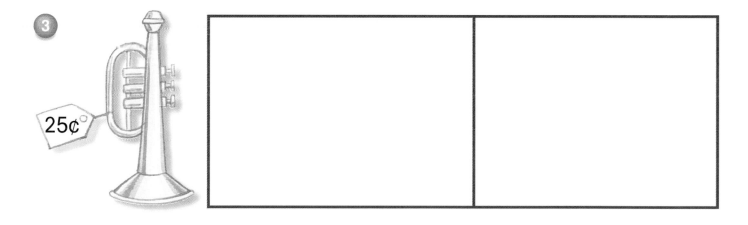

Circle the coins you need.
Use the fewest coins.

4 

Name ______________________________

Cumulative Chapters 1–17

TAAS Prep

Mark the best answer.

1. How many?

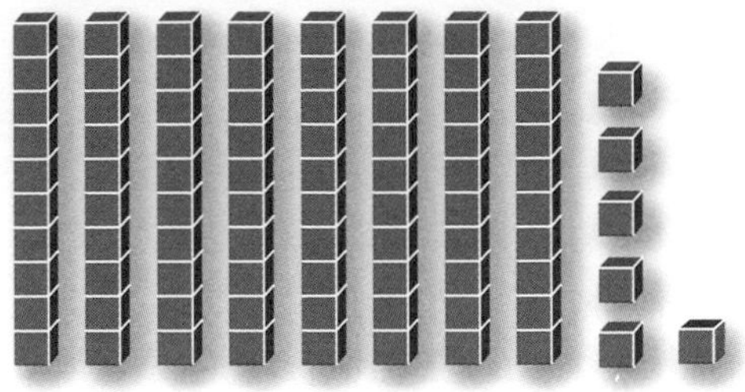

- ⬭ 68
- ⬭ 86
- ⬭ 46
- ⬭ 91

2. Which comes between?

29, ____ ,31

- ⬭ 28
- ⬭ 30
- ⬭ 32
- ⬭ 34

3. How much money?

- ⬭ 1¢
- ⬭ 5¢
- ⬭ 10¢
- ⬭ 25¢

4. How much money?

- ⬭ 10¢
- ⬭ 25¢
- ⬭ 1¢
- ⬭ 5¢

5. How many in all?

9 + 3 = ____

- ⬭ 12
- ⬭ 6
- ⬭ 93
- ⬭ 8

6. Which comes next?

10, 20, 30, 40, ____, ____

- ⬭ 45, 50
- ⬭ 41, 42
- ⬭ 60, 80
- ⬭ 50, 60

CHAPTER 18

Using a Calendar

Name a month that goes with each picture.

Home Note **In this chapter, your child is learning to order days and months and read a calendar.**
ACTIVITY **Have your child use the picture to name the seasons of the year and tell about different activities that people do in each season.**

SCHOOL-HOME CONNECTION

Dear Family,

Today we started Chapter 18. In this chapter we will learn to order months and days and to read a calendar. Here are the new vocabulary words and an activity for us to do together at home.

Love,

____

Vocabulary

days — month — week

OCTOBER						
Sunday	Monday	Tuesday	Wednesday	Thursday	Friday	Saturday
					1	2
3	4	5	6	7	8	9
10	11	12	13	14	15	16
17	18	19	20	21	22	23
24	25	26	27	28	29	30
31						

ACTIVITY

Write the days of the week and the months of the year on slips of paper. Give your child one group of names to put in order. Take away some of the names while your child is not looking, and have him or her tell you the missing days or months.

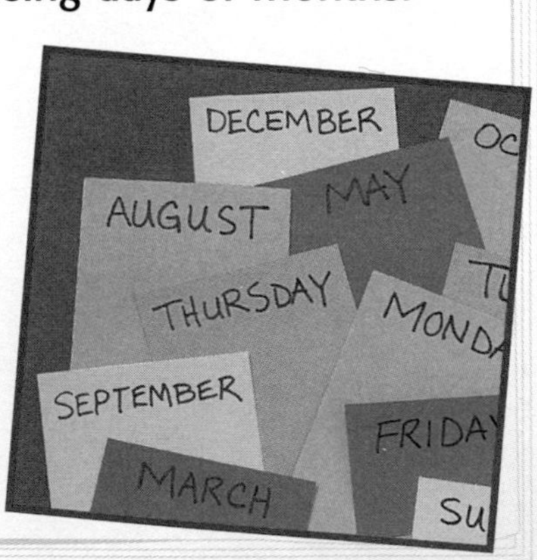

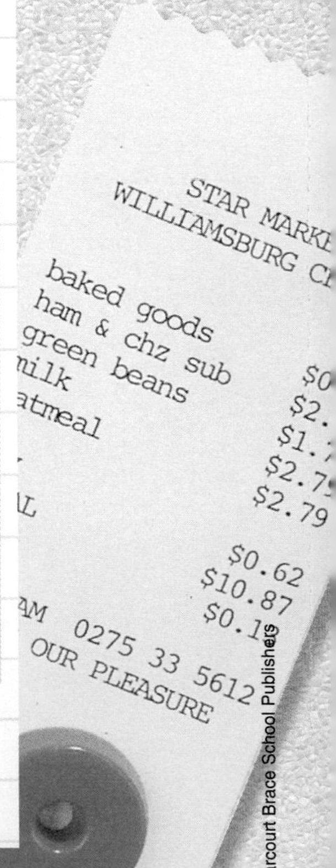

Visit our Web site for additional ideas and activities.
http://www.hbschool.com

Name ________________________________

Ordering Months and Days

January is the first month.

January						
Sunday	Monday	Tuesday	Wednesday	Thursday	Friday	Saturday
					1	2
3	4	5	6	7	8	9
10	11	12	13	14	15	16
17	18	19	20	21	22	23
24/31	25	26	27	28	29	30

February						
Sunday	Monday	Tuesday	Wednesday	Thursday	Friday	Saturday
	1	2	3	4	5	6
7	8	9	10	11	12	13
14	15	16	17	18	19	20
21	22	23	24	26	26	27
28						

March						
Sunday	Monday	Tuesday	Wednesday	Thursday	Friday	Saturday
	1	2	3	4	5	6
7	8	9	10	11	12	13
14	15	16	17	18	19	20
21	22	23	24	25	26	27
28	29	30	31			

April						
Sunday	Monday	Tuesday	Wednesday	Thursday	Friday	Saturday
				1	2	3
4	5	6	7	8	9	10
11	12	13	14	15	16	17
18	19	20	21	22	23	24
25	26	27	28	29	30	

May						
Sunday	Monday	Tuesday	Wednesday	Thursday	Friday	Saturday
						1
2	3	4	5	6	7	8
9	10	11	12	13	14	15
16	17	18	19	20	21	22
23/30	24/31	25	26	27	28	29

June						
Sunday	Monday	Tuesday	Wednesday	Thursday	Friday	Saturday
		1	2	3	4	5
6	7	8	9	10	11	12
13	14	15	16	17	18	19
20	21	22	23	24	25	26
27	28	29	30			

July						
Sunday	Monday	Tuesday	Wednesday	Thursday	Friday	Saturday
				1	2	3
4	5	6	7	8	9	10
11	12	13	14	15	16	17
18	19	20	21	22	23	24
25	26	27	28	29	30	31

August						
Sunday	Monday	Tuesday	Wednesday	Thursday	Friday	Saturday
1	2	3	4	5	6	7
8	9	10	11	12	13	14
15	16	17	18	19	20	21
22	23	24	25	26	27	28
29	30	31				

September						
Sunday	Monday	Tuesday	Wednesday	Thursday	Friday	Saturday
			1	2	3	4
5	6	7	8	9	10	11
12	13	14	15	16	17	18
19	20	21	22	23	24	25
26	27	28	29	30		

October						
Sunday	Monday	Tuesday	Wednesday	Thursday	Friday	Saturday
					1	2
3	4	5	6	7	8	9
10	11	12	13	14	15	16
17	18	19	20	21	22	23
24/31	25	26	27	28	29	30

November						
Sunday	Monday	Tuesday	Wednesday	Thursday	Friday	Saturday
	1	2	3	4	5	6
7	8	9	10	11	12	13
14	15	16	17	18	19	20
21	22	23	24	25	26	27
28	29	30				

December						
Sunday	Monday	Tuesday	Wednesday	Thursday	Friday	Saturday
			1	2	3	4
5	6	7	8	9	10	11
12	13	14	15	16	17	18
19	20	21	22	23	24	25
26	27	28	29	30	31	

Write **1** to **12** in front of the names of the months to show their order.

1. ______ March
2. ______ February
3. ______ May
4. __1__ January
5. ______ August
6. ______ June
7. ______ July
8. ______ September
9. ______ December
10. ______ October
11. ______ November
12. ______ April

Talk About It • Critical Thinking

What month comes after December? How do you know?

Practice

MARCH

Sunday	Monday	Tuesday	Wednesday	Thursday	Friday	Saturday
	1	2	3	4	5	6
7	8	9	10	11	12	13
14	15	16	17	18	19	20
21	22	23	24	25	26	27
28	29	30	31			

Write the days of the week in order.

1 ____________________ 2 ____________________

3 ____________________ 4 ____________________

5 ____________________ 6 ____________________

7 ____________________

8 Color the Sundays

.

9 Color the Wednesdays

.

10 Color the Fridays

.

Home Note Your child ordered the months of the year and the days of the week.
ACTIVITY Have your child draw pictures of events and label each with the day of the week or month of the year when it might happen.

Name ______________________

Understand • Plan • Solve • Look Back

Problem Solving
Using a Calendar

Fill in the calendar for this month.
Use the calendar to
answer the questions.

There are 7 days in 1 week.

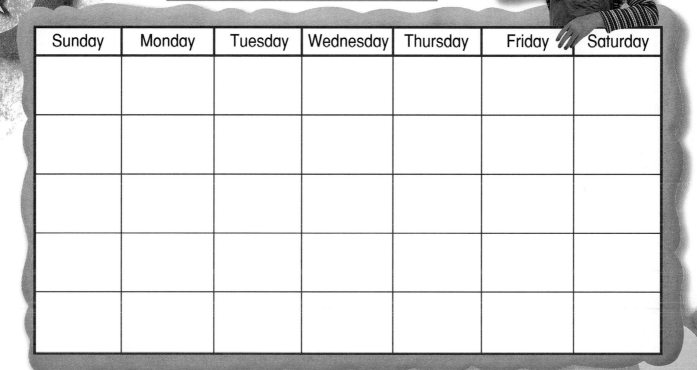

Sunday	Monday	Tuesday	Wednesday	Thursday	Friday	Saturday

1. How many days are in the month? ______

2. What day is it today? ______

3. What day was it yesterday? ______

4. What day will it be tomorrow? ______

Talk About It • Critical Thinking

How many days are there in two weeks?

Practice

Fill in the calendar for next month.
Use the calendar to answer the questions.

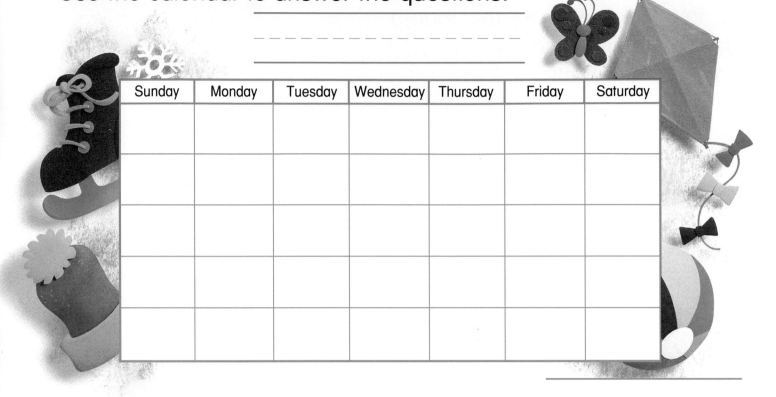

Sunday	Monday	Tuesday	Wednesday	Thursday	Friday	Saturday

1. On what day does the month begin?

2. On what day does the month end?

3. How many days are in the month?

4. What is the date of the first Monday?

Mixed Review

Write the amount.

5.  ¢

6.  ¢

Home Note **Your child made and used a calendar.**
ACTIVITY **Point out a date on a calendar, and have your child tell which dates come just before and just after it.**

Name ____________________

Ordering Events

Draw pictures of things you do each day.

Practice

Draw something special you do on

Write About It

Draw a picture of something you like to do.
Write the month and day of the week you do it.

Home Note **Your child ordered events.**
ACTIVITY Have your child tell you some events of his or her school day in the order in which they happened.

Name ______________________________

Understand • Plan • Solve • Look Back

Problem Solving
Estimating Time

How much time do these things take to do?
Number them from the shortest time to the longest time.

1.

1

3

2

2.

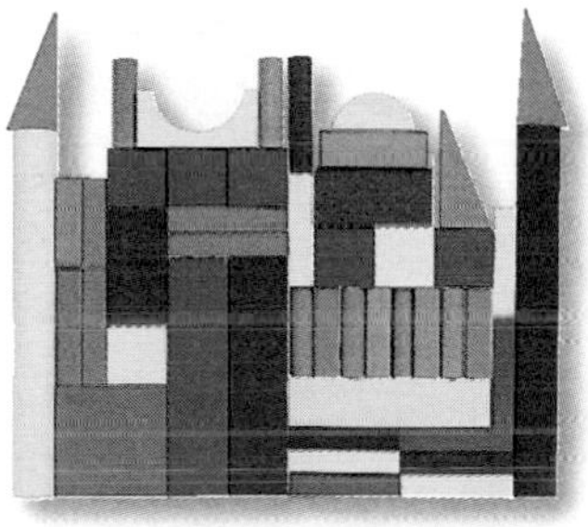

____

____

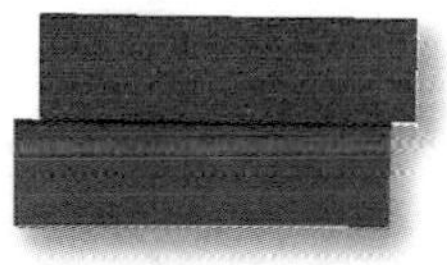

____

3.

____

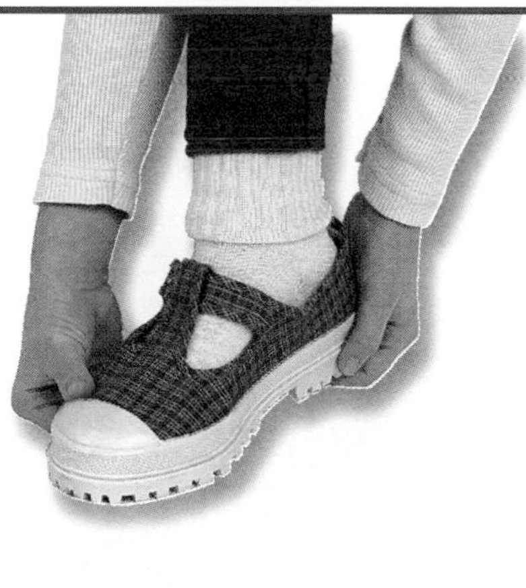

____

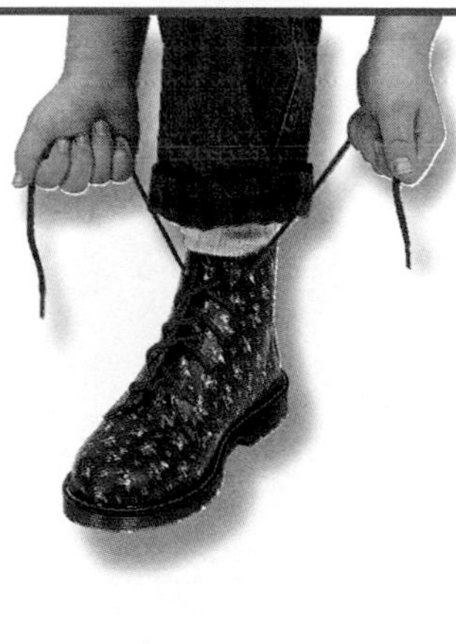

____

Talk About It • Critical Thinking

How did you decide which thing takes the longest time to do?

Practice

How much time do these things take to do? Number them from the shortest time to the longest time.

1

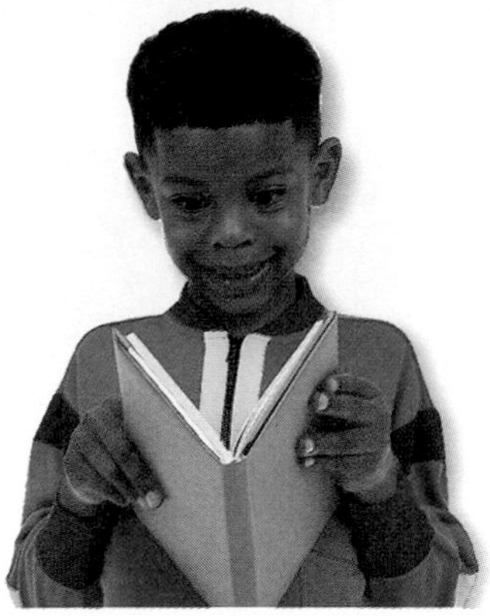

2 ____

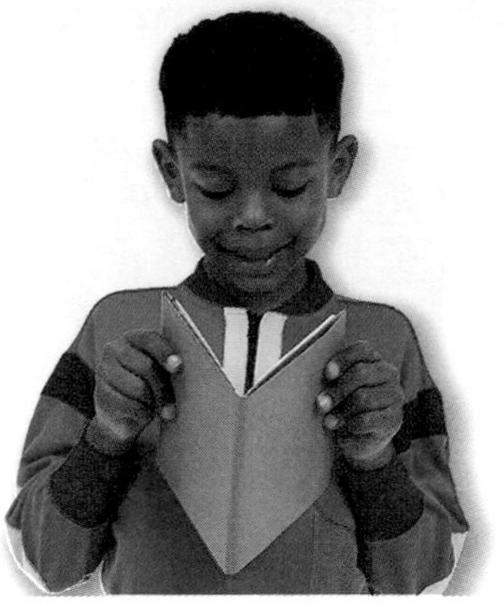

1 ____

3 ____

2

____

____

____

3

____

____

____

Home Note **Your child ordered events according to the time they would take to do.**
ACTIVITY **Have your child take note of everyday activities and order the activities according to the time they take.**

Name ______________________________

Chapter 18

Review/Test

Concepts and Skills

April

Sunday	Monday	Tuesday	Wednesday	Thursday	Friday	Saturday
				1	2	3
4	5	6	7	8	9	10

Use the calendar to answer these questions.

1. What is the name of this month? ____________
2. On what day does the month begin? ____________
3. What is the date of the first Friday? ____________
4. What is the date of the second Thursday? ____________

Draw pictures of things you do each day.

5.

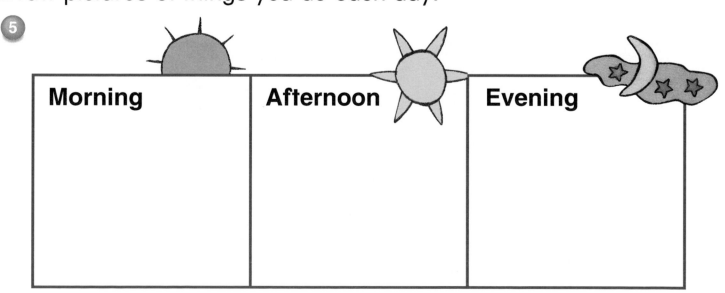

Morning	Afternoon	Evening

Name ________________________________

Cumulative Chapters 1–18

TAAS Prep

Mark the best answer.

1 Which thing takes longest to do?

- ⬭ 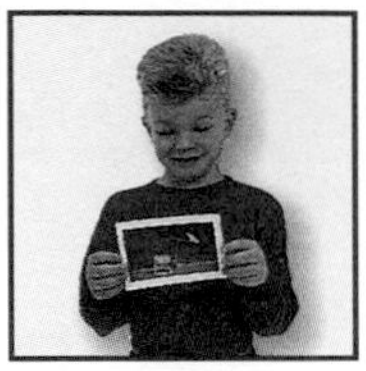
- ⬭ 
- ⬭ 

2 Which shows the numbers in order from least to greatest?

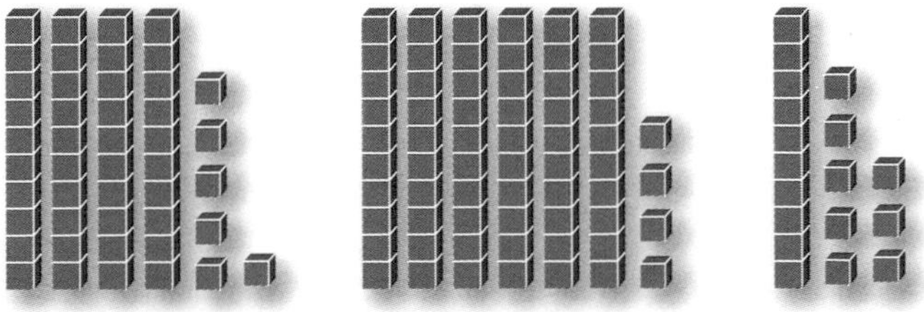

- ⬭ 18, 46, 64
- ⬭ 18, 64, 46
- ⬭ 64, 46, 187
- ⬭ 46, 18, 64

3 Which coin can you use to buy the pencil?

- ⬭ 1 penny
- ⬭ 1 nickel
- ⬭ 1 dime
- ⬭ 1 quarter

4 Count by fives. How many fingers in all?

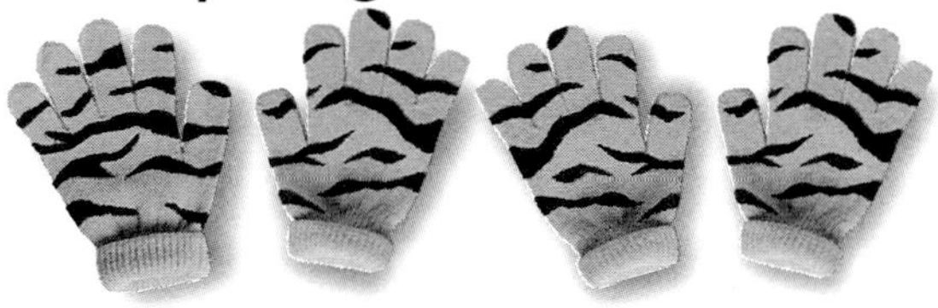

- ⬭ 4
- ⬭ 5
- ⬭ 10
- ⬭ 20

5 How many tens and ones?

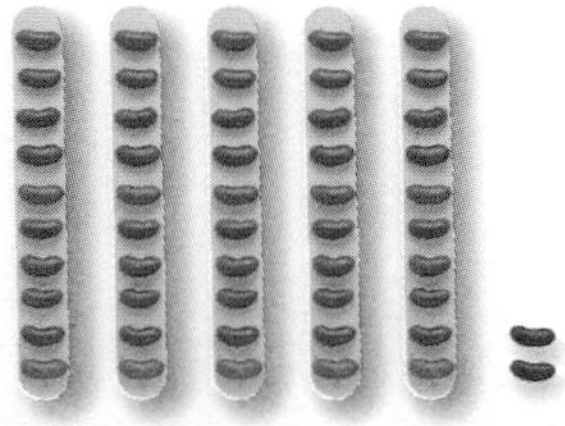

- ⬭ 4 tens 2 ones
- ⬭ 5 tens 2 ones
- ⬭ 2 tens 5 ones
- ⬭ 8 tens 3 ones

CHAPTER 19

Telling Time

What time of the day do you think it is?
What clues from the picture tell you this?

Home Note **In this chapter, your child is learning to tell time to the hour and half hour.**
ACTIVITY Discuss with your child what is happening in this picture. Have your child name the time of the day when the picture might happen.

SCHOOL-HOME CONNECTION

Dear Family,

Today we started Chapter 19. We will learn to tell time to the hour and the half hour. Here are the new vocabulary words and an activity for us to do together at home.

Love,

____

Vocabulary

The time is four **o'clock.**

The time is three thirty.

ACTIVITY

Give your child magazine or newspaper pictures of people doing everyday activities. Have your child tell a time of day (to the hour) when each activity could happen.

Visit our Web site for additional ideas and activities.
http://www.hbschool.com

Name ________________________

Reading the Clock

Write the missing numbers.

Use a clock. Show each time.
Write the time two ways.

1

1 o'clock

1:00

2

____ o'clock

___:___

3

____ o'clock

___:___

Talk About It • **Critical Thinking**

How are the minute hand and hour hand alike?
How are they different?

Practice

Use a clock. Show each time.
Write the time two ways.

1 

3 o'clock

3:00

2 

______ o'clock

___:___

3 

______ o'clock

___:___

4 

______ o'clock

___:___

5 

______ o'clock

___:___

6 

______ o'clock

___:___

7 

______ o'clock

___:___

8 

______ o'clock

___:___

9 

______ o'clock

___:___

Home Note Your child wrote the time to the hour.
ACTIVITY At times on the hour, have your child show you the minute hand and the hour hand on a clock and tell what numbers they are pointing to.

Name ______________________________

Hour

The hour is the same on both clocks. It is 6 o'clock.

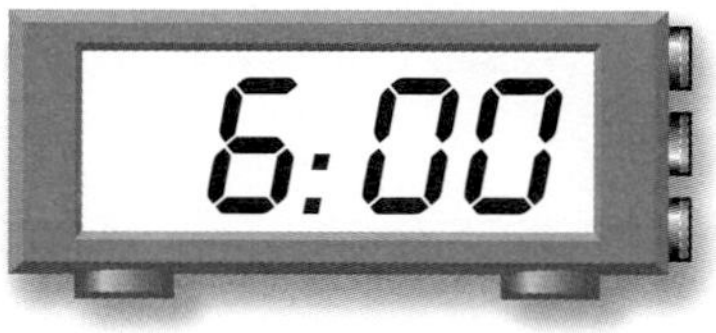

Read the clock. Write the time.

1

2

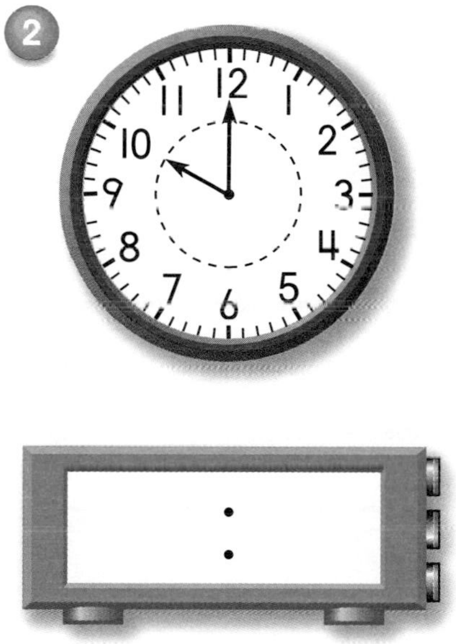

3

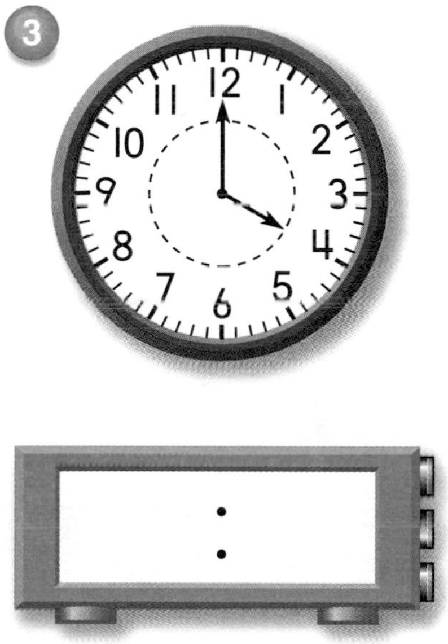

4

5

6

Practice

Write the time.

1

2

3
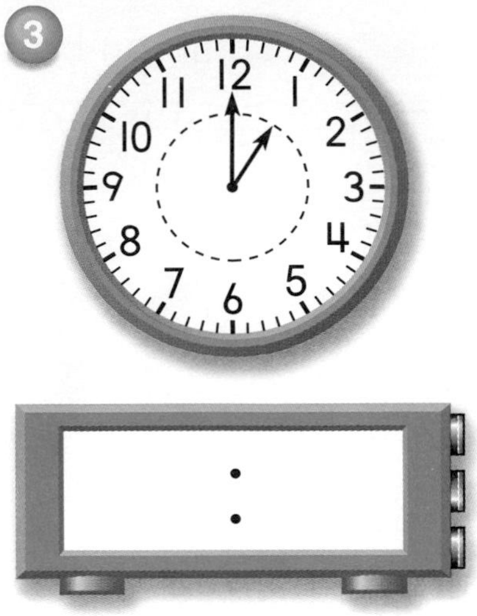

4

5

6

Mixed Review

How much money?

7

_____ ¢

8

_____ ¢

9
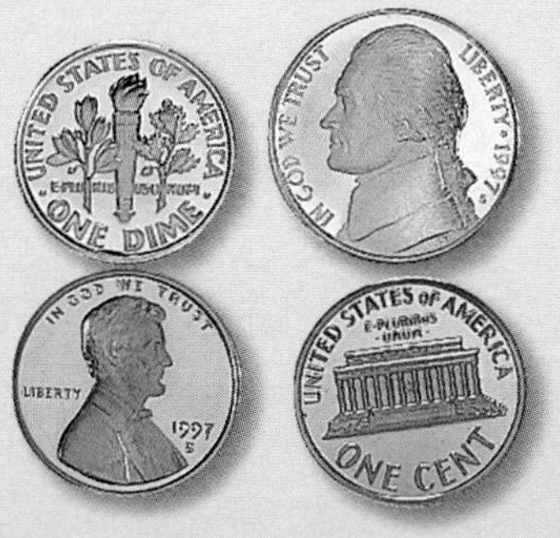

_____ ¢

Home Note **Your child learned to tell time on an analog and digital clock.**
ACTIVITY On the hour, have your child read the time on a digital clock.

Name ______________________________

Time to the Hour

Draw the hour hand and the minute hand.

1

2

3

4

5

6

7

8

9

Practice

Draw the hour hand and the minute hand.

1

2
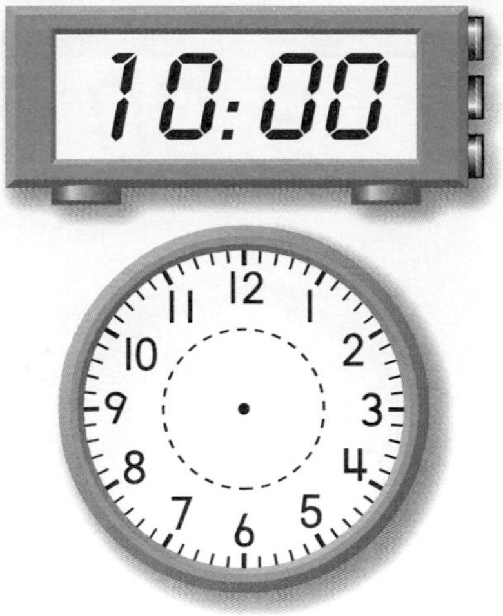

3

4

5

6

Problem Solving

Draw a clock to show your answer.

7 Matt gets up at 6 o'clock.
Mike gets up 3 hours later.
What time does Mike get up?

______ o'clock

Home Note **Your child drew hands on an analog clock.**
ACTIVITY **On the hour, have your child read the time on an analog clock.**

Name ______________________________

Half-Hour

Where are the hands?
Write the numbers. Write the time.

1. The hour hand is between 1 and 2.

   The minute hand is at 6.

1:30

---

2. The hour hand is between ______ and ______.

   The minute hand is at ______.

___ : ___

---

3. The hour hand is between ______ and ______.

   The minute hand is at ______.

___ : ___

Talk About It • Critical Thinking

How does the hour hand move when a half-hour passes?
How does the minute hand move?

Practice

Write the time.

1

____ : ____

2

____ : ____

3

____ : ____

4

____ : ____

5

____ : ____

6

____ : ____

Write About It

7 Write a sentence telling about something you do at 30 minutes after an hour.

Home Note Your child learned to read time to the half-hour.
ACTIVITY On the half-hour, have your child show you the minute hand and the hour hand on a clock and tell what numbers they are pointing to.

Name ______________________________

Understand • Plan • Solve • Look Back

Problem Solving
Act It Out

How long is a minute? You can estimate it.

Close your eyes.

Guess when 1 minute has passed. Raise your hand.

Was your guess too long or too short?
Try again. Was your guess better this time?
Name some things that you can do in 1 minute.

About how long would it take? Circle your estimate.
Then act it out to see if you are right.

1. sing a song

more than a minute
less than a minute

2. say hello

more than a minute
less than a minute

3. write a letter

more than a minute
less than a minute

Practice

About how long would it take?
Circle your estimate.
Then act it out to see
if you are right.

1. write the alphabet

more than a minute

less than a minute

2. write 1 to 10

more than a minute

less than a minute

3. paint a picture

more than a minute

less than a minute

4. say your name

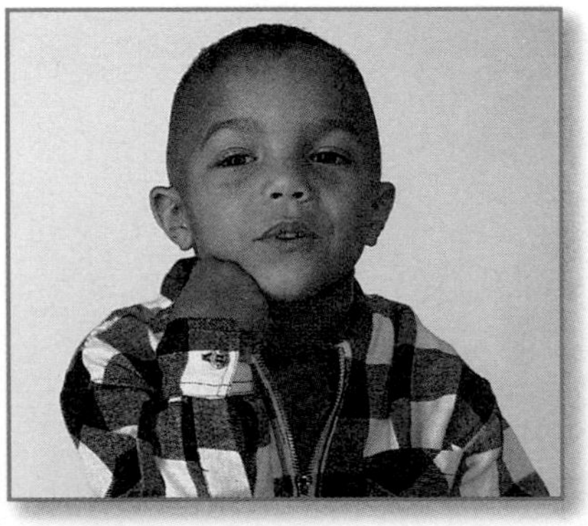

more than a minute

less than a minute

5. count to 100

more than a minute

less than a minute

6. write your name

more than a minute

less than a minute

Home Note Your child estimated time and acted out activities.
ACTIVITY Have your child name an activity that he or she thinks will take about one minute. Time the activity to see if it takes one minute, or more, or less.

Name ______________________

Chapter 19

Review/Test

Concepts and Skills

Write the time two ways.

1 

_____ o'clock

_____ : _____

2 

_____ o'clock

_____ : _____

3 

_____ o'clock

_____ : _____

Write the time.

4 

_____ : _____

5 

_____ : _____

6 

_____ : _____

Problem Solving

About how long would it take? Circle your estimate.

7 write to 100

more than a minute

less than a minute

8 say good morning

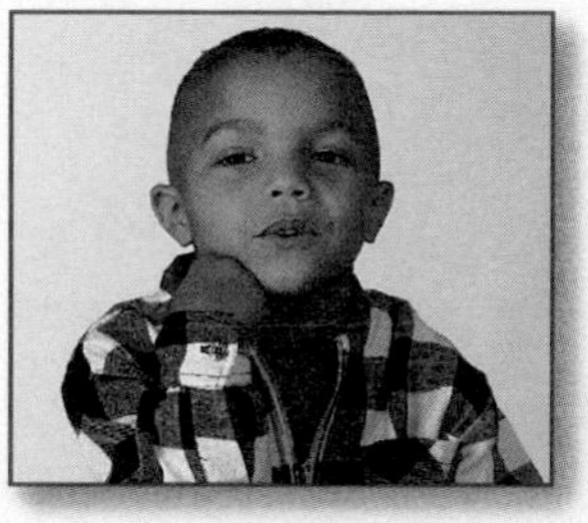

more than a minute

less than a minute

Name ______________________________

TAAS Prep

Cumulative Chapters 1–19

Mark the best answer.

1. What time is it?

- ⬭ 1:30
- ⬭ 2:30
- ⬭ 5:00
- ⬭ 5:30

2. Which clock shows 7:00?

- ⬭ 
- ⬭ 
- ⬭ 
- ⬭ 

3. Karly put her pennies in groups of five. She wants to trade them for nickels. How many nickels will she get?

- ⬭ 4
- ⬭ 5
- ⬭ 15
- ⬭ 20

4. Count by tens. What is the amount?

- ⬭ 7¢
- ⬭ 20¢
- ⬭ 65¢
- ⬭ 70¢

5. Is 15 even or odd?

- ⬭ even
- ⬭ odd

6. Which number is greater?

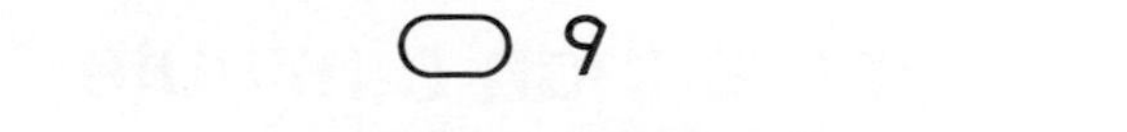

16

9

- ⬭ 9
- ⬭ 16

Name ______________________________

MATH FUN

Use:

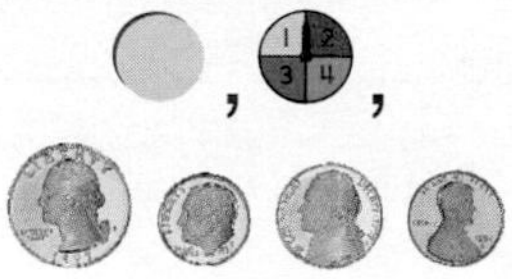

Time for a Game!

1. Play with a partner.
2. Spin to find out how many spaces to move.
3. You earn 1 penny when you land on a red space.
4. You earn 1 nickel if you name the month when you land on a green space.
5. You earn 1 dime if you name the time when you land on a blue space.
6. Count all your coins at the end of the game.
7. The winner is the player who earns the most money.

Home Note **Your child played a game in which he or she identified months and times and counted money.**
ACTIVITY **Play the game on this page with your child. Work together to add up the money each of you earned.**

Name ____________________

Calculator | Computer

Use a .
What is the total?
Write the keys you press.
Write what you see.

1

AC 5 + 5 + 1 + 1 + 1 = 13

2

AC ___ + ___ + ___ + ___ + ___ = ___

3

AC ___ + ___ + ___ + ___ + ___ = ___

4

AC ___ + ___ + ___ + ___ + ___ = ___

5

AC ___ + ___ + ___ + ___ + ___ = ___

This book will help me review trades for nickels and dimes.

This book belongs to ______________________.

A

"I helped Dad wash the dog," said Sarah.
"He gave me all the coins in his pocket.
It is a lot of coins."

"Can I trade some pennies for a nickel?" asked Sarah.

"Yes, Sarah. How many pennies equal one nickel?" asked Mom.

_______________ = 

Can I trade some pennies for a dime?" asked Sarah.

"Yes, Sarah. How many pennies equal one dime?" asked Mom.

____

= 

____

Can I trade some nickels for a dime?"
asked Sarah.

"Yes, Sarah. How many nickels equal
one dime?" asked Mom.

_________________ = 

"Can I trade a nickel and some pennies for a dime?" asked Sarah.

"Yes, Sarah. A nickel and how many pennies equal a dime?" asked Mom.

= 

____

“Now your purse is full of coins, Mom.
I can help,” said Sarah.

“How?” asked Mom.

"I can wash your car. Then you can give me all the coins in your purse!" said Sarah.

Name ______________________________

Chapters 16-19

Review/Test

Concepts and Skills

Write the amount.

1 

_______¢

2 

_______¢

3 

_______¢

4 

_______¢

Show the amount in two ways. Draw the coins.
Circle the way that uses fewer coins.

5 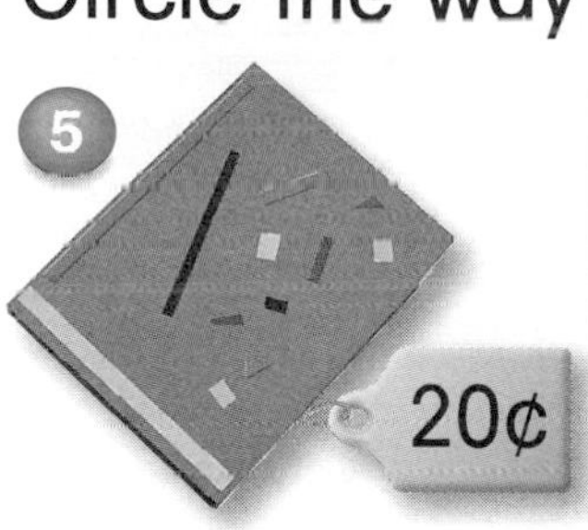

6 Circle the coins you need.
Use the fewest coins.

Use the calendar to answer questions 7 and 8.

MAY						
Sunday	Monday	Tuesday	Wednesday	Thursday	Friday	Saturday
		1	2	3	4	5
6	7	8	9	10	11	12

7. On which day does the month begin?

____

8. On which day is May 7?

____

Write the time.

9. 

___:___

10. 

___:___

Problem Solving

About how long would it take? Circle your estimate.

11. Water a plant.

more than a minute
less than a minute

12. Water a garden.

more than a minute
less than a minute

Name ______________________________

Performance Assessment

Use .

Put 10 coins in a bag.
Take 3 coins. Draw the coins.
Write the amount. Do this 3 times.

1 ______¢

2 ______¢

3 ______¢

4 Show the amount in two ways.
Draw the coins.
Circle the way that uses fewer coins.

Write About It

5 Write about something you do at school that takes more than one minute.

Name ______________________________

Chapters 1–19 Cumulative Review

Fill in the ⬭ for the correct answer.

1 Which number sentence tells the story?

⬭ 3 + 3 = 6
⬭ 3 + 2 = 5
⬭ 3 − 2 = 1
⬭ 5 − 1 = 4

2 Which is the same shape as this tent?

⬭ 
⬭ 
⬭ 
⬭ 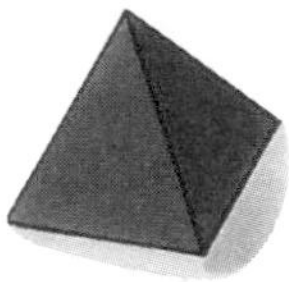

3 Which shape is next in the pattern?

⬭ 
⬭ 

4 Count by twos. What number comes after 8?

⬭ 9
⬭ 10
⬭ 11
⬭ 12

2, 4, 6, 8, ______

5

$$\begin{array}{r} 5 \\ +5 \\ \hline \end{array}$$

⬭ 0
⬭ 5
⬭ 9
⬭ 10

6

$$\begin{array}{r} 11 \\ -\ 3 \\ \hline \end{array}$$

⬭ 7
⬭ 8
⬭ 9
⬭ 10

7 How many?

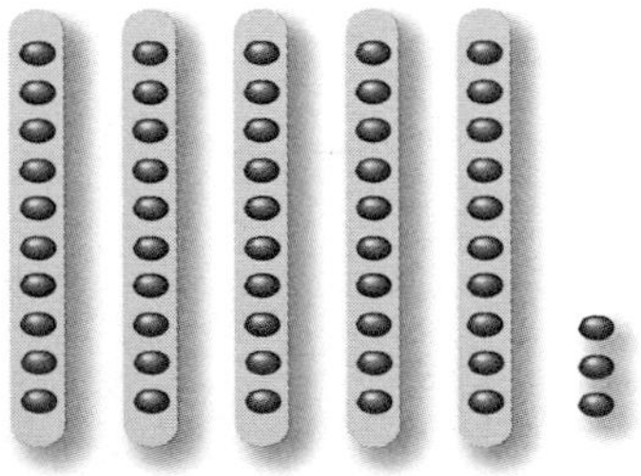

⬭ 8
⬭ 35
⬭ 50
⬭ 53

8 What is the amount?

⬭ 3¢
⬭ 15¢
⬭ 30¢
⬭ not here

9 What is the time?

⬭ 9:00
⬭ 10:00
⬭ 11:00
⬭ not here

CHAPTER 20

Measuring Length

What math tool is he using to measure?

Home Note In this chapter, your child is estimating and measuring objects using inches and centimeters.
ACTIVITY Have your child use an inch or centimeter ruler to measure things in the picture.

SCHOOL-HOME CONNECTION

Dear Family,

Today we started chapter 20. We will learn to estimate and measure how long things are in inches and centimeters. Here are the new vocabulary words and an activity for us to do together at home.

Love,

____

Vocabulary

inch a customary unit for measuring

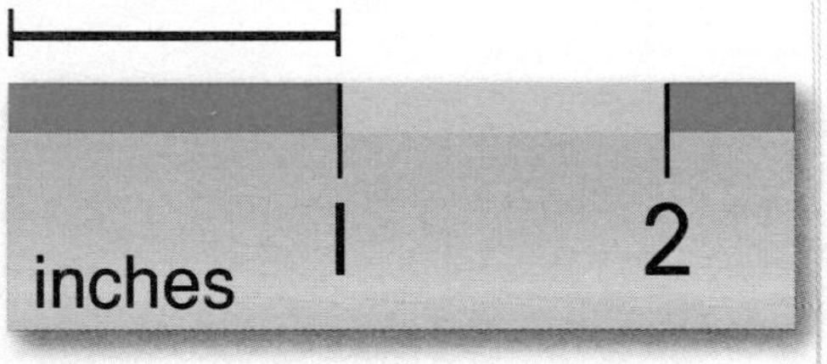

centimeter a metric unit for measuring

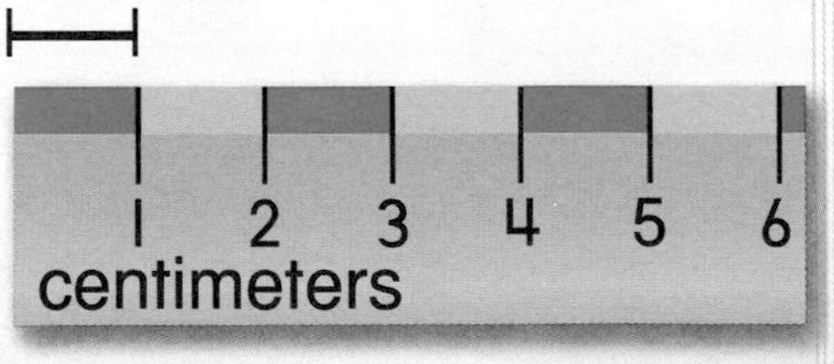

ACTIVITY

Invite your child to estimate in inches or in centimeters the length of some objects that are less than 12 inches or 30 centimeters. Help your child use a ruler to check the estimates.

Visit our Web site for additional ideas and activities.
http://www.hbschool.com

Name ______________________

Using Nonstandard Units

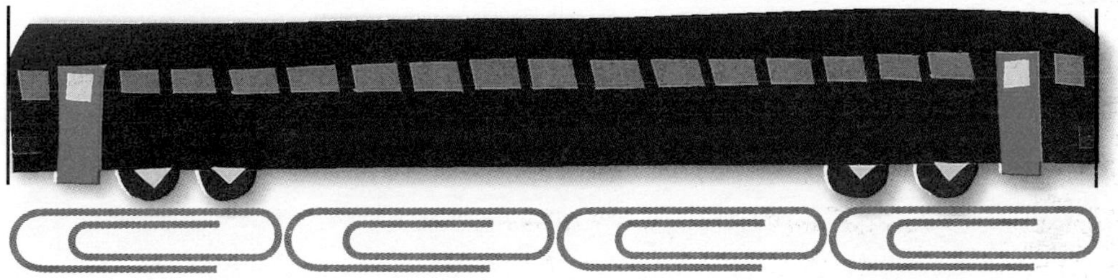

about 4

Use to measure.
About how many long?

1

about ____

2
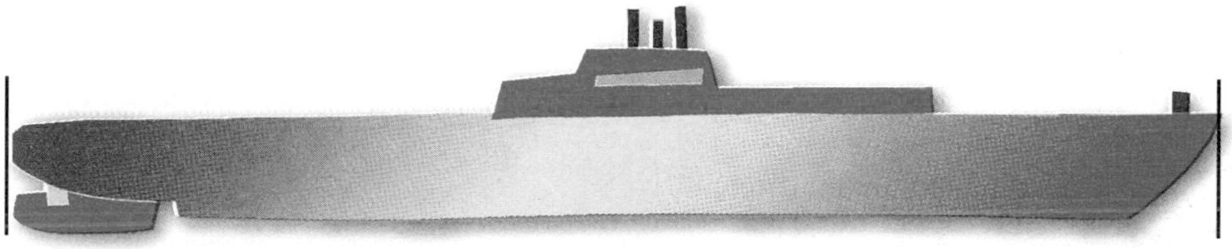

about ____

3

about ____

4

about ____

5

about ____

Talk About It • Critical Thinking

How would your measures change if you used a smaller ?

Practice

Estimate. Then use  to measure.

1

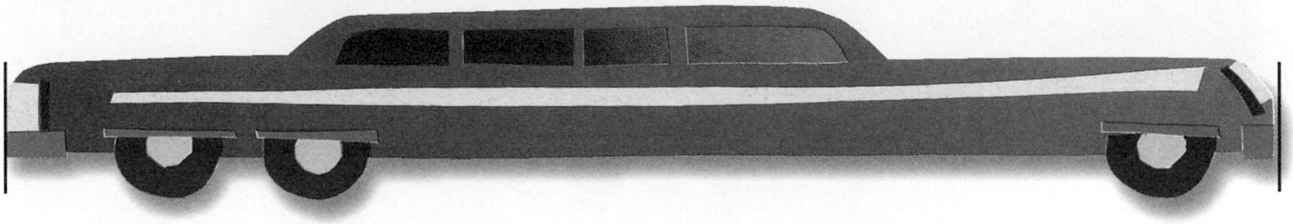

Estimate about ____  Measure about ____ 

2

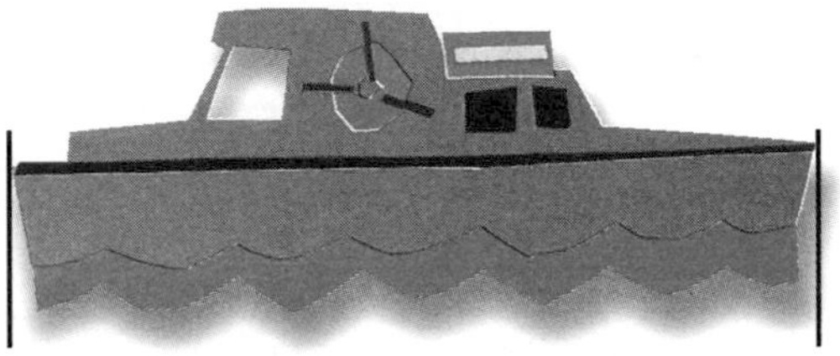

Estimate about ____  Measure about ____ 

3

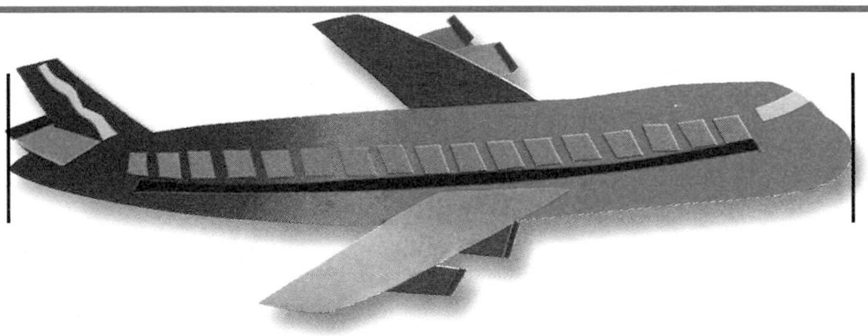

Estimate about ____ 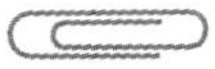 Measure about ____ 

4

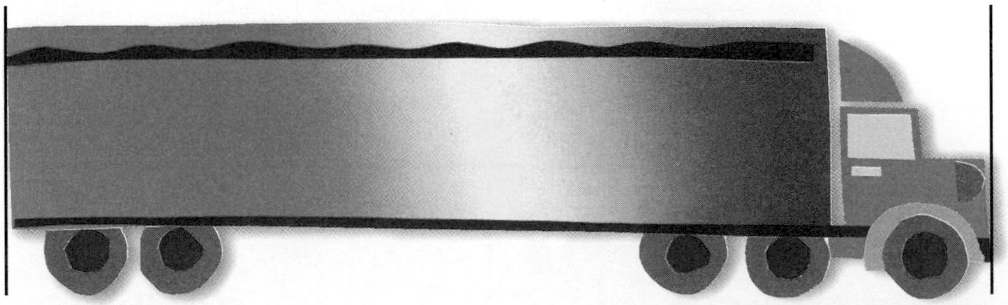

Estimate about ____  Measure about ____ 

Home Note Your child used paper clips to measure length.
ACTIVITY Give your child some paper clips or other objects that are all the same length. Have him or her measure things around the house such as spoons and forks.

Name ______________________________

Measuring in Inch Units

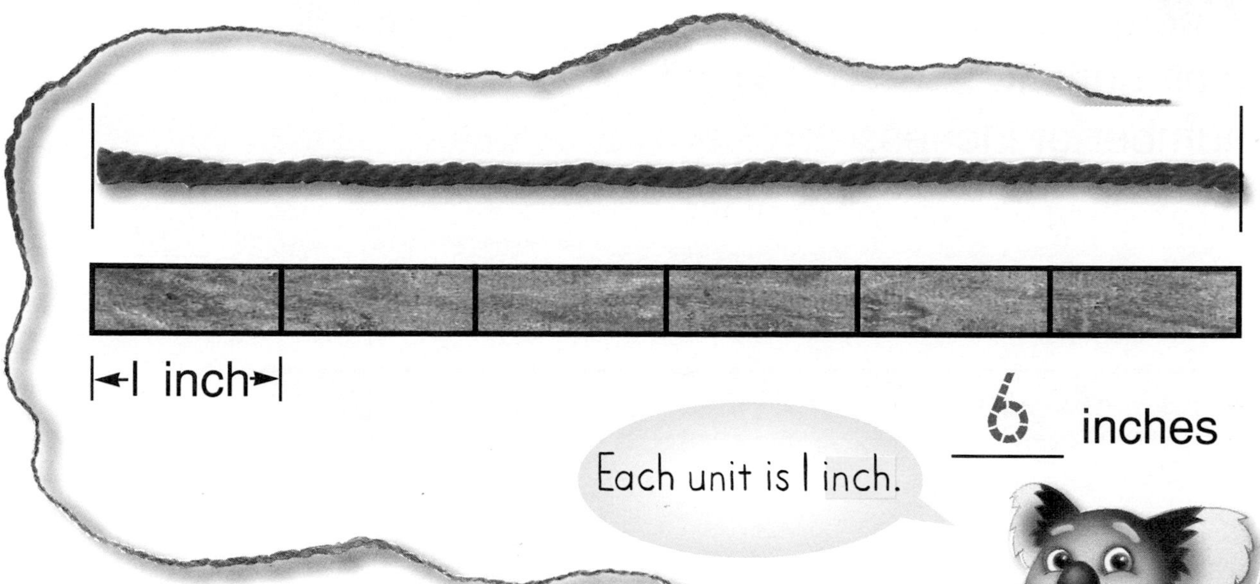

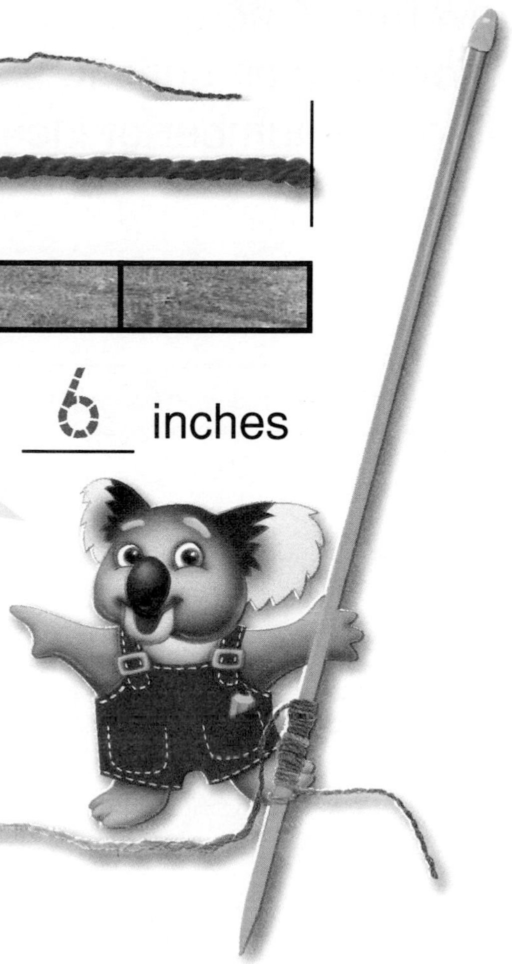

How many inches long?
Color the inch units to show.
Write the number of inches.

1

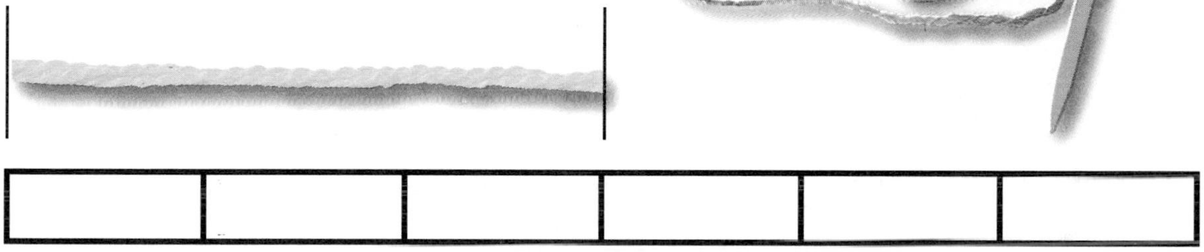

____ inches

2

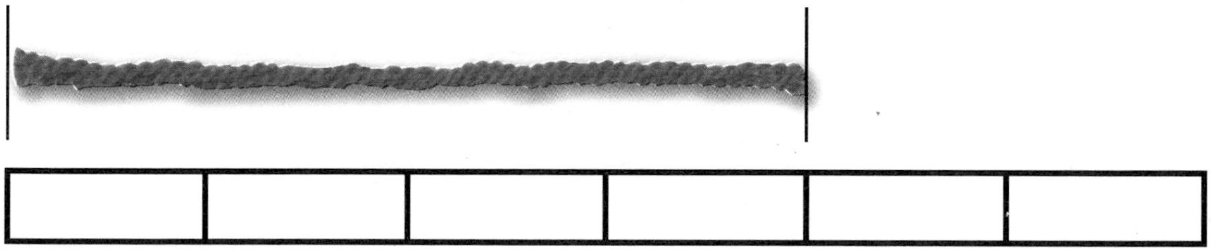

____ inches

3

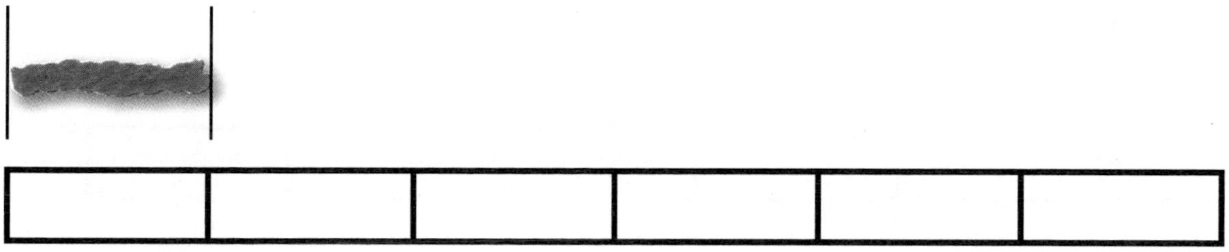

____ inch

Practice

How many inches long?
Color the inch units to show.
Write the number of inches.

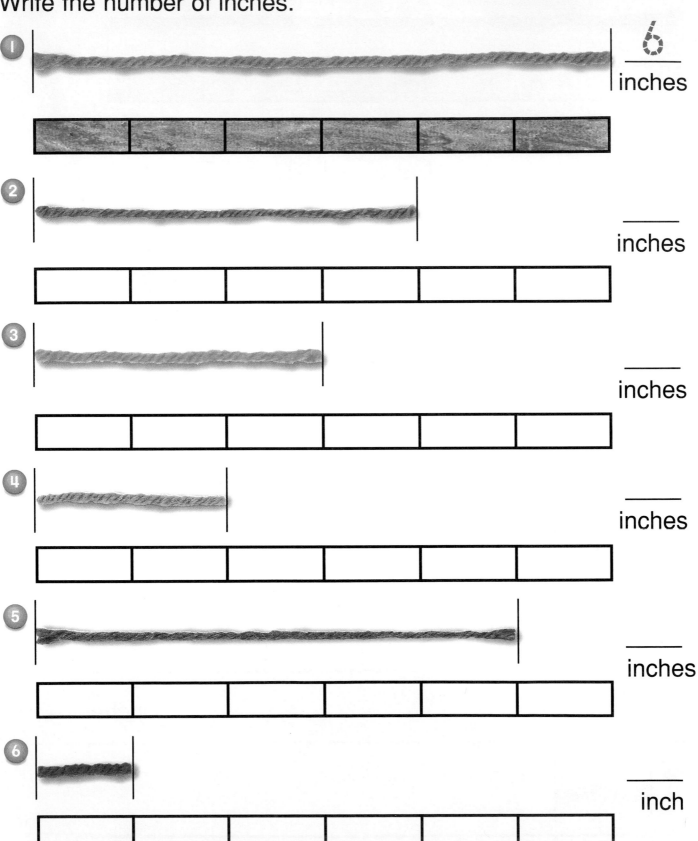

1 

6 inches

2 

____ inches

3 

____ inches

4 

____ inches

5 

____ inches

6 

____ inch

Home Note **Your child used inch units to measure.**
ACTIVITY Cut a 6-inch-long string for your child. Have him or her find objects that are the same length as the string.

Name ______________________________

Using an Inch Ruler

You can use an inch ruler to measure.

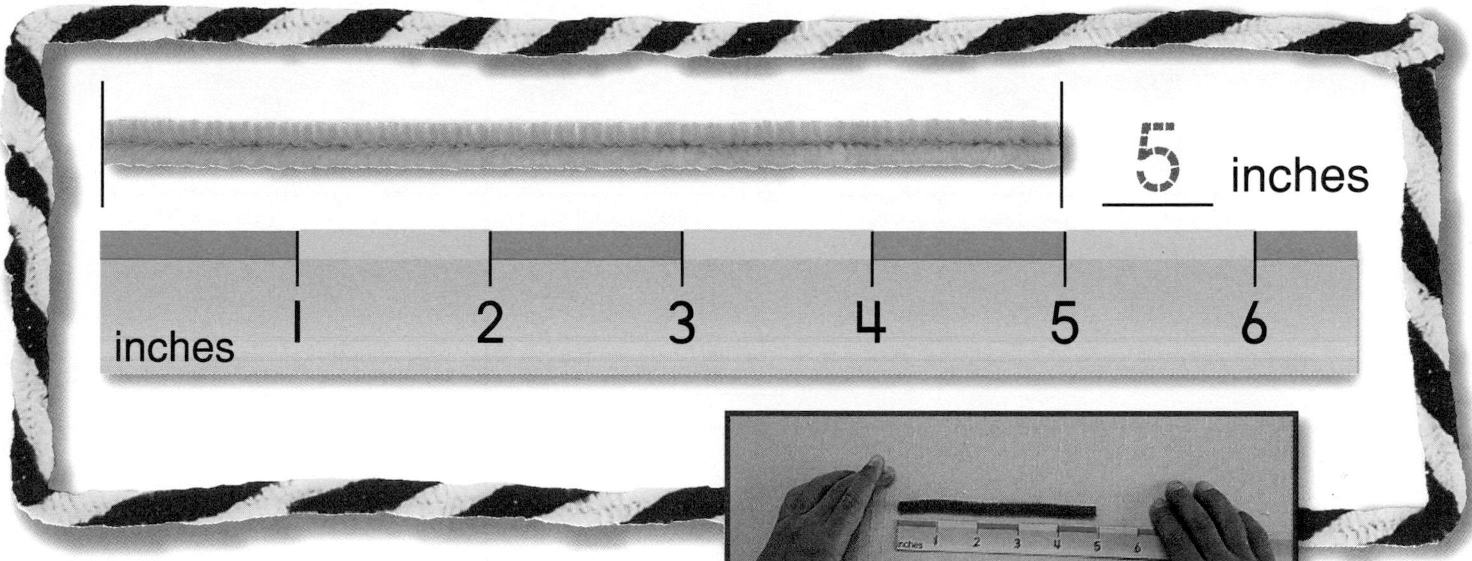

Use an inch ruler to measure.
Write how many inches long.

1. 
____ inches

2. 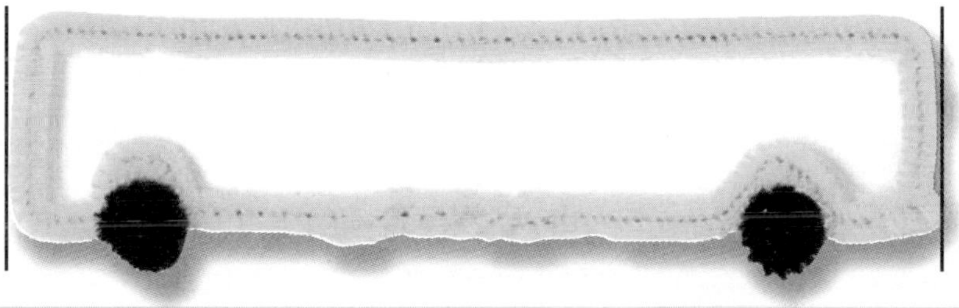
____ inches

3. 
____ inches

4. 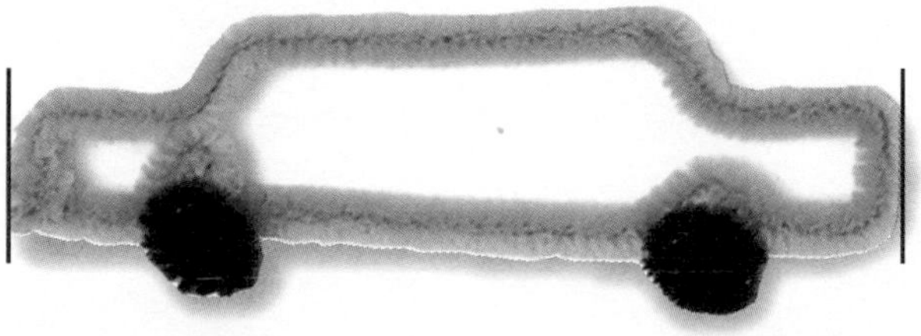
____ inches

Talk About It • Critical Thinking

How should you place the ruler?

Practice

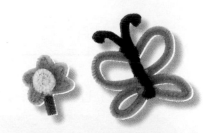

Use real objects and an inch ruler.
Estimate. Then measure.

Objects	Estimate	Measure
1 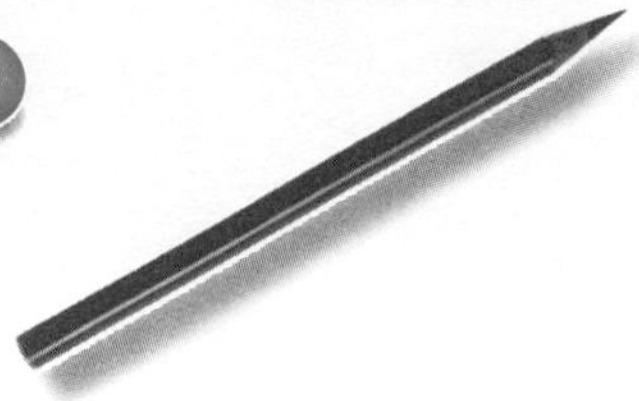	about ____ inches	about ____ inches
2 	about ____ inches	about ____ inches
3 	about ____ inches	about ____ inches
4 	about ____ inches	about ____ inches

Problem Solving • **Visual Thinking**

Cut a string 6 inches long.
Circle the strings that are as long as your string.

5

Home Note **Your child estimated and used an inch ruler to measure.**
ACTIVITY **Have your child estimate the length in inches of some small objects. Together, use a ruler to check the estimates.**

Name ______________________________

Measuring in Centimeter Units

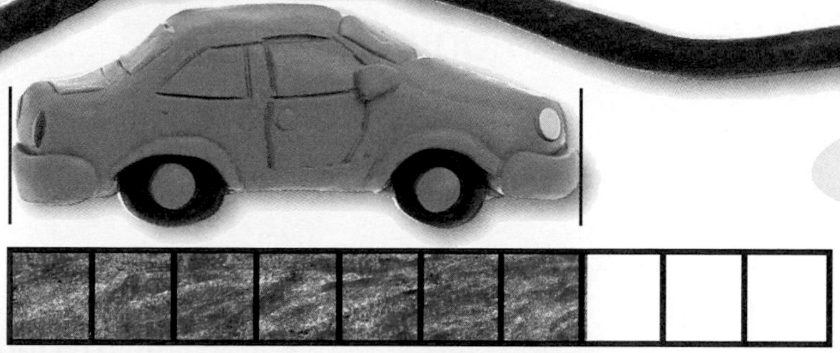

This toy car is
7 centimeters long.

7 centimeters

How many centimeters long?
Color the centimeter units. Write how many.

1 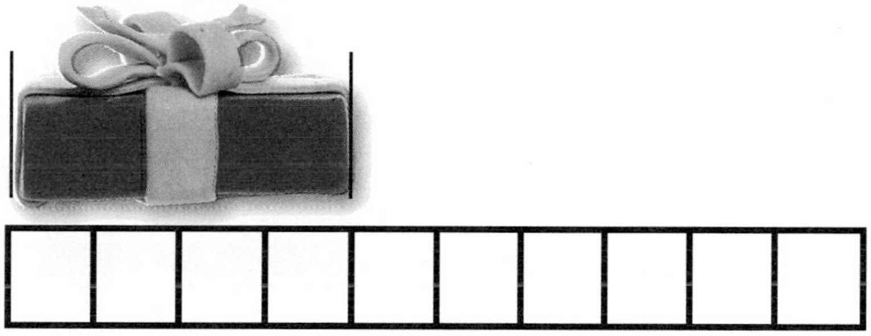

____ centimeters

2 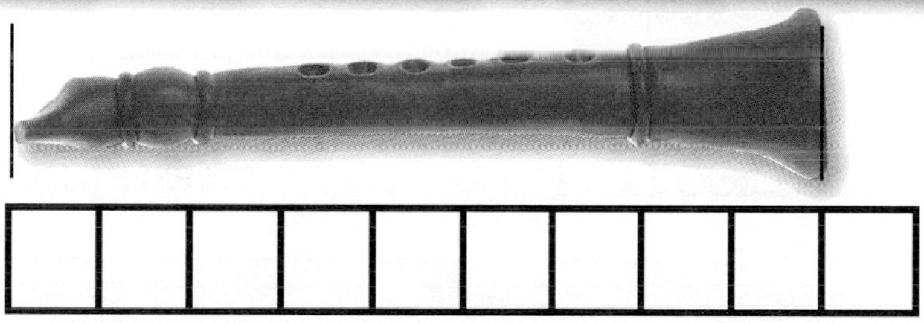

____ centimeters

3 

____ centimeters

4 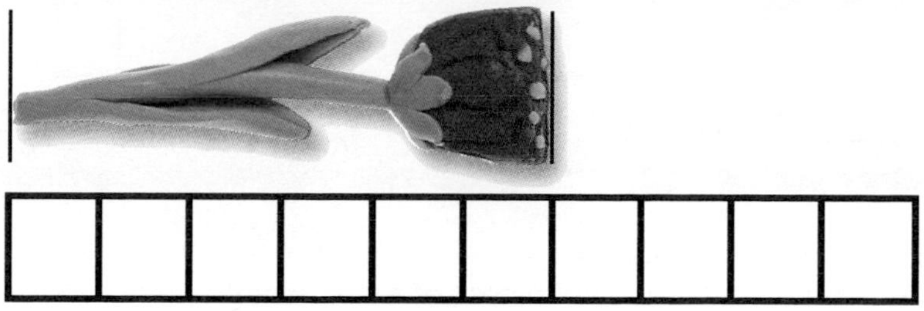

____ centimeters

Practice

How many centimeters long?
Color the centimeter units. Write how many.

1 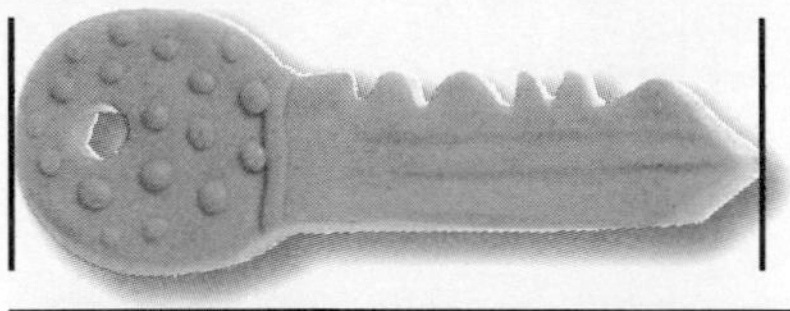

5 centimeters

2 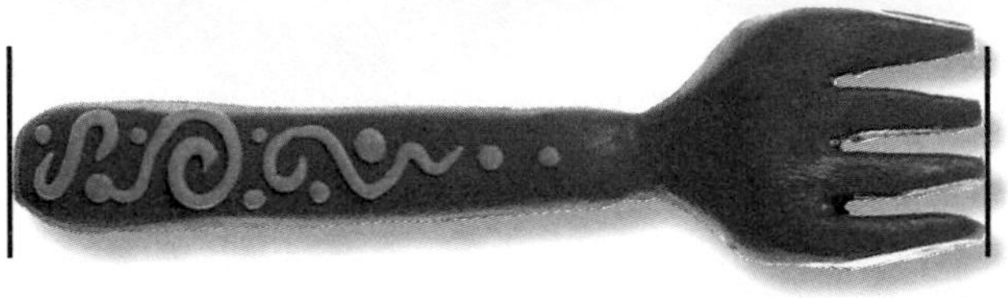

____ centimeters

3 

____ centimeters

Mixed Review

Circle the shape that comes next in the pattern.

4 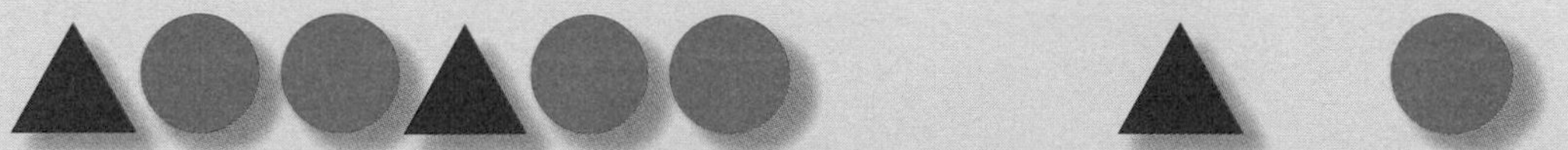

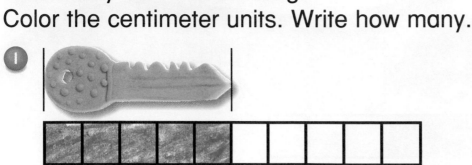

Count. Write the amount.

5 

____ ¢, ____ ¢, ____ ¢, ____ ¢, ____ ¢ ____ ¢

Home Note Your child used centimeter units to measure.
ACTIVITY Invite your child to estimate the length in centimeters of some small objects. Together, use a centimeter ruler to check.

Name ________________________

Using a Centimeter Ruler

You can use a centimeter ruler to measure.

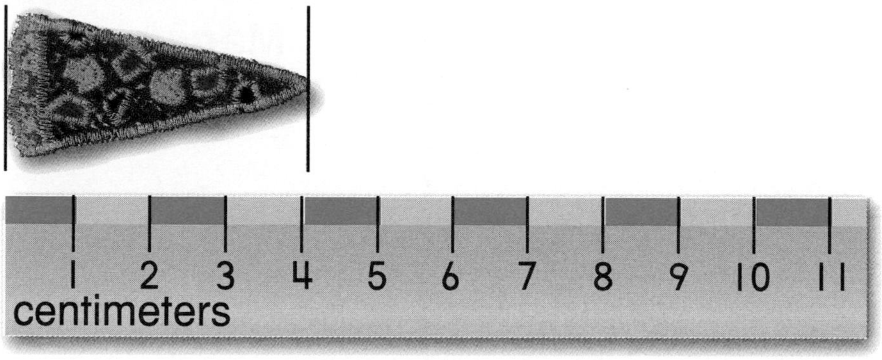

__4__ centimeters

Use a centimeter ruler to measure.
Write how many centimeters long.

1. 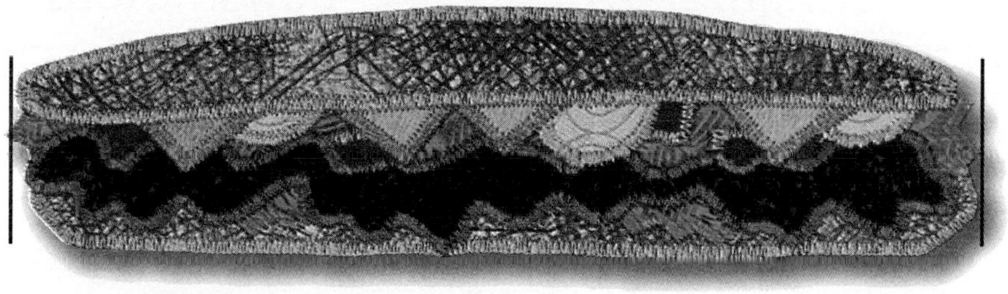

____ centimeters

2. 

____ centimeters

3. 

____ centimeters

4. 

____ centimeters

5. 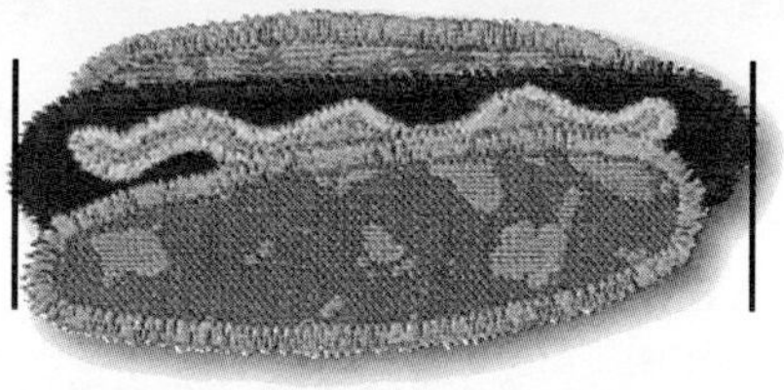

____ centimeters

Practice

Use real objects and a centimeter ruler.
Estimate. Then measure.

Objects	Estimate	Measure
1 	about ____ centimeters	about ____ centimeters
2 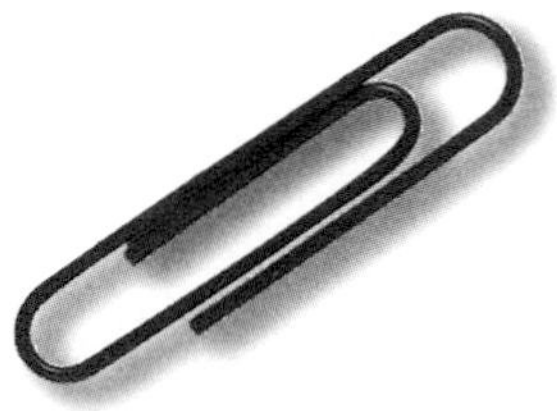	about ____ centimeters	about ____ centimeters
3 	about ____ centimeters	about ____ centimeters
4 	about ____ centimeters	about ____ centimeters

Write About It

5 Draw a picture of an object 7 inches long.
Draw another picture of an object
7 centimeters long.
Tell how the objects are different.

Home Note **Your child estimated and used a centimeter ruler to measure.**
ACTIVITY **Give your child two objects of different lengths. Tell him or her how long in centimeters one of them is. Have your child guess which one.**

Name ______________________________

Chapter 20

Review/Test

Concepts and Skills

Use 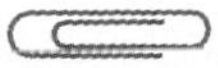 to measure.

1. 

about ____ 

2. 

about ____ 

Use an inch ruler to measure.

3. 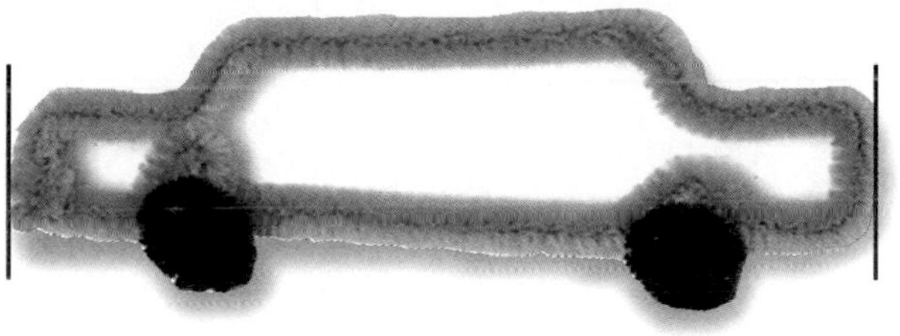

____ inches

4. 

____ inches

Use a centimeter ruler to measure.

5. 

____ centimeters

6. 

____ centimeters

Name ______________________________

Cumulative Chapters 1–20

TAAS Prep

Mark the best answer.

1. Which number tells the sum or difference?

   Ann has 12 crayons.
   She gives away 3.
   How many are left?

   - 6
   - 7
   - 8
   - 9

2. Who has more money?

   Rosa has 1 dime.
   Jill has 3 nickels.

   - Rosa
   - Jill

3. Which is the same size and shape as this triangle?

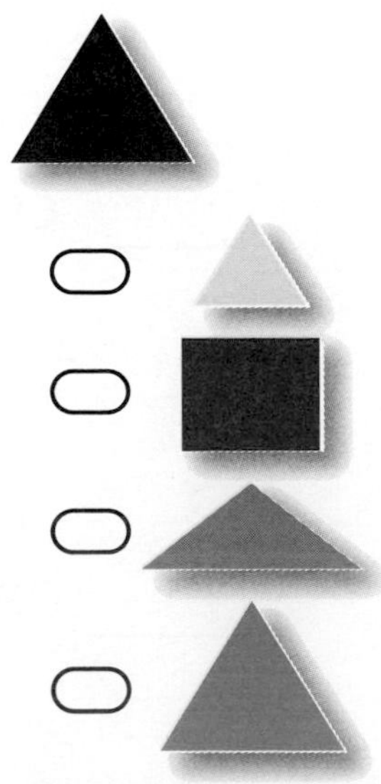

4. Which number tells the sum or difference?

   Phil has 9 pennies.
   He finds 3 more.
   How many does he have now?

   - 9
   - 10
   - 11
   - 12

5. Which number comes next?

   2, 4, 6, ____

   - 7
   - 8
   - 9
   - 10

6. How long is the chalk?

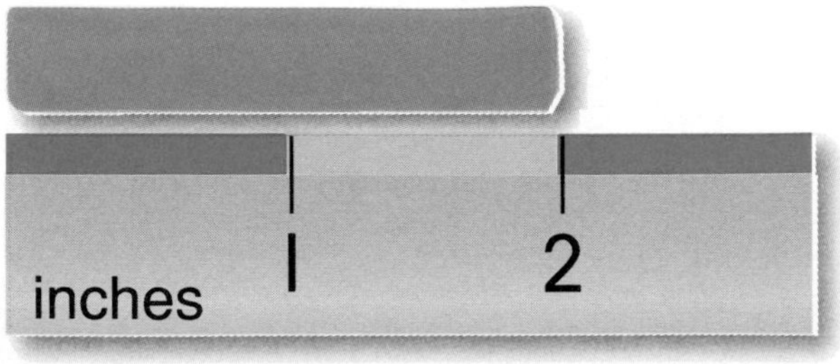

   - 1 inch
   - 2 inches
   - 3 inches
   - 4 inches

CHAPTER 21

Measuring Mass, Capacity, and Temperature

Name some things that are heavier than the fish.

Home Note In this chapter, your child will learn about mass (weight), capacity, and temperature.
ACTIVITY Have your child compare two of the items in the picture and tell which is heavier or which is lighter.

SCHOOL-HOME CONNECTION

Dear Family,

Today we started Chapter 21. We will learn about measuring mass, capacity, and temperature. Here are the new vocabulary words and an activity for us to do together at home.

Love,

____

Vocabulary

Heavier and **lighter** are words that compare two objects of different weights.

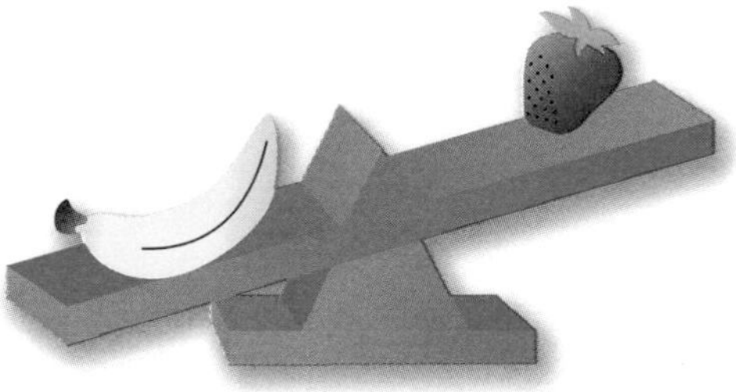

ACTIVITY

Invite your child to estimate the number of cups of water it would take to fill some of the pots and pans you use for cooking. Together, use a cup to measure how much water the pots and pans hold. Compare the measurements with the estimates.

Visit our Web site for additional ideas and activities.
http://www.hbschool.com

Name ______________________________

Using a Balance

Find each object and compare it to a box of crayons.
Estimate. Write **H** for heavier. Write **L** for lighter.
Then use a  to measure.

Object	Estimate	Measure
1 	H	H
2 		
3 		
4 		

Talk About It ● **Critical Thinking**

How would your estimate change if you used a bigger box of crayons?

Practice

Find each object and compare it to a box of crayons.
Estimate. Write **H** for heavier. Write **L** for lighter.
Then use a  to measure.

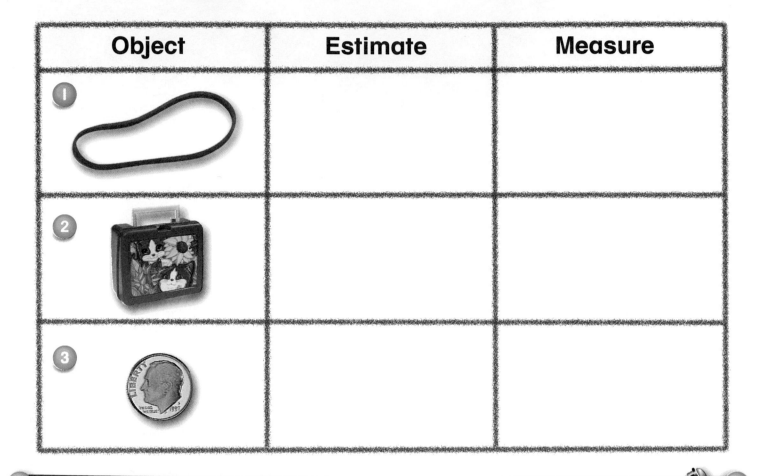

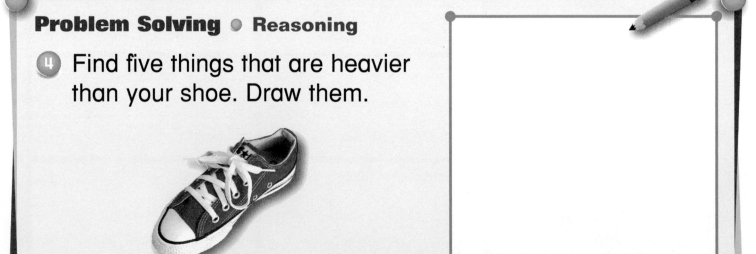

Object	Estimate	Measure
1 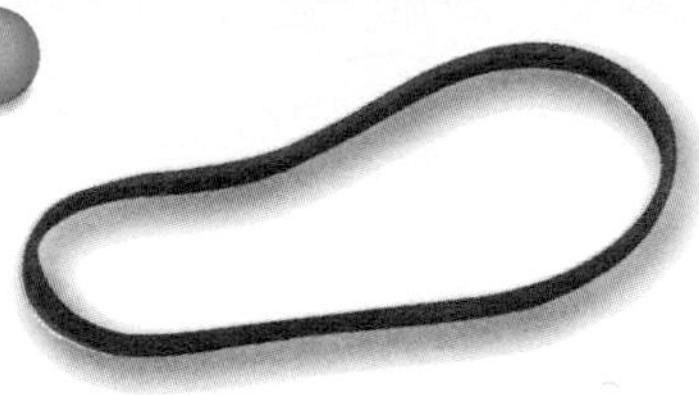		
2 		
3 		

Problem Solving • Reasoning

4 Find five things that are heavier than your shoe. Draw them.

Home Note Your child compared real objects to find which was heavier and which was lighter.
ACTIVITY Give your child an object to hold. Then have him or her find one object that is heavier and one that is lighter.

Name ______________________________

Understand • Plan • Solve • Look Back

Problem Solving
Guess and Check

Use real objects, a (scale), and (cube).
Find each object.
About how many (cube) does it take to balance the scale?
Estimate. Then measure.

Object	Estimate	Measure
1 	about ______	about ______
2 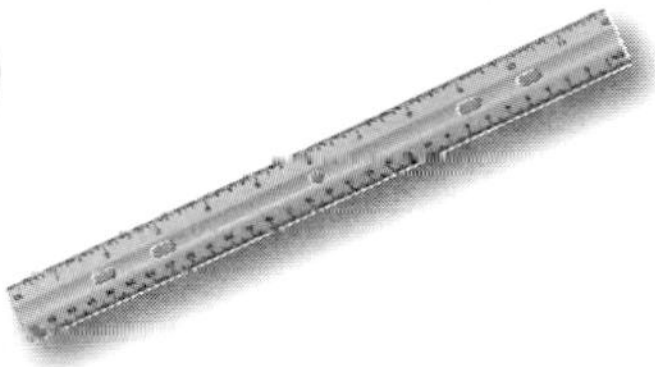	about ______	about ______
3 	about ______	about ______
4 	about ______	about ______
5 	about ______	about ______
6 	about ______	about ______

Practice

Use real objects, a balance, and cubes.
Find each object.
About how many cubes does it take to balance the scale?
Estimate. Then measure.

Object	Estimate	Measure
1 	about ____	about ____
2 	about ____	about ____
3 	about ____	about ____
4 	about ____	about ____
5 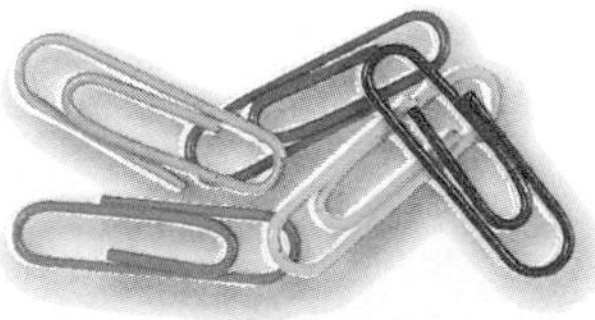	about ____	about ____
6 	about ____	about ____

Home Note Your child estimated the weight of objects and weighed them on a balance.
ACTIVITY Ask your child to draw a picture to show the way a balance looks when the object on one side is heavier than the object on the other.

Name ______________________________

Measuring with Cups

Use a  and containers.
About how many cups of rice does each container hold?
Estimate. Then measure.

Container	Estimate	Measure
1 	about _____ cups	about _____ cups
2 	about _____ cups	about _____ cups
3 	about _____ cups	about _____ cups
4 	about _____ cups	about _____ cups
5 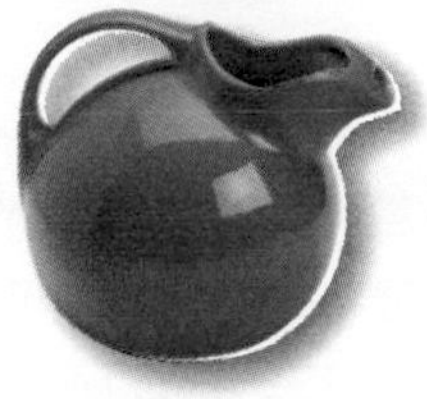	about _____ cups	about _____ cups

Practice

Use a  and containers.
About how many cups of rice does each container hold? Estimate. Then measure.

Container	Estimate	Measure
1 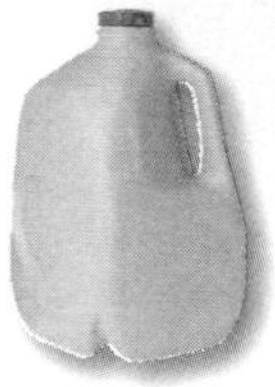	about _____ cups	about _____ cups
2 	about _____ cups	about _____ cups
3 	about _____ cups	about _____ cups
4 	about _____ cups	about _____ cups

Mixed Review

Write the time.

5 

___ : ___

6 

___ : ___

7 

___ : ___

Home Note Your child estimated the capacity of containers and then measured their capacity using a cup.
ACTIVITY Have your child estimate how many cups of rice, cereal, or water some different containers hold. Then have him or her measure to check.

Name ________________________________

Temperature Hot and Cold

Circle the picture that shows something hot.

1

2

3

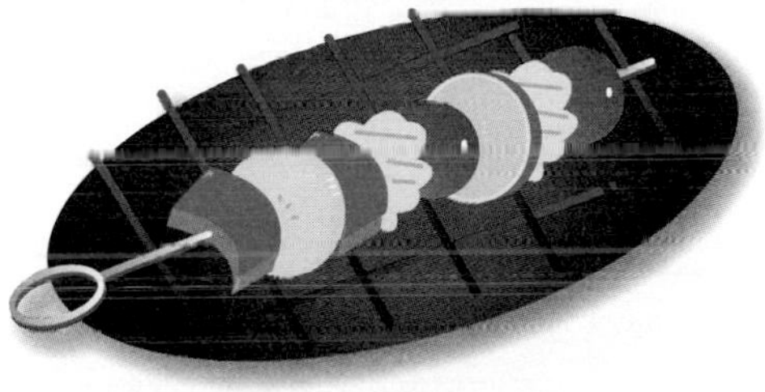

4

5

Talk About It • Critical Thinking

How can you tell if something is hot or cold?

Practice

Circle the picture that shows a time of the year that is cold.

1  

2  

3  

4  

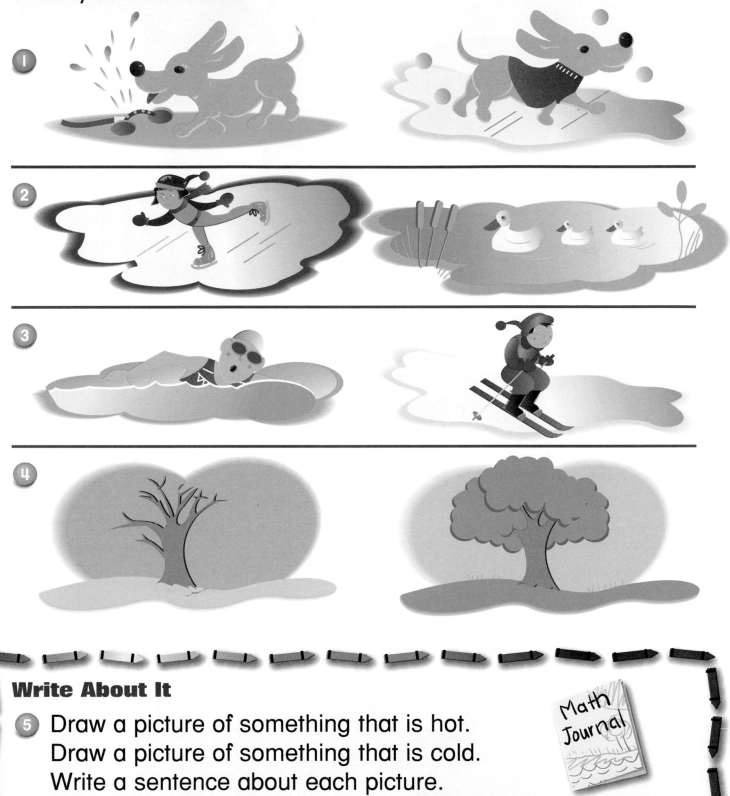

Write About It

5 Draw a picture of something that is hot.
Draw a picture of something that is cold.
Write a sentence about each picture.

Home Note **Your child compared hot and cold objects and times of the year.**
ACTIVITY **Have your child cut out magazine pictures that show hot or cold objects or times of the year. Ask him or her to sort the pictures into two groups—hot and cold.**

Name ______________________________

Concepts and Skills

Use a . Find each object.
Write **H** if the object is heavier than .
Write **L** if the object is lighter than .

1.  ______  ______ 

______

---

2. Use a  and .

Find a pencil. Write how many  it takes to balance the scale.

 about ______ 

Use a  and a container.

3. Write how many cups of rice the container holds.

about ______ cups

---

4. Circle in  the picture that shows something cold.

Name ______________________________

Cumulative Chapters 1–21

TAAS Prep

Mark the best answer.

1. Which picture shows $6 - 3 = 3$?

- 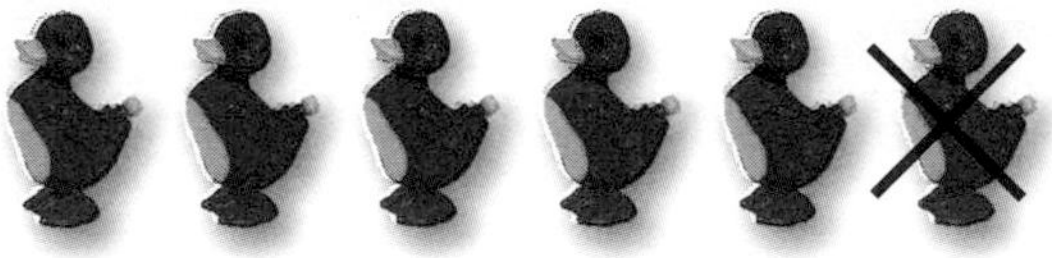
- 
- 
- 

2. Which picture shows a hot day?

- 
- 
- 
- 

3. Count by tens. Then count on by ones. How many pennies in all?

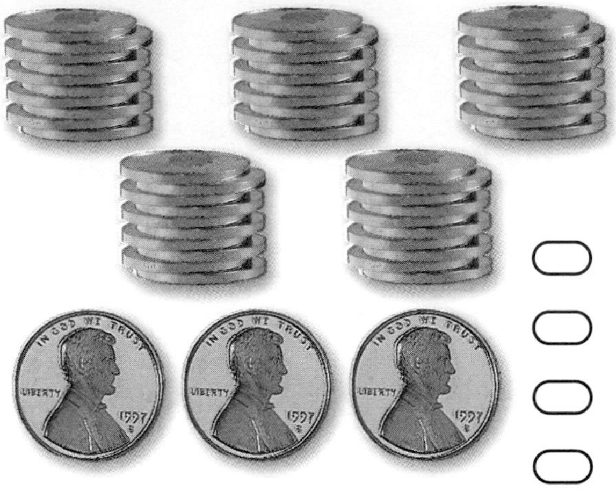

- 13
- 43
- 53
- 63

4. Which weighs about the same as an apple?

- 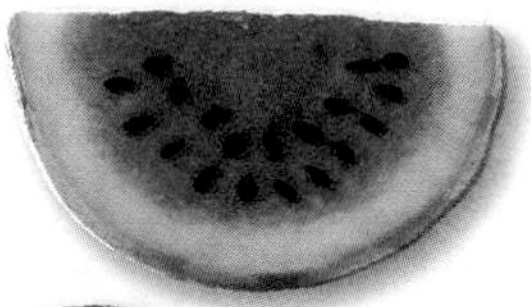
- 
- 
- 

5. Which subtraction sentence tells how many are left?

- $9 + 4 = 13$
- $4 + 5 = 9$
- $9 - 4 = 5$
- $9 - 5 = 4$

CHAPTER 22

Fractions

How many equal parts do you see? Tell about them.

Home Note In this chapter, your child will learn about equal parts of a whole, including halves, fourths, and thirds.
ACTIVITY Fold three sheets of paper to show halves, thirds, and fourths. Have your child tell you about each sheet.

SCHOOL-HOME CONNECTION

Dear Family,

Today we started Chapter 22. We will learn about fractions. We will learn about equal parts, halves, thirds, and fourths. Here are the new vocabulary words and an activity for us to do together at home.

Love,

____

Vocabulary

whole — two equal parts, halves

 **one half** $\frac{1}{2}$ → 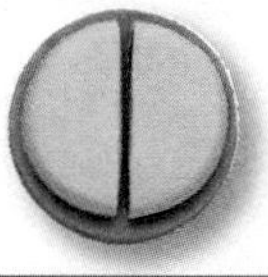

whole — four equal parts, fourths

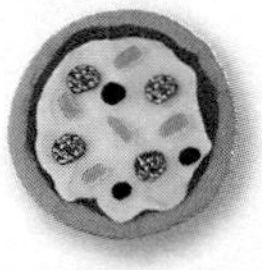 **one fourth** $\frac{1}{4}$ → 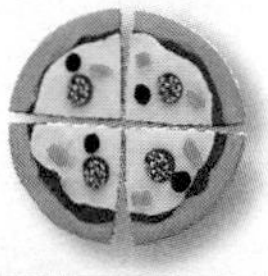

whole — three equal parts, thirds

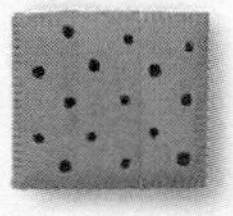 **one third** $\frac{1}{3}$ → 

ACTIVITY

Provide opportunities for your child to show how to divide food items into equal parts. Oranges, apples, bread, and pizza are good items to use.

Visit our Web site for additional ideas and activities.
http://www.hbschool.com

Name ______________________________

Equal and Unequal Parts of Wholes

Circle the pictures that show equal parts.

Equal parts are
the same size.

1

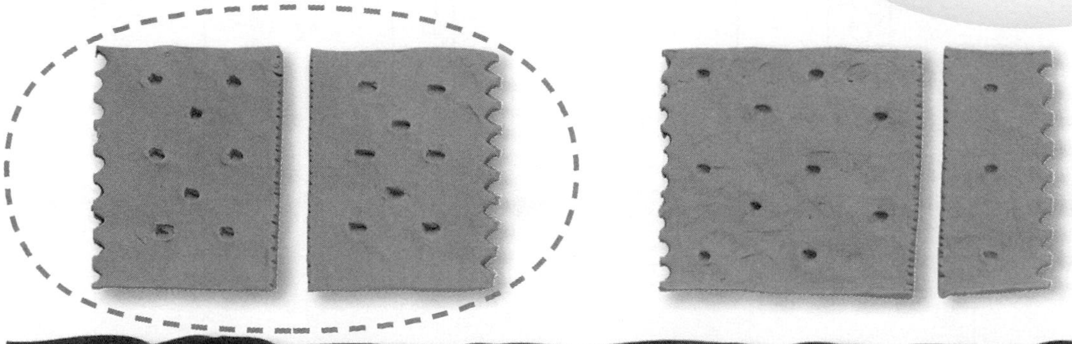

2

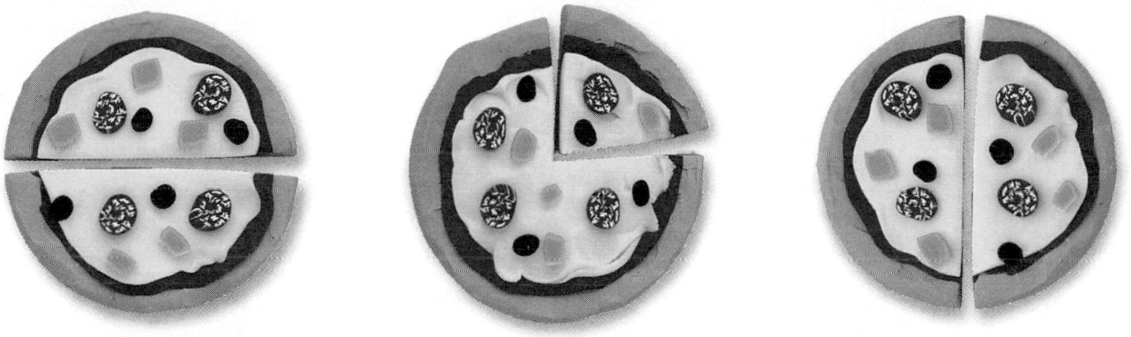

3

4

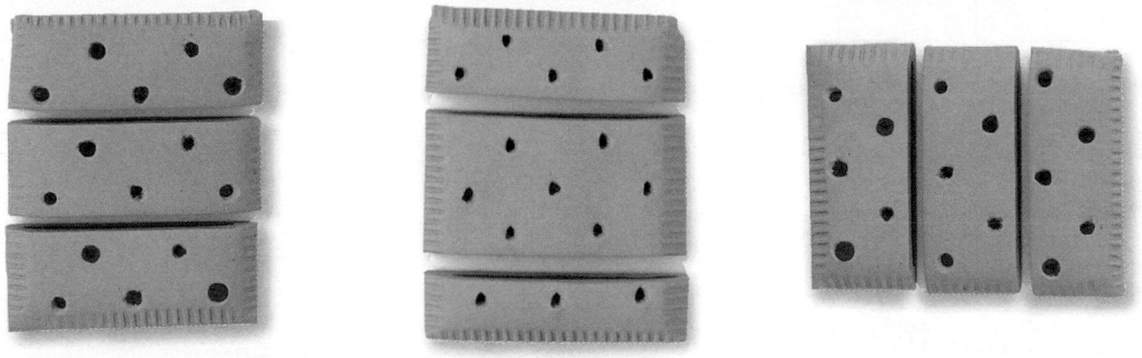

Talk About It • **Critical Thinking**

Which pictures show ways for two friends to have equal parts? How can you tell?

Practice

Circle the figures that show equal parts.

1

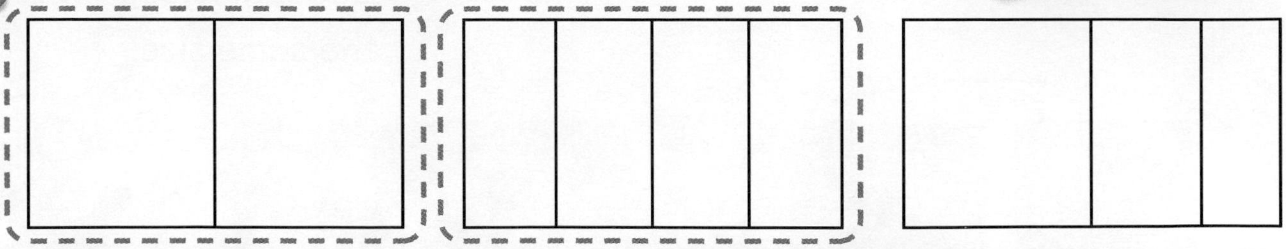

2

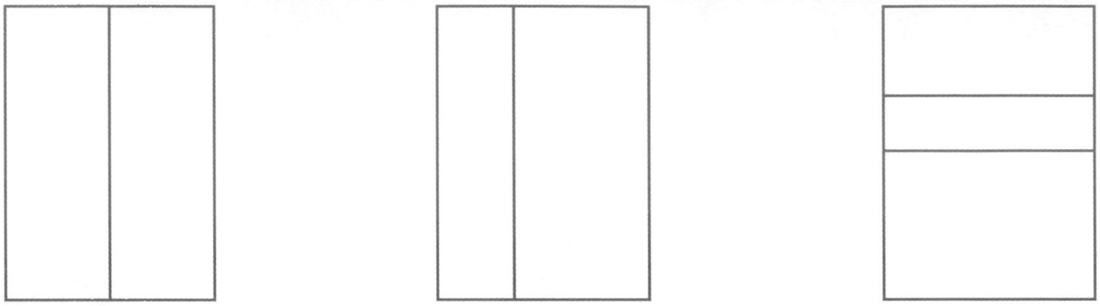

3

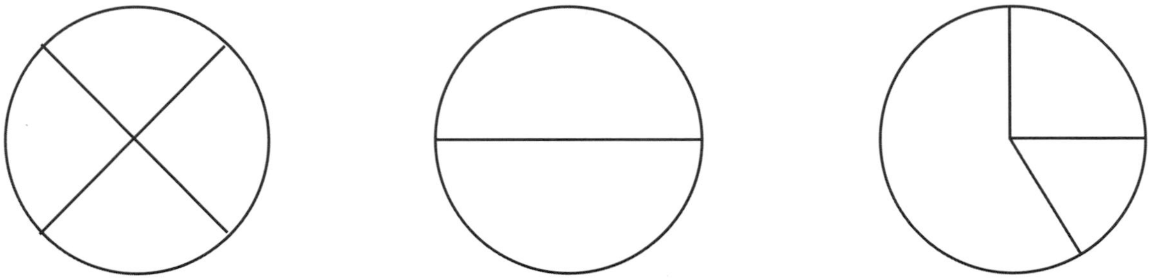

4

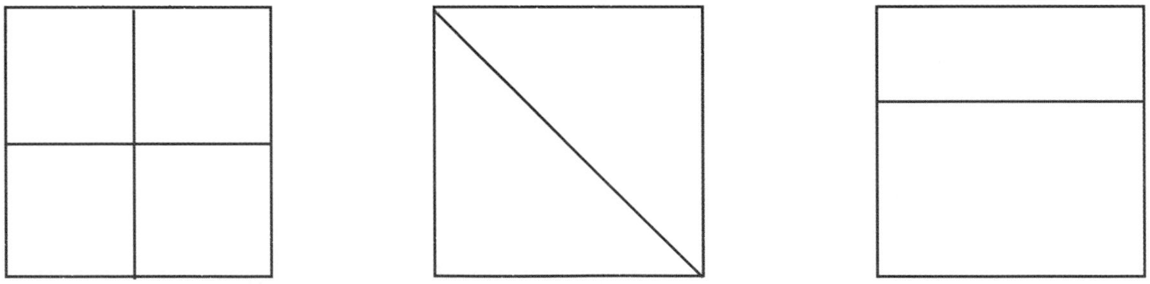

Problem Solving

5 Two children want equal parts. Draw a line to show where you would cut this sandwich.

Home Note **Your child identified equal parts of a whole.**
ACTIVITY **Invite your child to help you divide a food item into equal parts for 2, 3, or 4 family members to share.**

Halves

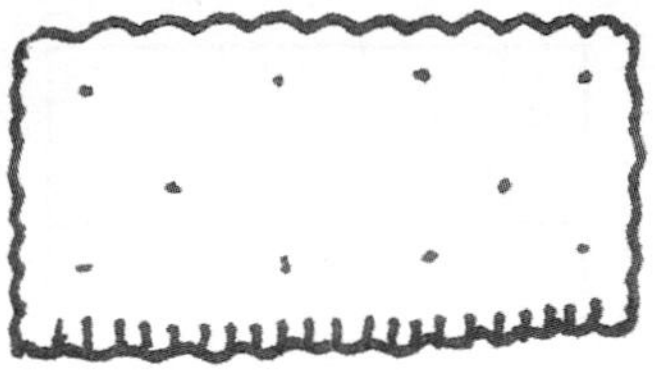

whole

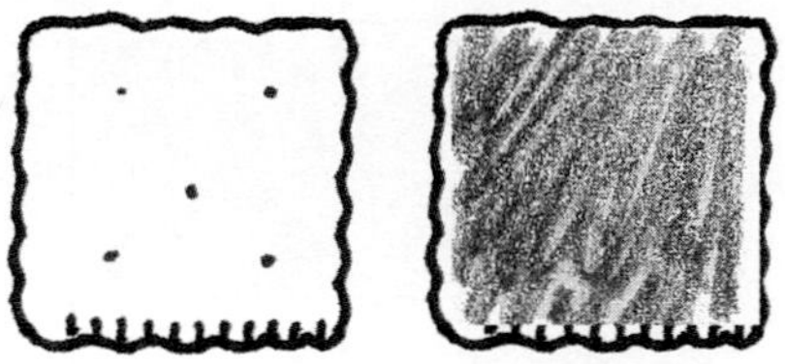

I out of 2 equal parts is $\frac{1}{2}$, or one half.
Two halves make one whole.

Find the figures that show halves. Color $\frac{1}{2}$ .

1

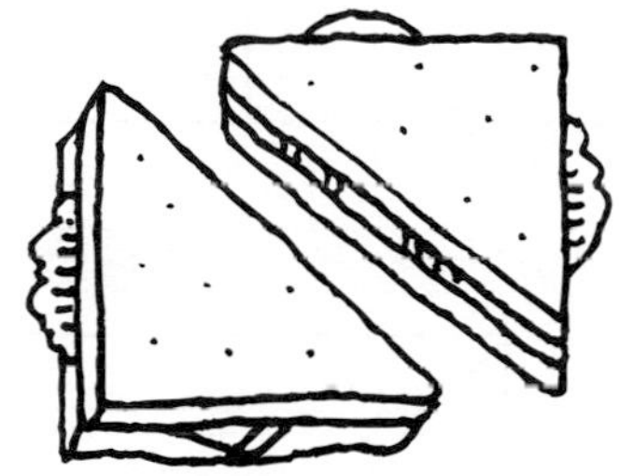

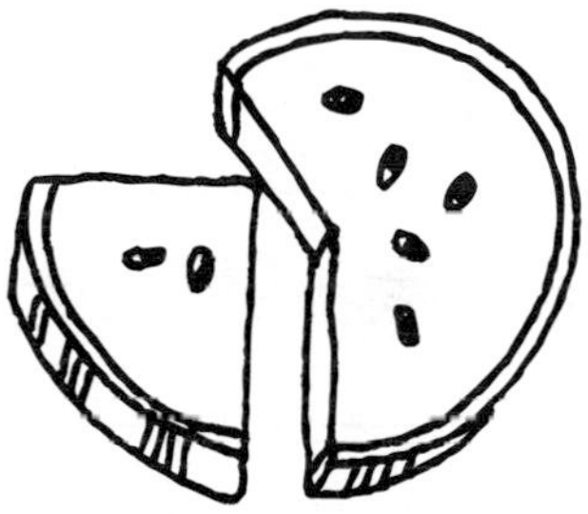

2

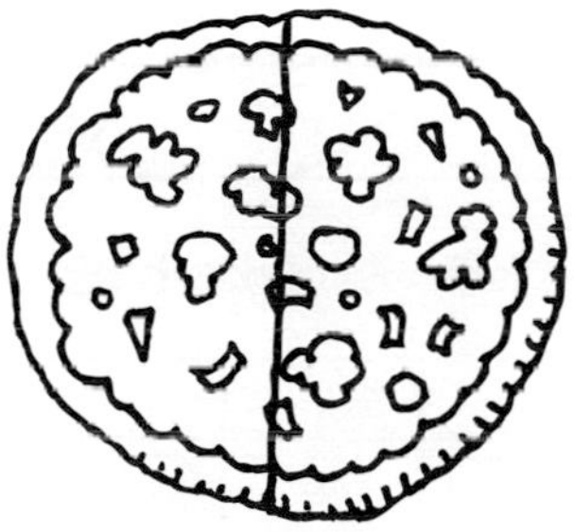
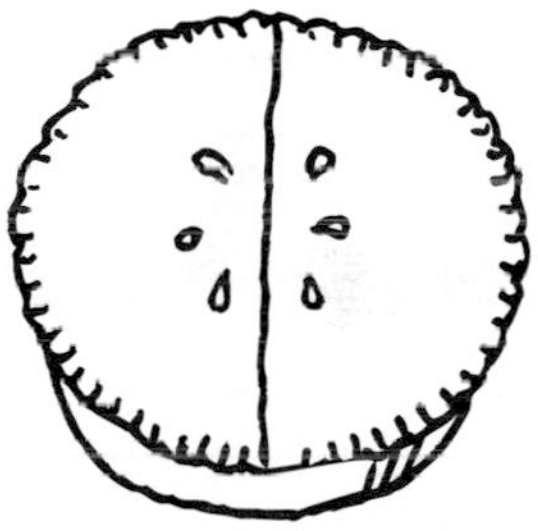

3

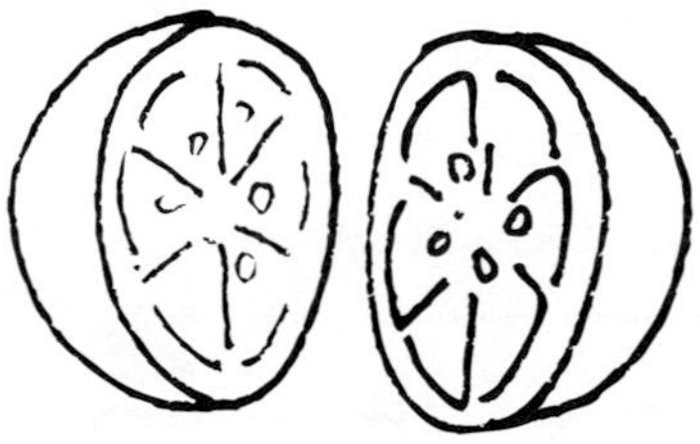
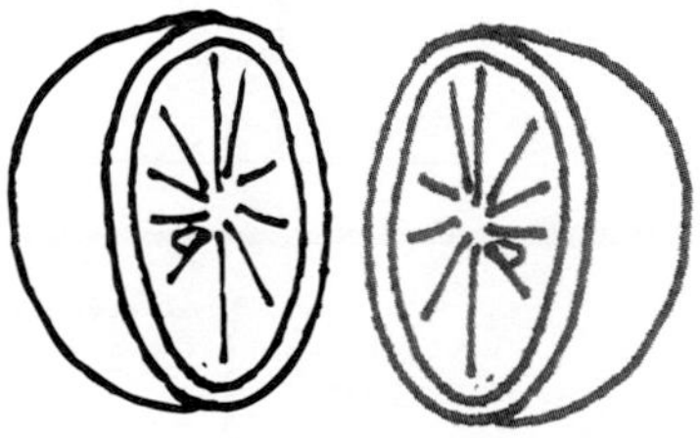

4

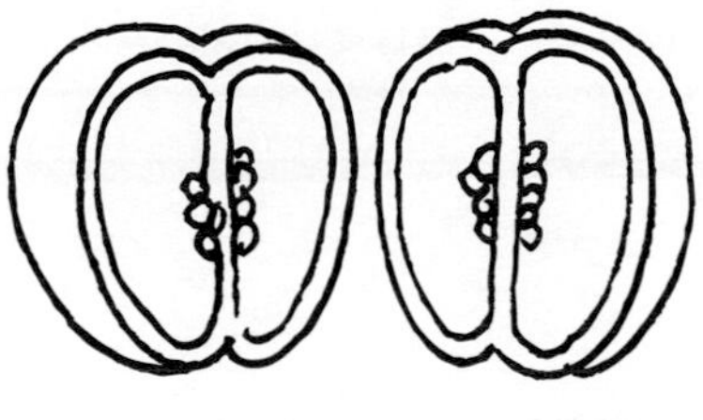
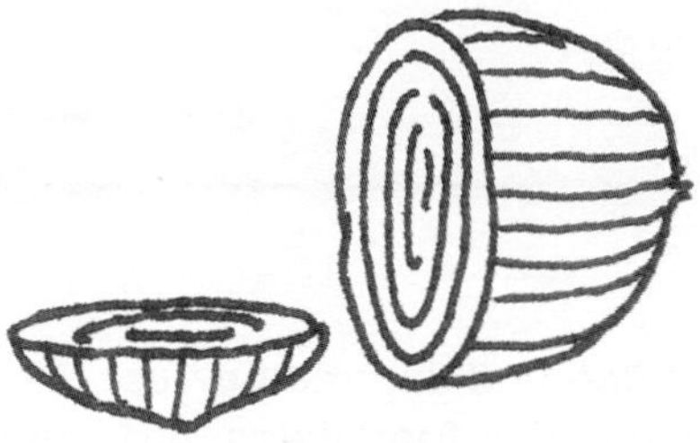
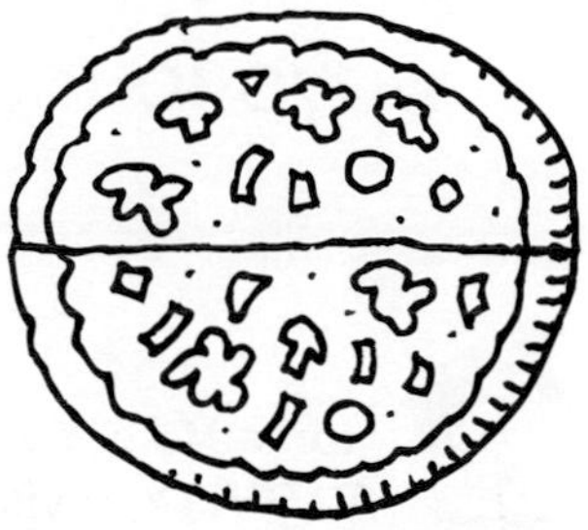

Practice

Find the figures that show halves. Color $\frac{1}{2}$.

1

2

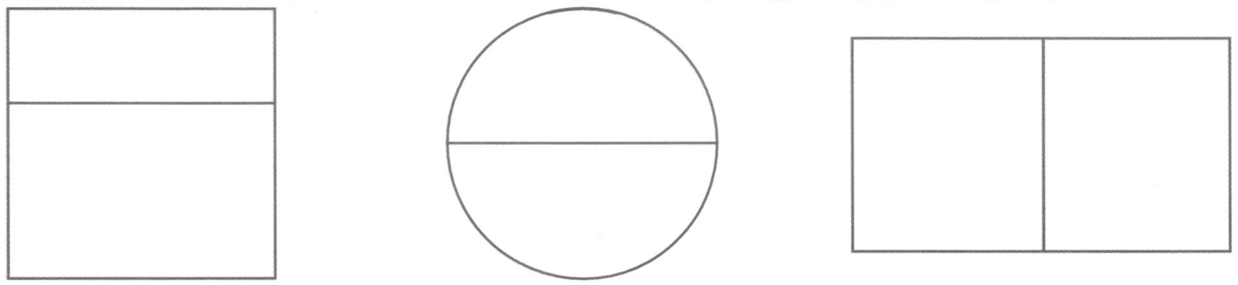

3

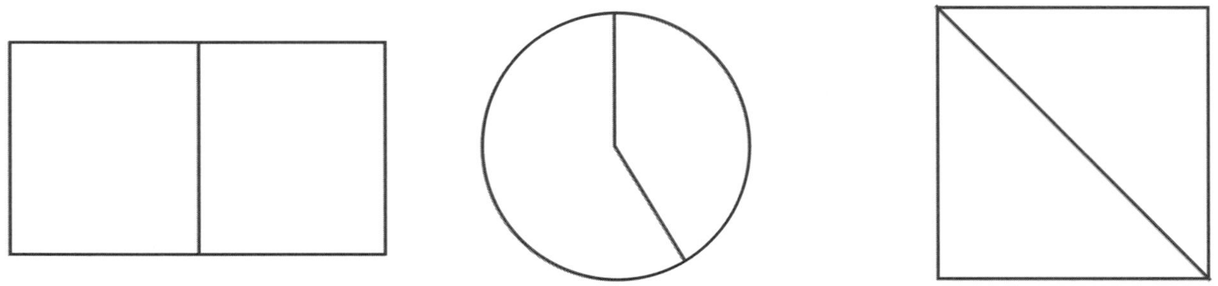

4

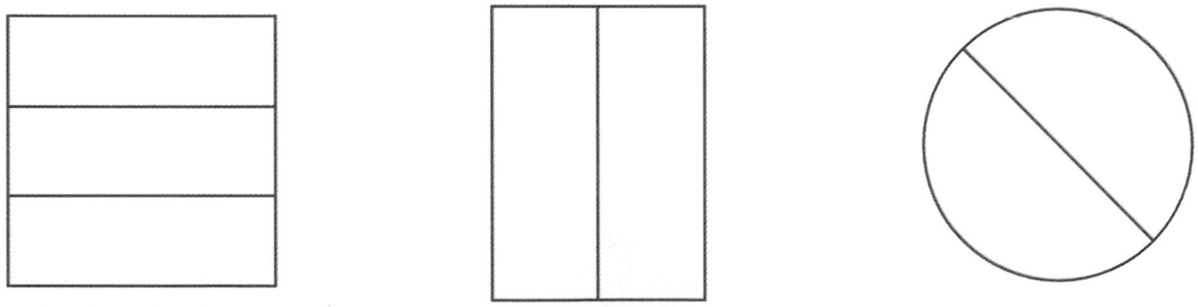

Problem Solving

5 Draw a line on each cracker to show different ways to make halves.

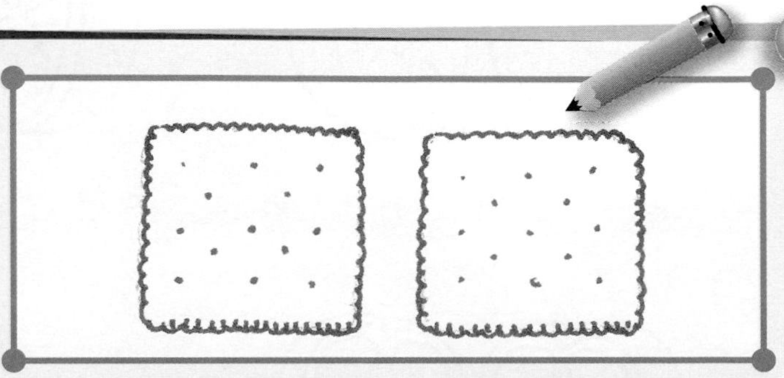

Home Note Your child identified and colored halves.
ACTIVITY Invite your child to divide food items into halves and to name each part.

Name ______________________________

Fourths

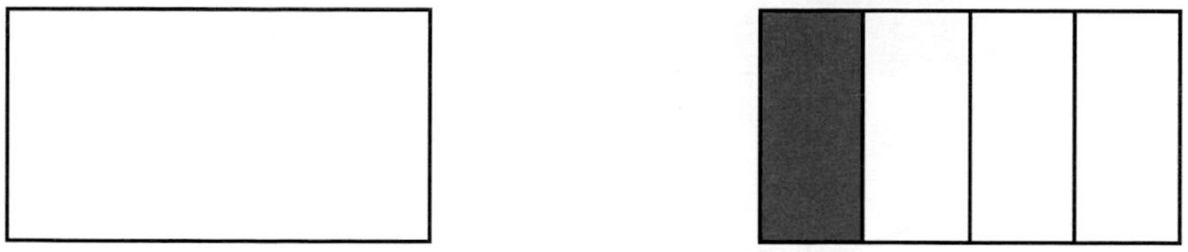

whole

I out of 4 equal parts is $\frac{1}{4}$, or one fourth.
Four fourths make one whole.

Find the figures that show fourths. Color $\frac{1}{4}$ with a crayon.

1

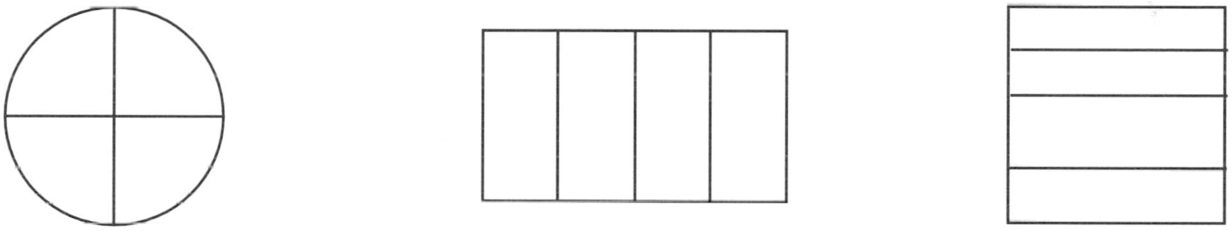

2

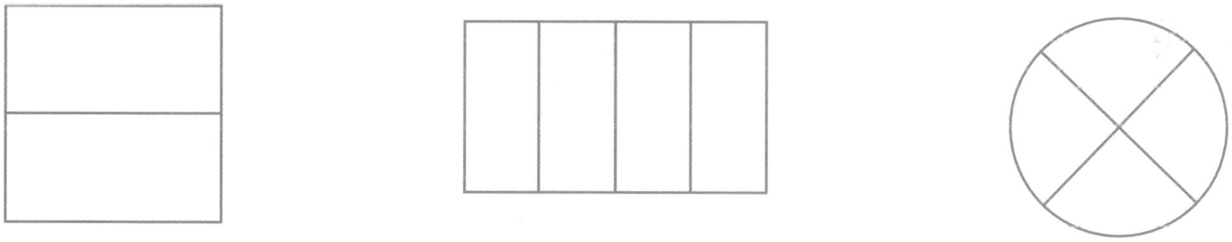

3

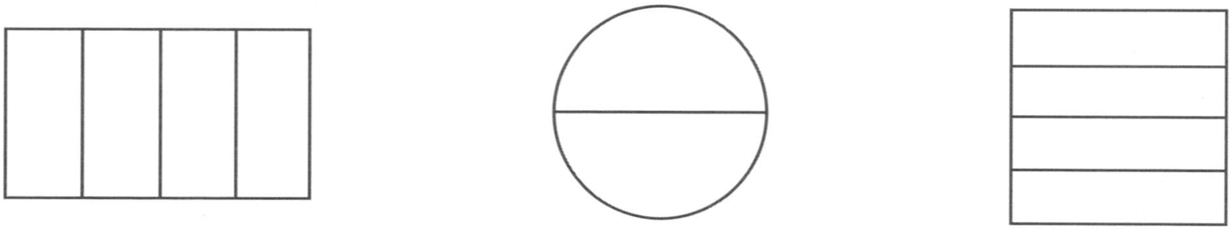

4

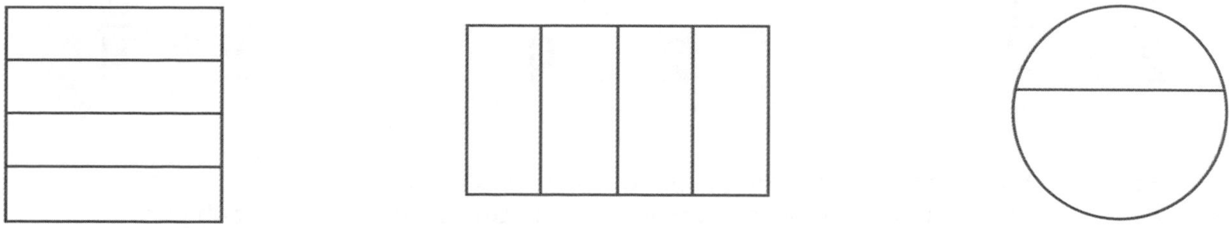

Practice

$\frac{1}{2}$ 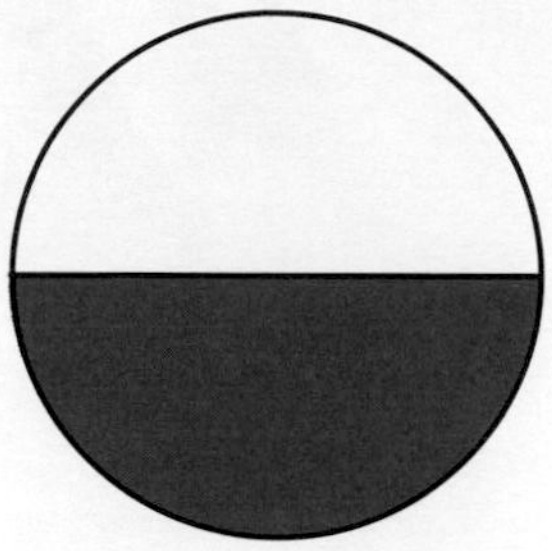 $\frac{1}{4}$ 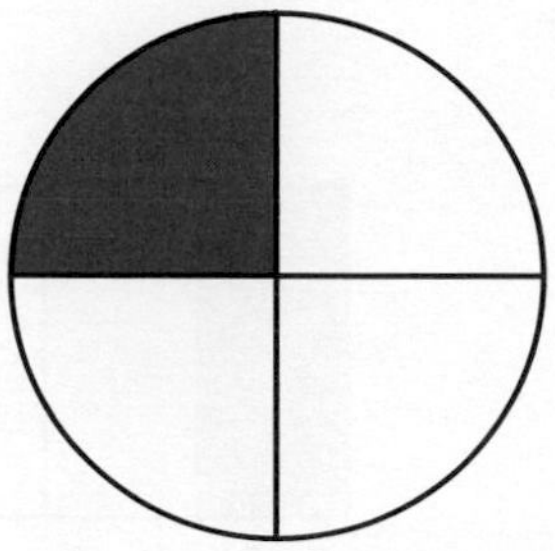

Color one part. Circle the fraction.

1 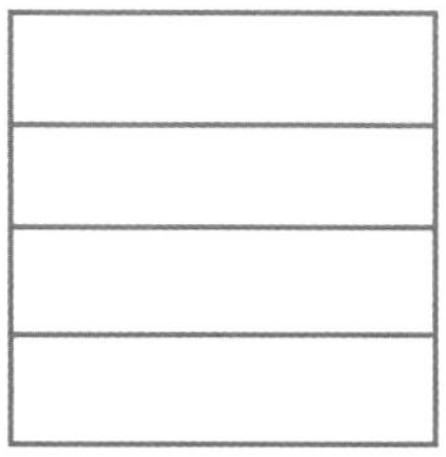

$\frac{1}{2}$ $\frac{1}{4}$

2 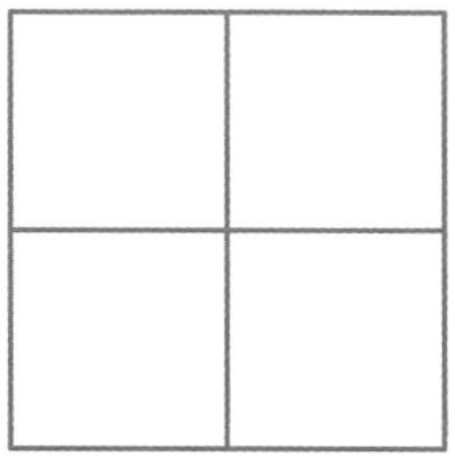

$\frac{1}{2}$ $\frac{1}{4}$

3 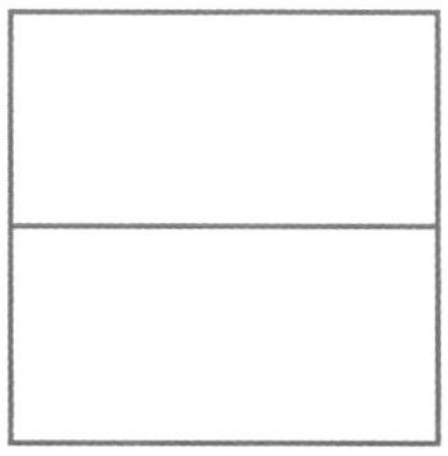

$\frac{1}{2}$ $\frac{1}{4}$

4 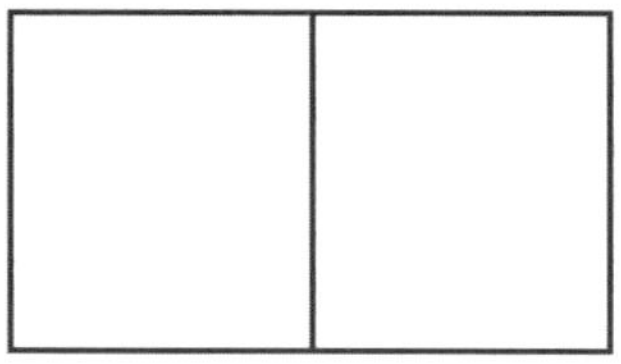

$\frac{1}{2}$ $\frac{1}{4}$

5 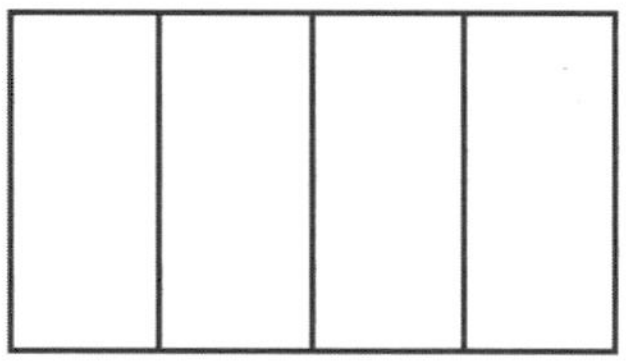

$\frac{1}{2}$ $\frac{1}{4}$

6 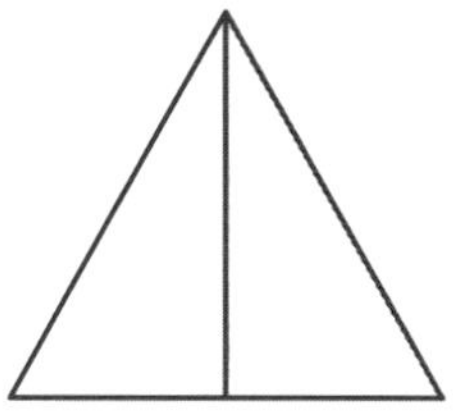

$\frac{1}{2}$ $\frac{1}{4}$

7 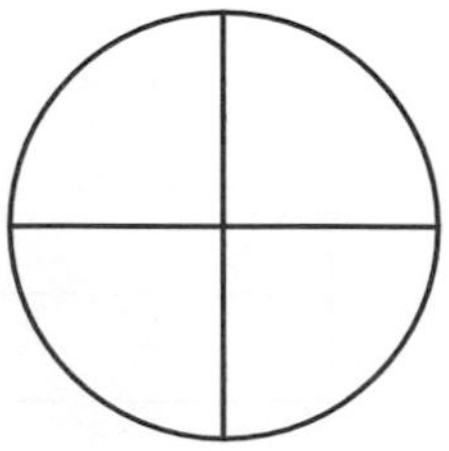

$\frac{1}{2}$ $\frac{1}{4}$

8 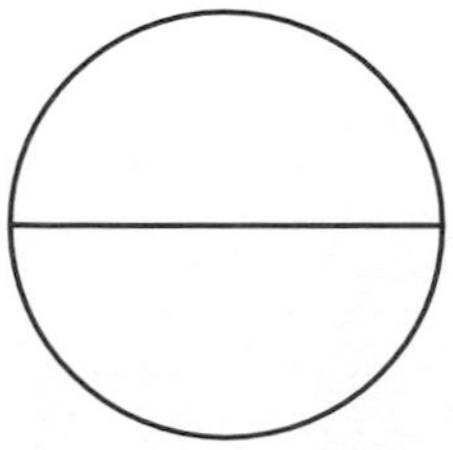

$\frac{1}{2}$ $\frac{1}{4}$

9 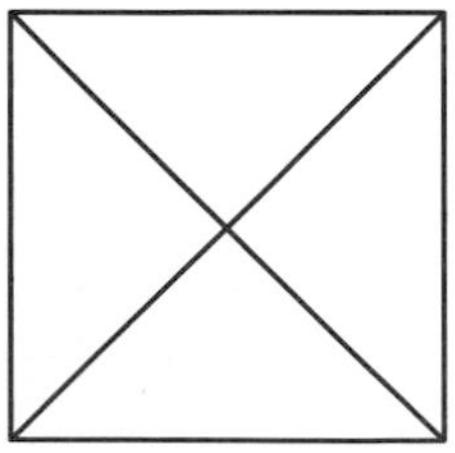

$\frac{1}{2}$ $\frac{1}{4}$

Home Note **Your child identified and colored halves and fourths.**
ACTIVITY **Invite your child to divide food items into fourths and to name each part.**

Name ______________________________

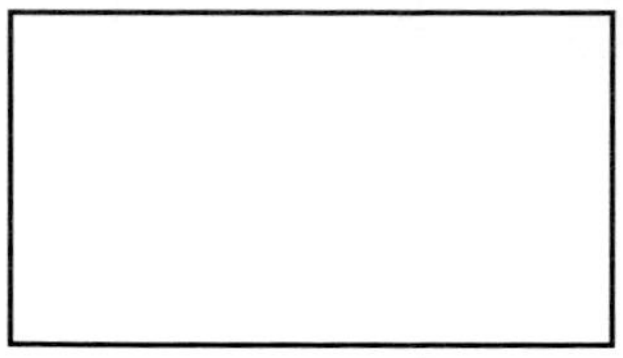

whole

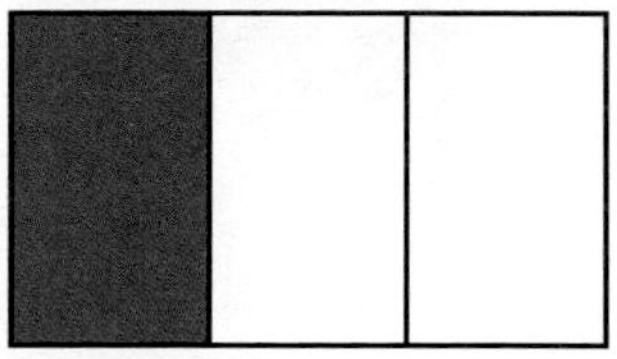

1 out of 3 equal parts is $\frac{1}{3}$, or one third.
Three thirds make one whole.

Find the figures that show thirds. Color $\frac{1}{3}$ .

1

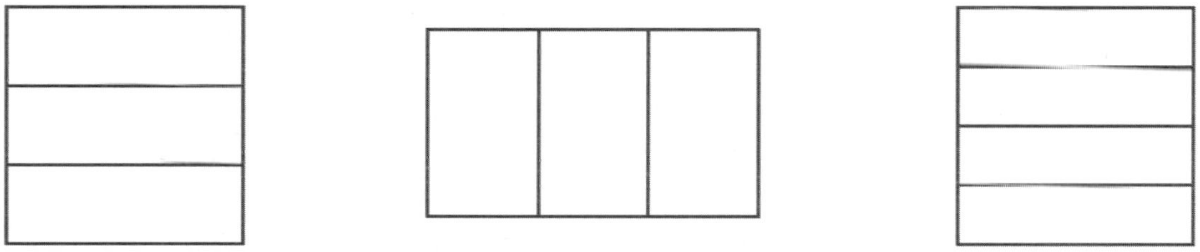

2

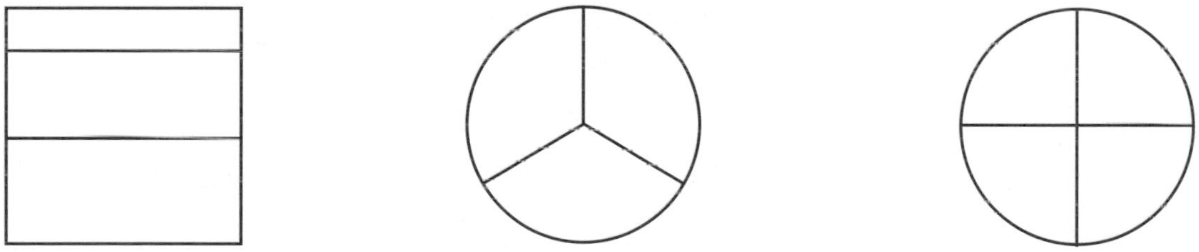

3

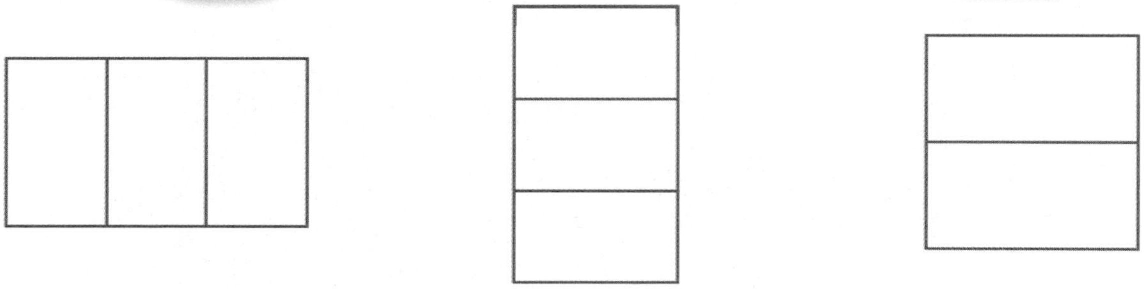

4

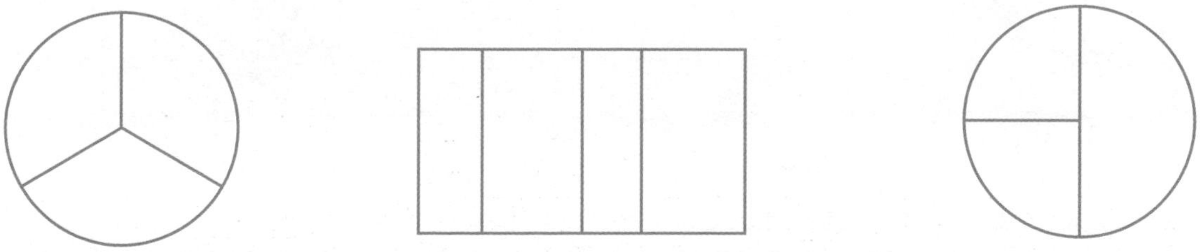

Practice

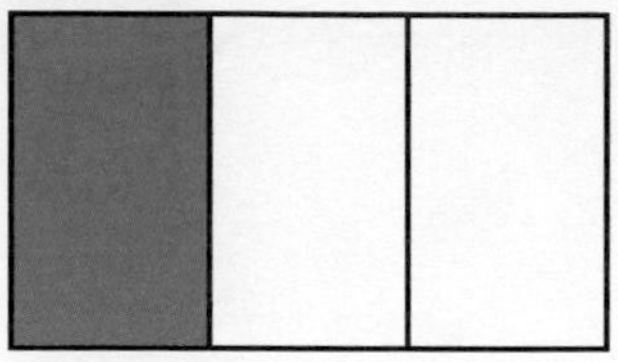

$\frac{1}{3}$

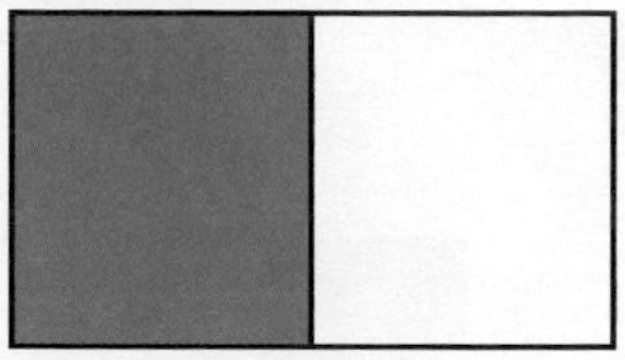

$\frac{1}{2}$

$\frac{1}{4}$

Color one part. Circle the fraction.

1. 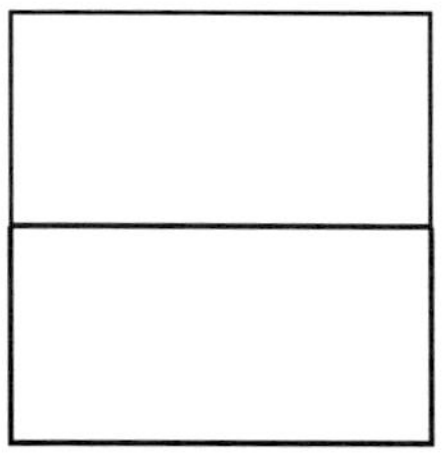

$\frac{1}{3}$ $\frac{1}{2}$ $\frac{1}{4}$

2. 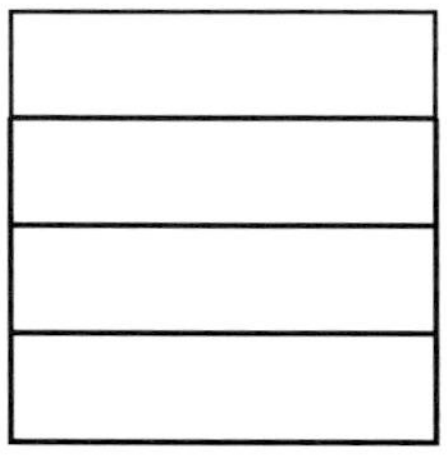

$\frac{1}{3}$ $\frac{1}{2}$ $\frac{1}{4}$

3. 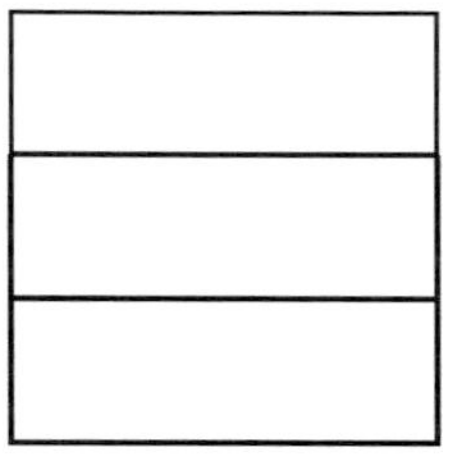

$\frac{1}{3}$ $\frac{1}{2}$ $\frac{1}{4}$

4. 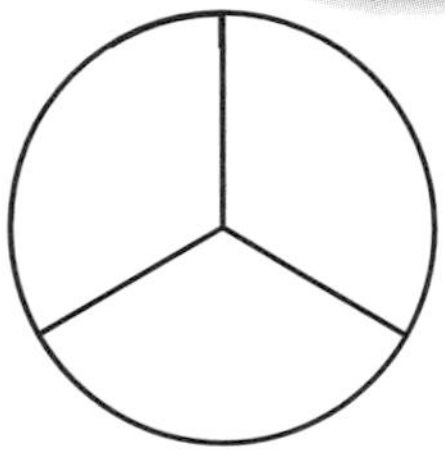

$\frac{1}{3}$ $\frac{1}{2}$ $\frac{1}{4}$

5. 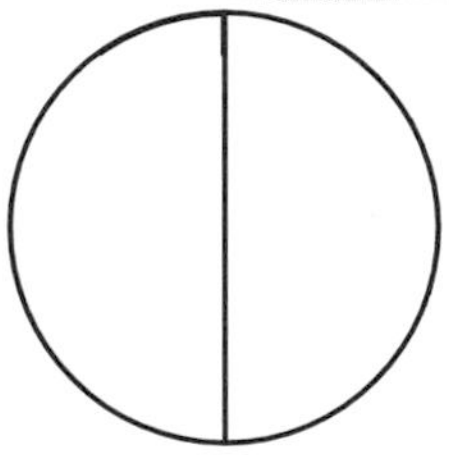

$\frac{1}{3}$ $\frac{1}{2}$ $\frac{1}{4}$

6. 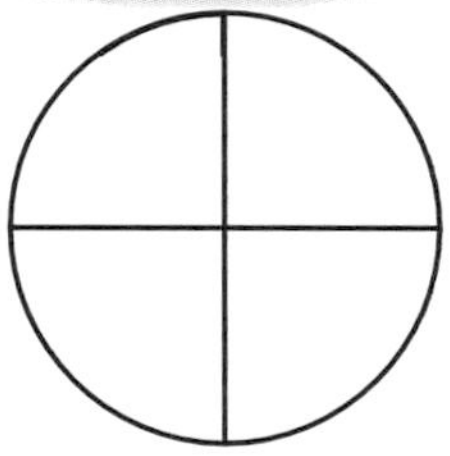

$\frac{1}{3}$ $\frac{1}{2}$ $\frac{1}{4}$

Mixed Review

7. Write the time.

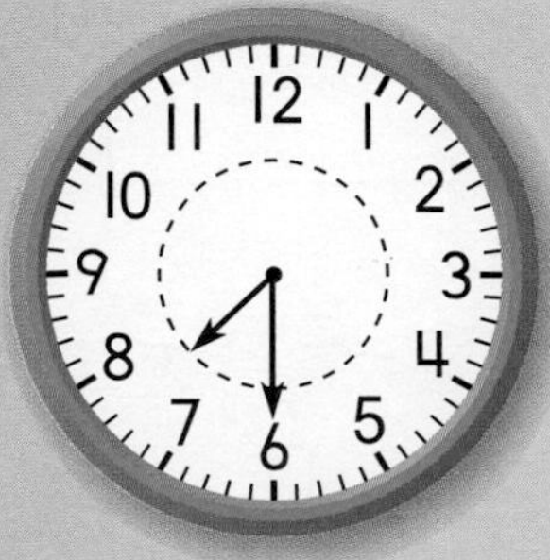

______

8. Count. Write the amount.

______¢

Home Note Your child identified and colored thirds.
ACTIVITY Invite your child to divide food items into thirds and to name each part.

Name ______________________________

Understand • Plan • Solve • Look Back

Problem Solving
Visualizing Results

Think about sharing this pizza.
Circle the picture that answers the question.

1. There are 2 children. Each one gets an equal part. How would you cut the pizza?

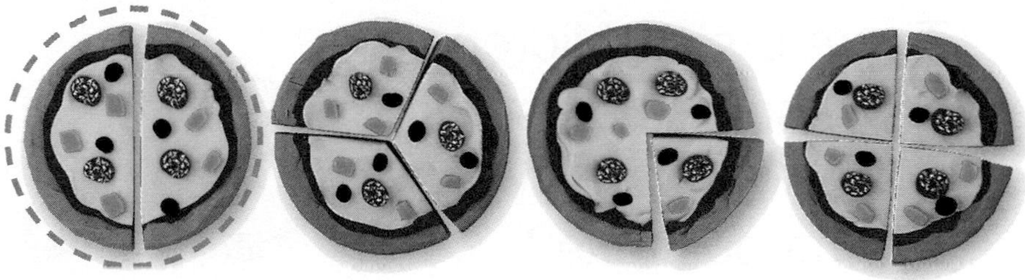

2. There are 3 children. Each one gets an equal part. How would you cut the pizza?

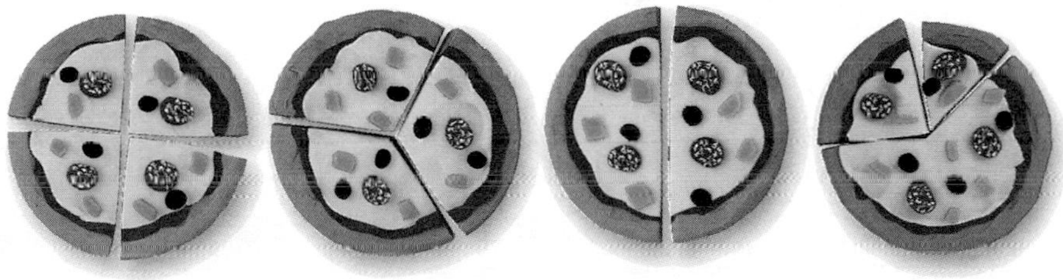

3. You want to share with 2 other children. You give an equal part to each of them. How much is left for you?

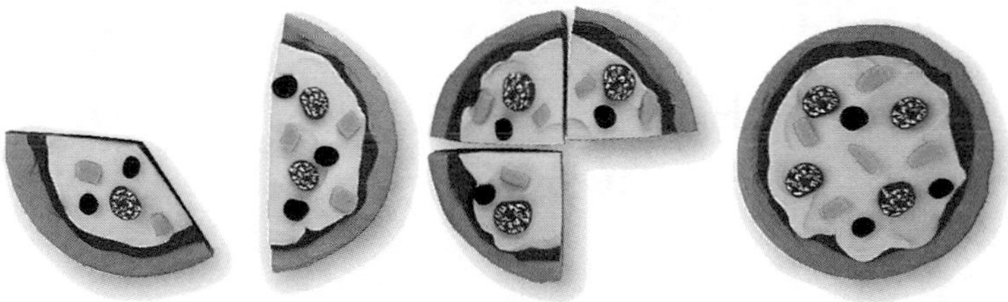

4. There are 4 children. Each one gets an equal part. How much is 1 share?

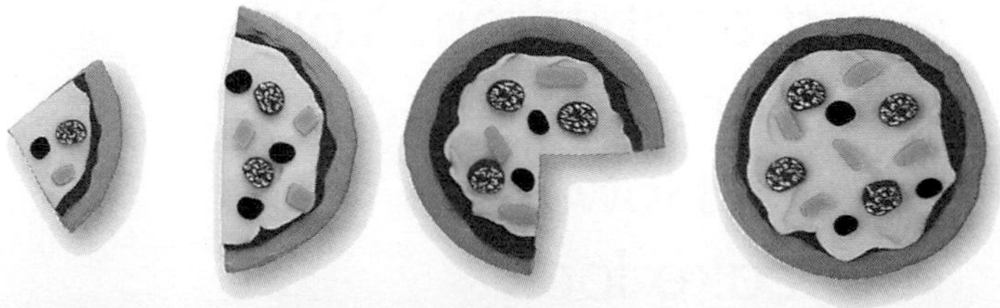

Practice

Think about sharing a giant pancake.
Circle the picture that answers the question.

1. There are 4 children. Each one gets an equal part. How would you cut the pancake?

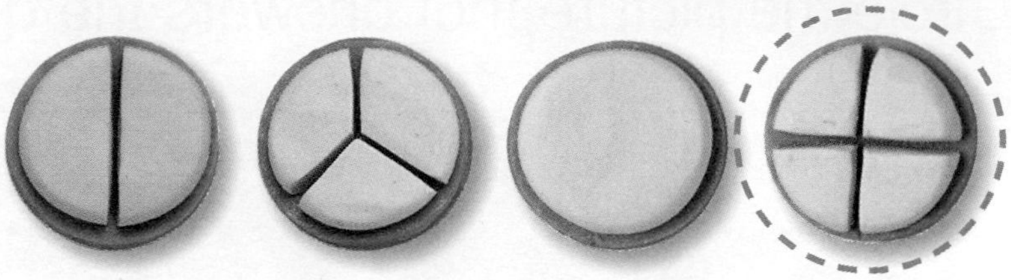

---

2. There are 3 children. Each one gets an equal part. How would you cut the pancake?

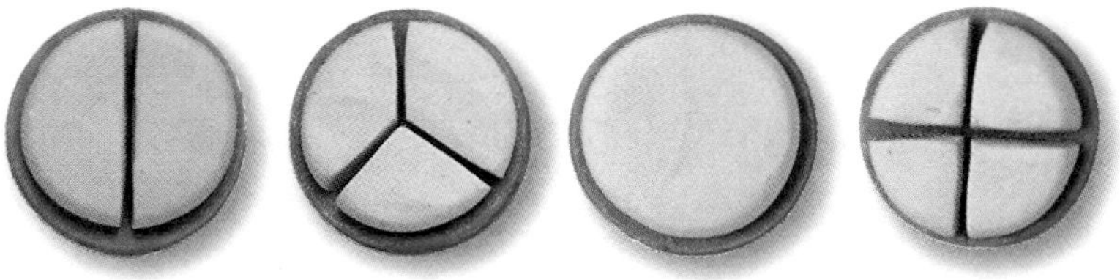

---

3. There are 2 children. Each one gets an equal part. Which pancake shows 2 equal parts?

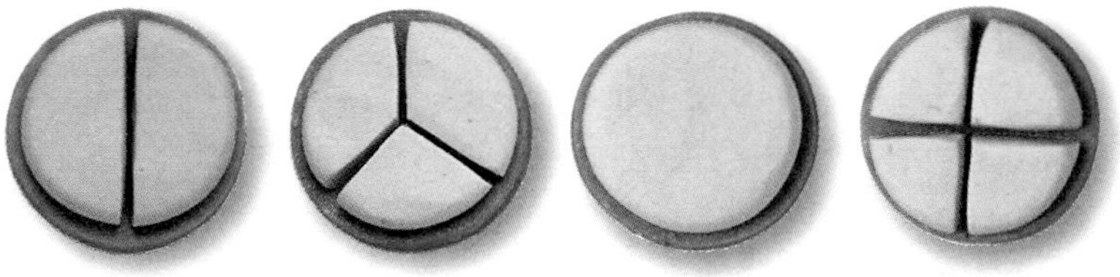

---

4. There are 3 children. Each one gets an equal part. How much is an equal part?

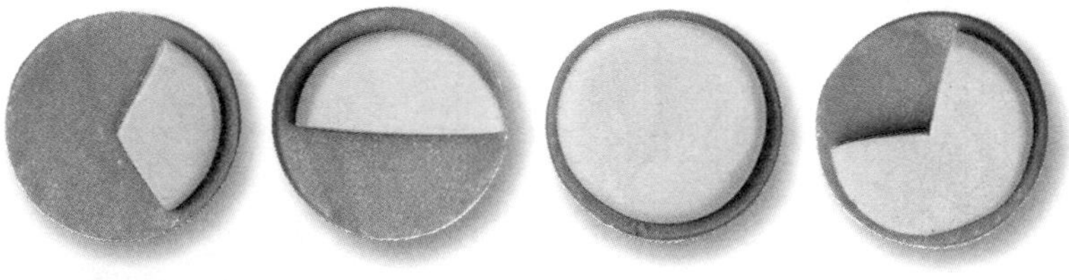

---

5. You want to share with 2 other children. You give one equal part away. How does the pancake look?

Home Note **Your child solved problems about equal parts.**
ACTIVITY Have your child draw to show equal parts of a food item for different numbers of people.

Name ______________________

Remember—
$\frac{1}{4}$ is one out of four
equal parts.

1 out of 4 of the apples is red.

$\frac{1}{4}$ of the apples are red.

Color $\frac{1}{3}$.

1

2

Color $\frac{1}{4}$.

3

4

Talk About It • Critical Thinking

How do you know if you have an equal part of a group?

Practice

$\frac{1}{2}$ are red.

$\frac{1}{3}$ are red.

$\frac{1}{4}$ are red.

Color to show each fraction.

1. 

$\frac{1}{2}$

2. 

$\frac{1}{3}$

3. 

$\frac{1}{4}$

4. 

$\frac{1}{4}$

Write About It

5. Draw a picture of a pie. Show how 4 friends can each get an equal part. Write about your picture.

Home Note Your child learned about equal parts of a group.
ACTIVITY Invite your child to identify $\frac{1}{2}$ of a group of 2 equal objects, $\frac{1}{4}$ of a group of 4 equal objects, and $\frac{1}{3}$ of a group of 3 equal objects.

Name ____________________

Concepts and Skills

Circle the pictures that show equal parts.

1

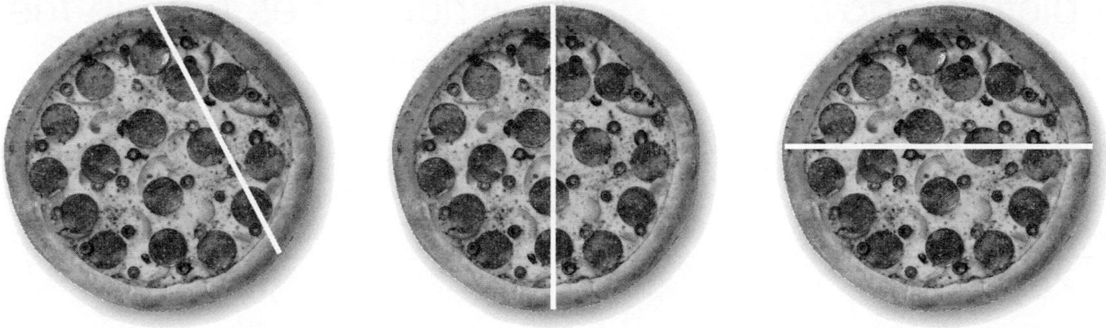

2 Color $\frac{1}{2}$.

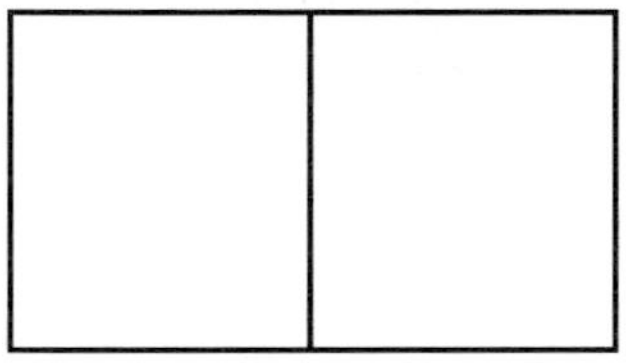

3 Color $\frac{1}{3}$.

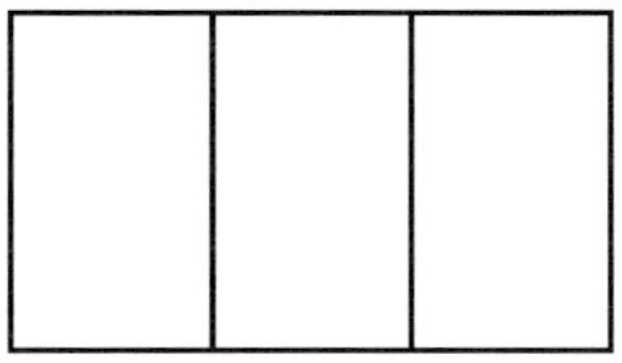

4 Color $\frac{1}{4}$.

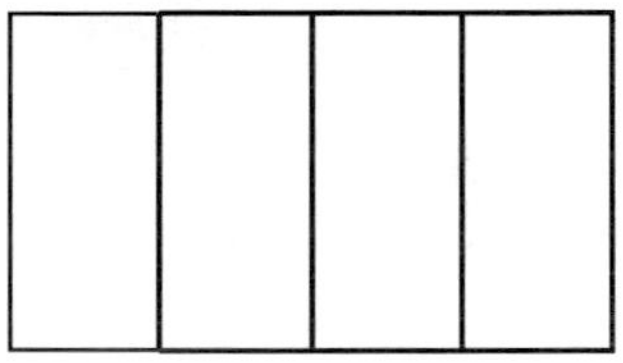

Color one part. Circle the fraction.

5

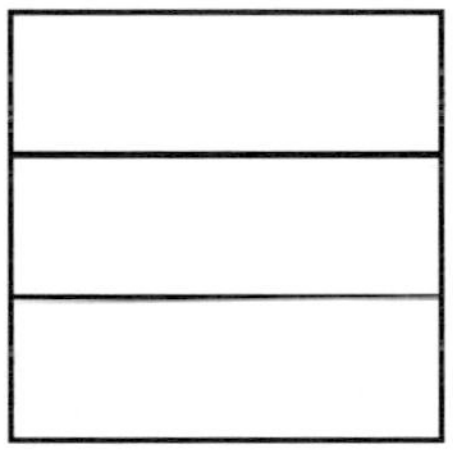

$\frac{1}{3}$ $\frac{1}{2}$ $\frac{1}{4}$

6

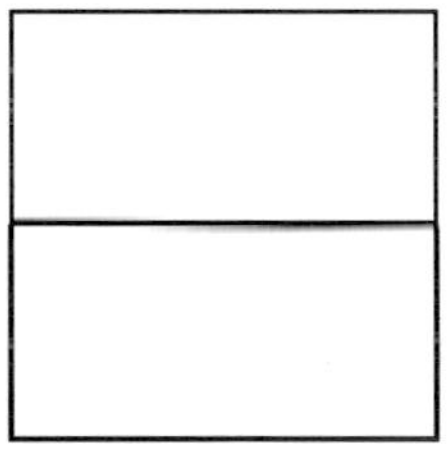

$\frac{1}{3}$ $\frac{1}{2}$ $\frac{1}{4}$

7

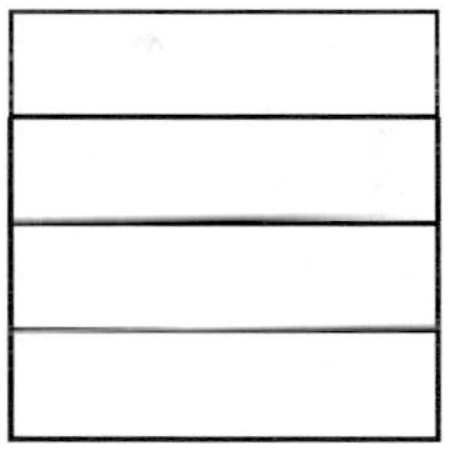

$\frac{1}{3}$ $\frac{1}{2}$ $\frac{1}{4}$

Circle the picture that answers the question.

8 There are 3 children.
Each one gets an equal part.
Which pizza would you use?

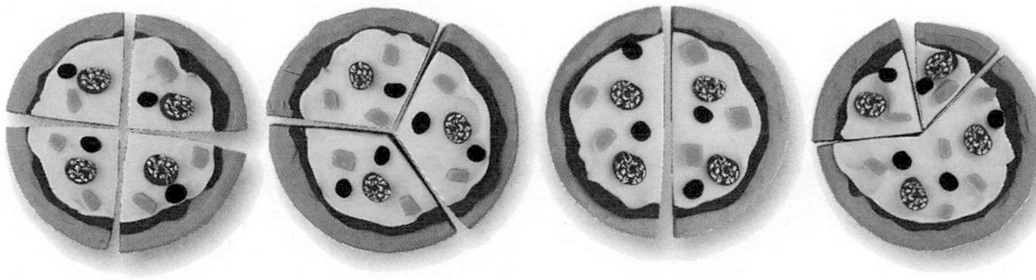

Name ______________________________

Cumulative Chapters 1–22

TAAS Prep

Mark the best answer.

1. Which picture shows 4 equal parts?

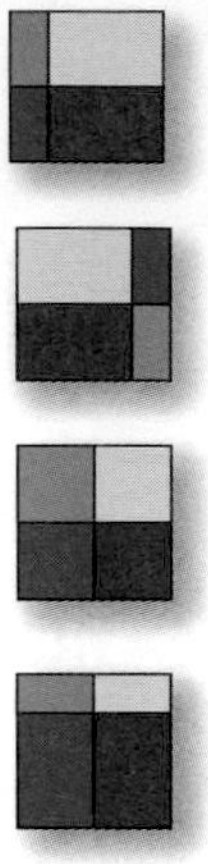

2. Which picture shows $\frac{1}{3}$ of the cats gray?

3. Which picture shows something hot?

4. What comes next in the pattern?

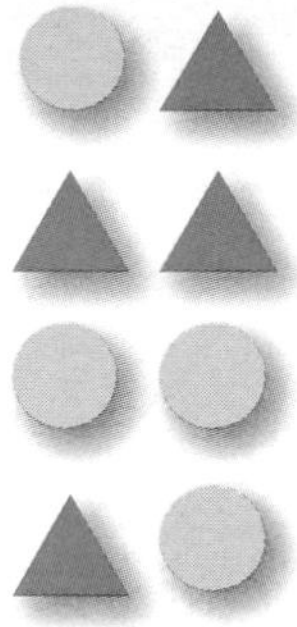

5. How many dimes can you trade for  ?

- 1
- 2
- 3
- 10

6. Which one would take longer to eat?

Name ________________________________

MATH FUN

Use:
number cube, and a crayon

The Fraction Race

Play with a partner.

1. Take turns. Toss the number cube and move that many spaces.
2. Name the fraction that tells what part is red.
3. Color part of your fraction strip to match the fraction you named.
4. The first one to color a whole fraction strip wins.

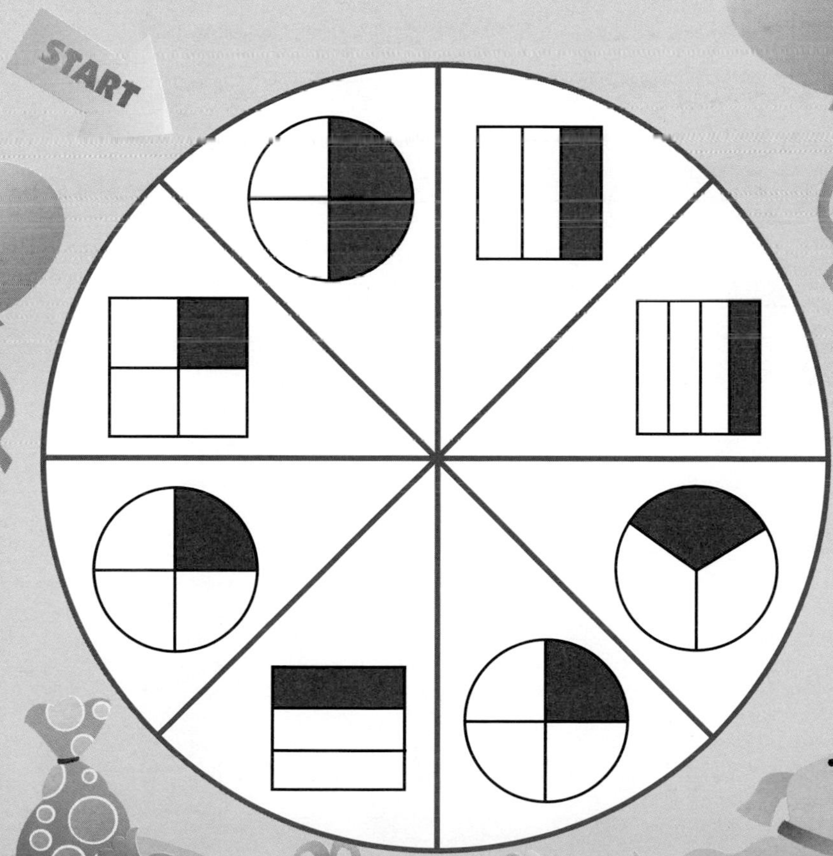

Player 1 **Player 2**

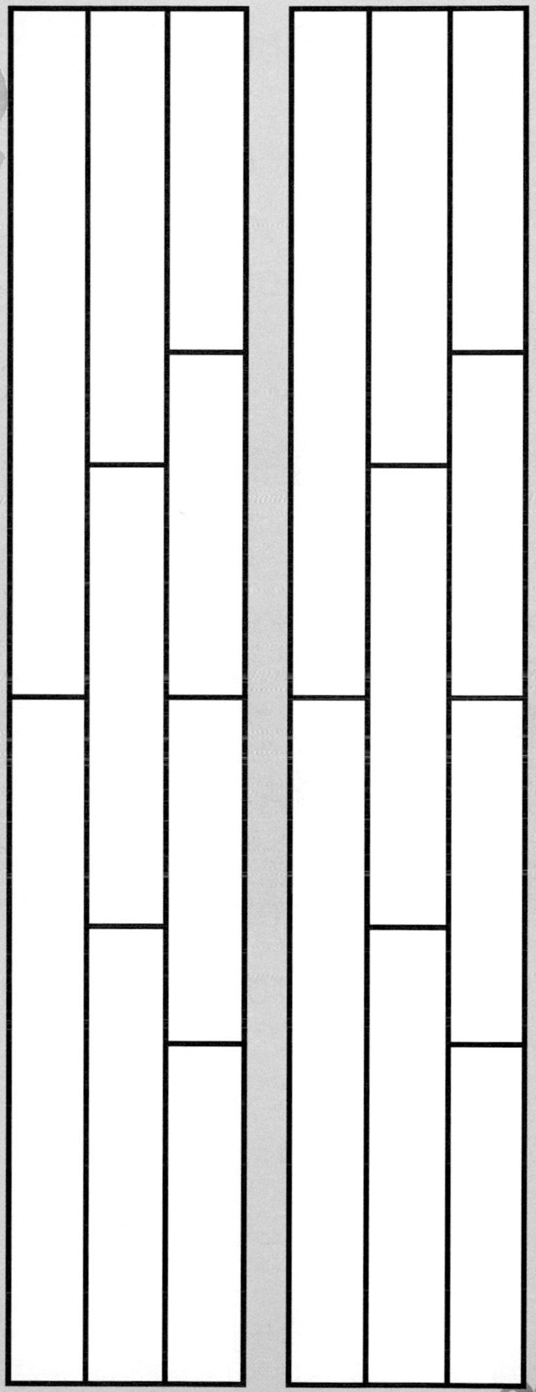

Home Note **Your child has learned the fractions $\frac{1}{2}$, $\frac{1}{3}$, and $\frac{1}{4}$.**
ACTIVITY **Play this game to give him or her practice with fractions.**

Name ______________________________

Calculator | Computer

You can use a to to show fractions.

1. Find the figure that shows $\frac{1}{2}$. Color $\frac{1}{2}$.

   Find the figure that show $\frac{1}{4}$. Color $\frac{1}{4}$.

   Find the figure that show $\frac{1}{3}$. Color $\frac{1}{3}$.

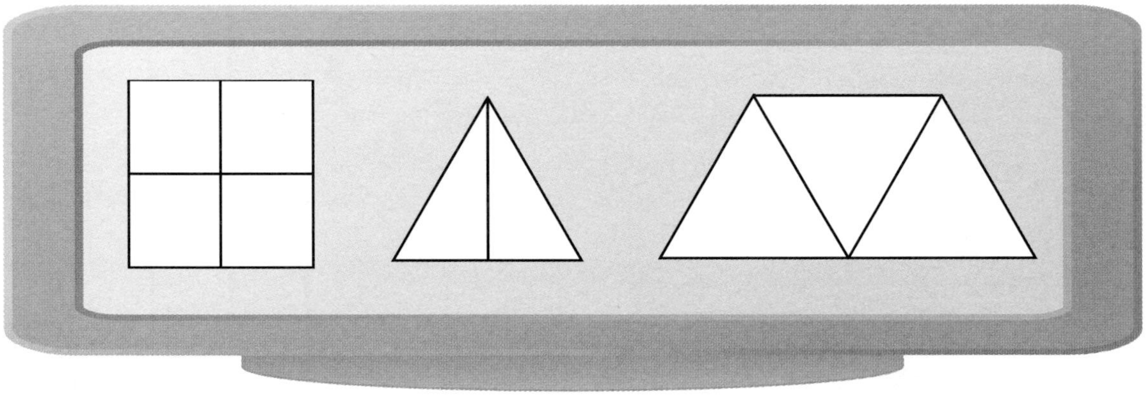

2. Use a , , , , , , or a to make figures that show $\frac{1}{2}$, $\frac{1}{3}$, and $\frac{1}{4}$. Draw your figures.

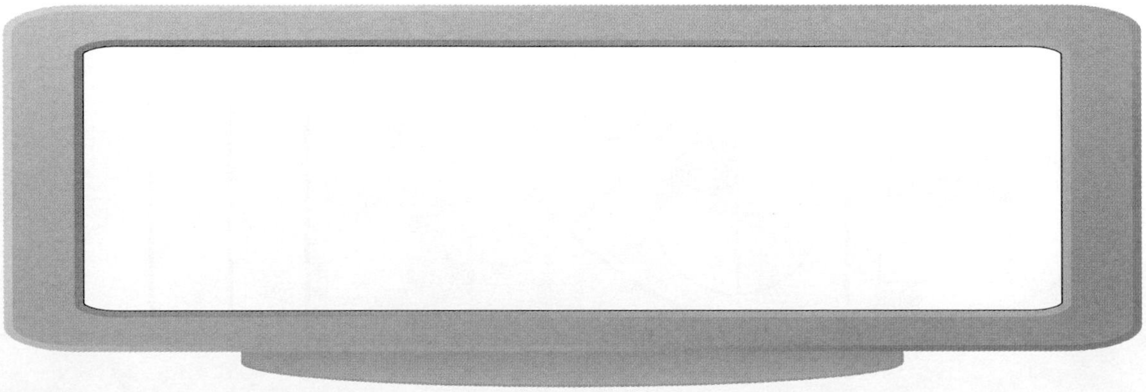

3. Have a classmate find and color each figure that shows $\frac{1}{2}$, $\frac{1}{3}$, and $\frac{1}{4}$.

Green bug,
Orange bug,
sitting on a leaf.

2 BUGS

written and illustrated by

Gerald McDermott

This book will help me review measurement concepts.

This book belongs to ______________________.

2 BUGS

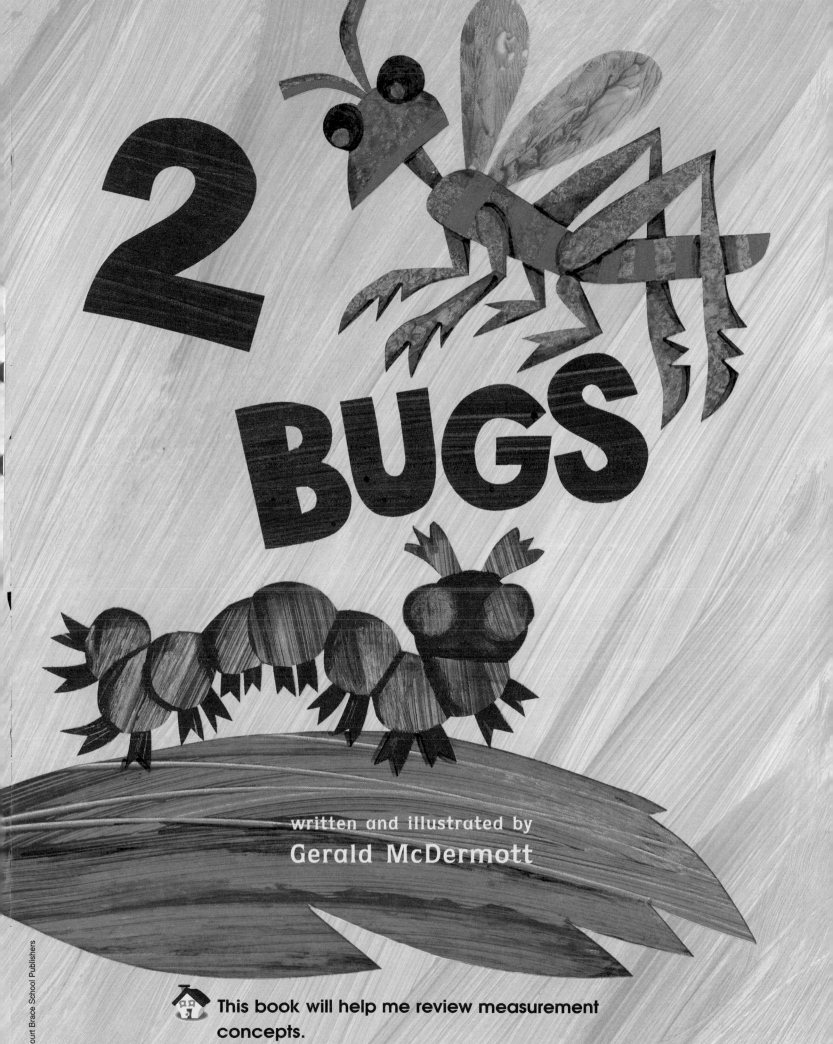

written and illustrated by
Gerald McDermott

This book will help me review measurement concepts.

This book belongs to ______________________.

Harcourt Brace School Publishers

Green bug,
Orange bug,
sitting on a leaf.

B

Who is **taller**?

Who is **l o n g e r ?**

F

Who is **heavier?**

Now, who is the tallest? Who is the longest? Who is the heaviest?

Name ______________________

Concepts and Skills

Measure.

1. About how many  long?

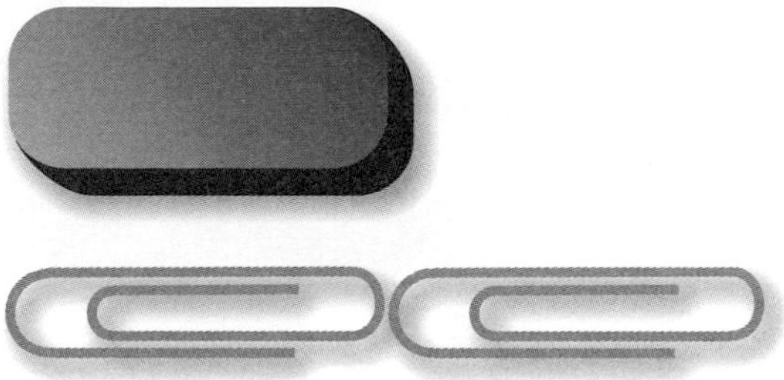

about ____ 

2. About how many centimeters long?

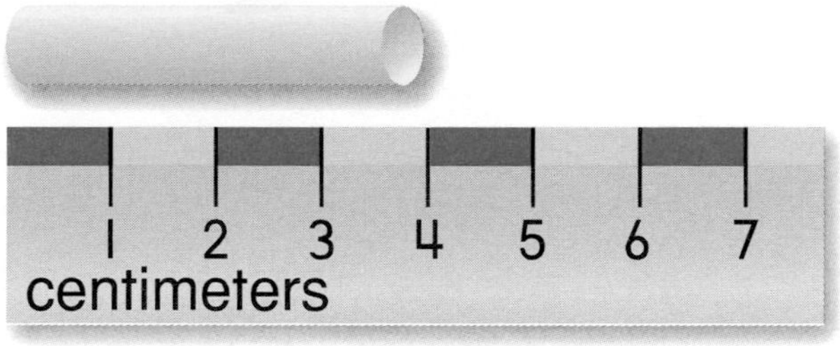

____ centimeters

3. How many inches long?

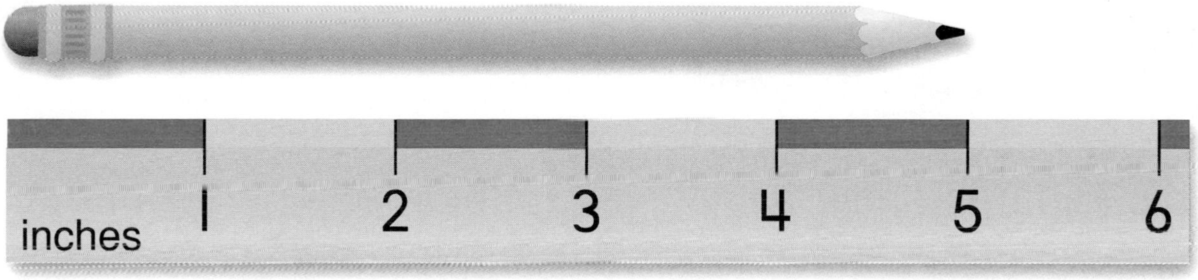

____ inches

Write **H** under the object that is heavier.
Write **L** under the object that is lighter.

4. 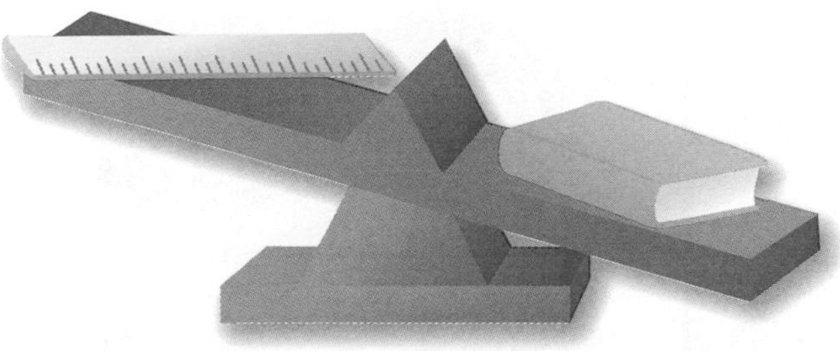

____ ____

5. 

____ ____

Circle the answers.

6 Use a cup and a milk carton. About how many cups of rice does the milk carton hold?

about 2  about 4 

7 Which pictures show something hot?

8 Which pictures show equal parts?

9 Which figures show $\frac{1}{2}$?

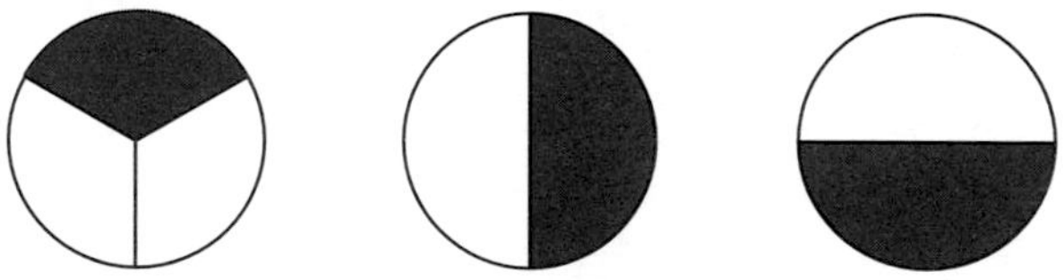

10 Which figures show $\frac{1}{3}$?

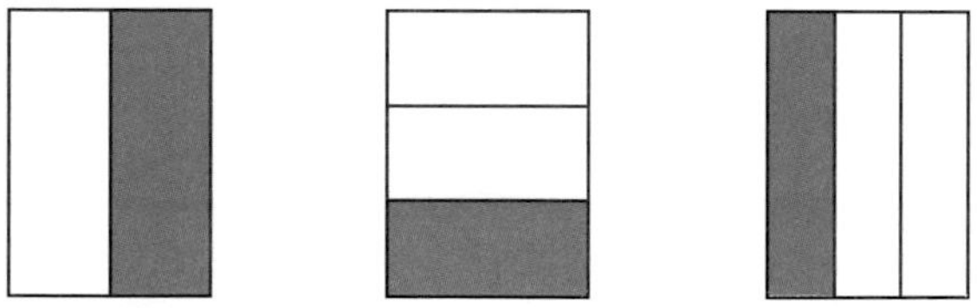

11 Color to show $\frac{1}{2}$ of the apples (crayon).

Problem Solving

12 There are 4 children. Each one gets an equal part. Which pizza would you use?

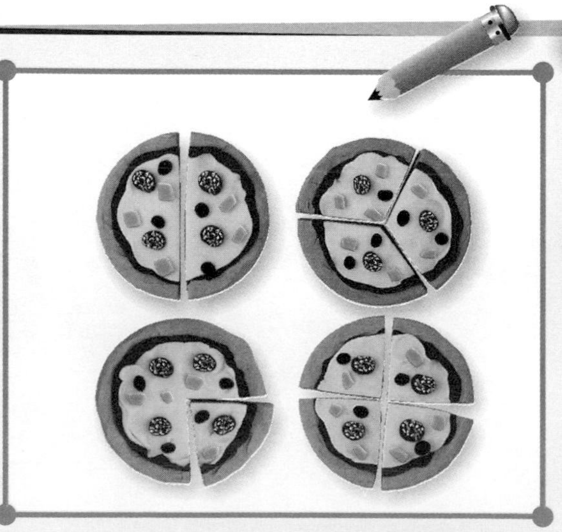

Name ______________________________

Performance Assessment

Use real objects and an inch ruler.

Find 2 objects to measure.
Draw them.
Estimate. Then measure.

	Objects	Estimate	Measure
1		about ____ inches	about ____ inches
2		about ____ inches	about ____ inches
3		about ____ inches	about ____ inches

Write About It

4. You share a large cookie with 2 friends. Each friend gets an equal part.
Draw the cookie.
Draw lines to show how you would cut the cookie.
Circle the part you would eat.

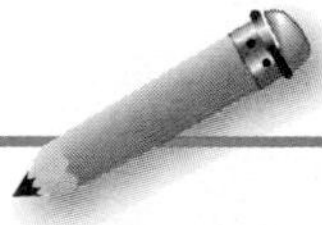

Name ______________________________

Fill in the ⬭ for the correct answer.

Chapters 1–22 Cumulative Review

1. Which figure has no faces?

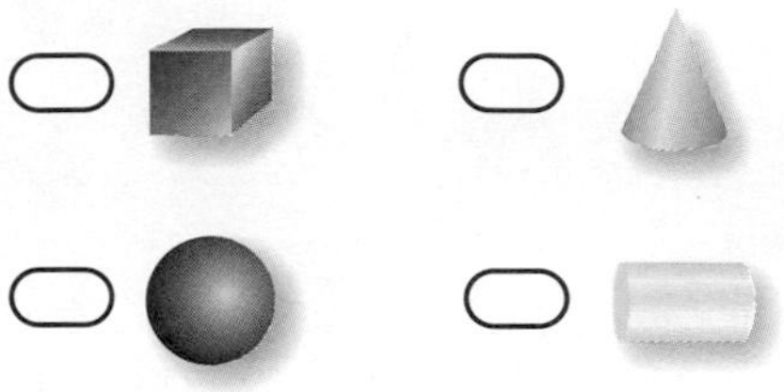

2. Which figure is a circle?

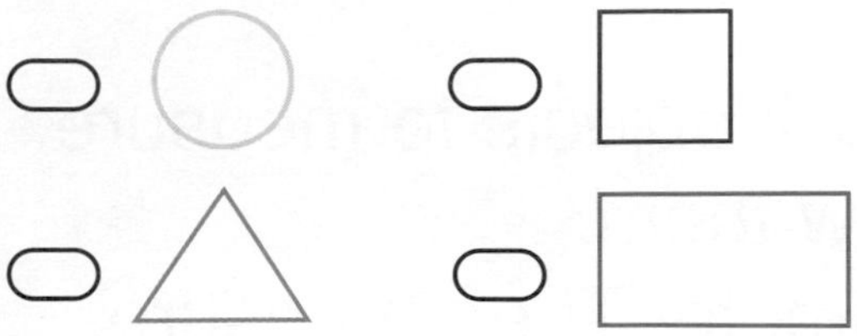

Which number sentence does the story show?

3. 7 books are red. 5 books are blue. How many books are there?

⬭ 7 − 5 = 2
⬭ 7 + 5 = 12
⬭ 5 + 2 = 7
⬭ not here

4. Bob had 11 toy cars. He lost 3 of them. How many toy cars are left?

⬭ 3 + 5 = 8
⬭ 11 + 3 = 14
⬭ 11 − 9 = 2
⬭ 11 − 3 = 8

5. Which tells the amount?

⬭ 5¢
⬭ 25¢
⬭ 50¢
⬭ 55¢

6. Is 11 even or odd?

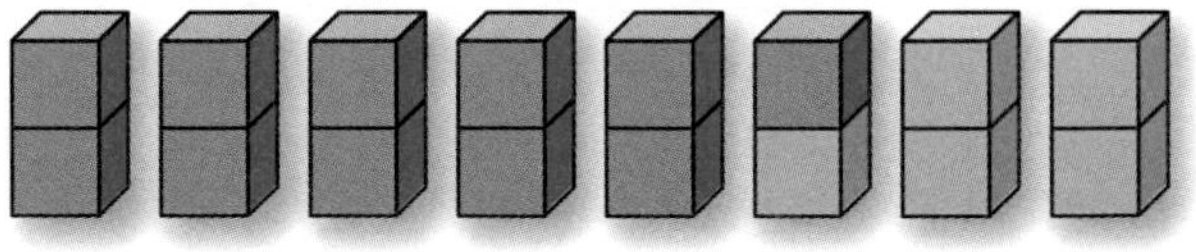

⬭ even ⬭ odd

7. Which takes longer to do?

8. Which shows equal parts?

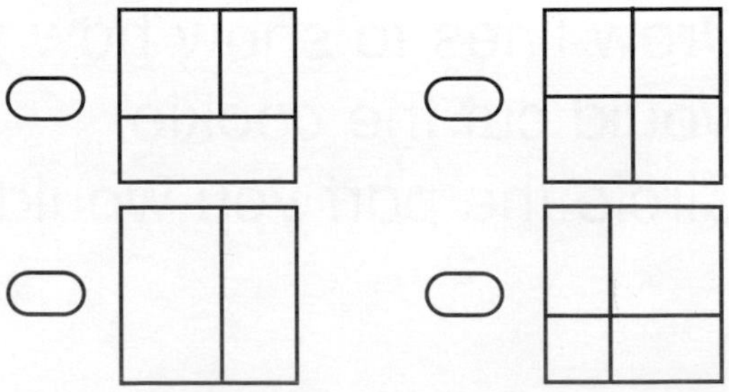

CHAPTER
23

Organizing Data

Make a table to show what is in the garden.

Home Note In this chapter, your child will learn how to organize data.
ACTIVITY Have your child make a table to show how many flowers of each kind are in the garden.

SCHOOL-HOME CONNECTION

Dear Family,

Today we started Chapter 23. We will sort things into groups and use tally marks to make tables. Here are the new vocabulary words and an activity we can do together at home.

Love,

____

Vocabulary

In this table the children are **sorted** into groups by the kind of pets they have.

Pets		
	Ms. Lee's Class	Ms. Miller's Class
cats	||||	卌
dogs	卌 卌 ||	卌 ||
hamsters	||	|||

Each **tally mark** stands for one child.

ACTIVITY

Have your child choose something that he or she would like to inventory, such as things in a closet, a toy box, or the refrigerator. Help your child make a table to show what he or she finds, using tally marks to show how many of each thing.

Visit our Web site for additional ideas and activities.
http://www.hbschool.com

Name ______________________________

Sort and Classify

Colors of My Butterflies	
red	III
blue	~~IIII~~

Each I stands for 1 butterfly.
~~IIII~~ stands for 5 butterflies.

This table shows one way to sort and count the butterflies. Each I is called a tally mark.

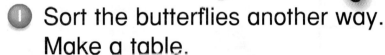

1. Sort the butterflies another way. Make a table.

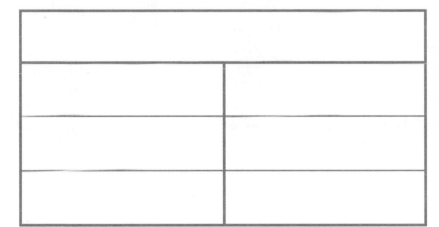

2. Write a question about your table. Give your question to a friend.

____

____

Talk About It • Critical Thinking

Explain to a classmate how you sorted the butterflies.

Practice

This table shows one way to sort and count the children.

Boys and Girls in Our Class	
girls	卌
boys	||||

Each | stands for 1 child.
卌 stands for 5 children.

1. Sort the children another way. Make a table.

2. Write a question about your table. Give your question to a friend.

____

Mixed Review

Write the numbers in each fact family.

3. $4 + 3 = $ ____ $3 + 4 = $ ____

$7 - 3 = $ ____ $7 - 4 = $ ____ ____, ____, ____

Home Note Your child sorted children into two groups and made a table.
ACTIVITY Have your child sort some objects in several ways. Make up questions about the objects for your child to answer.

Name ______________________________

Certain or Impossible

Look in the bag.
Circle the pictures that show what can come out of it.

1

2

3

Practice

Look in the box.
Circle the pictures that show what can come out of it.

1

2

3

Home Note Your child identified which events were certain and which were impossible.
ACTIVITY Ask your child more questions about what could and could not come out of this box.

Name ______________________________

Most Likely

Use a bag, 3 , 1 , 1 .

1. Put all the cubes into the bag.
Take out one cube.

2. Make a tally mark on the table to show which color you got.

3. Put the cube back into the bag. Shake.

4. Make a prediction.
If you do this 9 more times, which color do you think you will get most often?
Circle that color.

green   yellow   red

	Tally Marks	Total

5. Do this 9 more times.
Make a tally mark each time.
Count the tally marks for each color.
Write the totals.

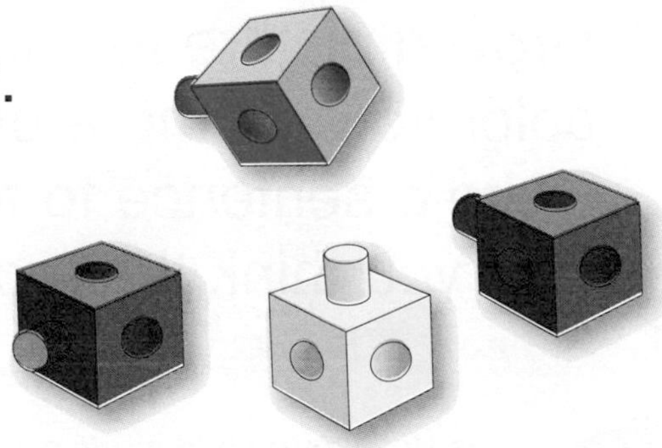

Talk About It • Critical Thinking

Which color did you get most often?
Why do you think that happened?

Practice

Use a plastic jar, 10 ■, 20 ■.

1. Put all the color tiles into the jar. Take out one color tile.

2. Make a tally mark on the table to show which color you got.

3. Put the color tile back into the jar. Shake.

4. Make a prediction.
   If you do this 9 more times, which color do you think you will get more often?
   Circle that color.

   red   blue

5. Do this 9 more times.
   Make a tally mark each time.
   Count the tally marks for each color.
   Write the totals.

	Tally Marks	Total
■		
■		

Write About It

6. Draw a picture to show which color you got more often.
   Write a sentence to tell why you think that happened.

Home Note Your child made a prediction, collected data, and made a table.
ACTIVITY Put 4 pennies and 2 dimes into a bag. Have your child predict which kind of coin he or she will get more often.

Name ______________________________

Understand • Plan • Solve • Look Back

Problem Solving
Use a Table

Use a  and a  to make a spinner.

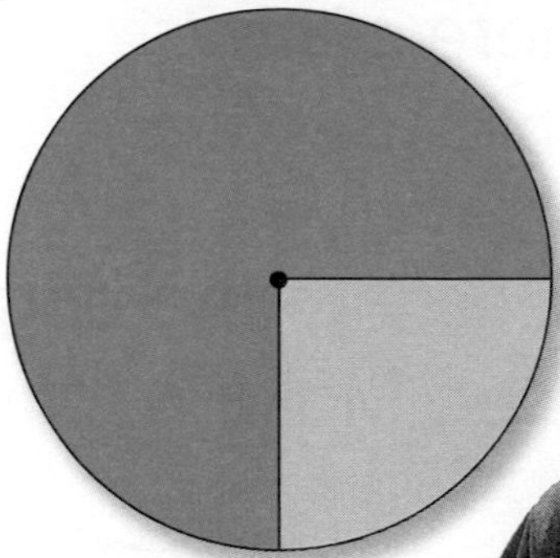

1. Which color do you predict the spinner will stop on more often?
   Circle that color.

   green   yellow

2. Try it.
   Spin 10 times.
   Make a tally mark after each spin.
   Write the totals.

	Tally Marks	Total
green		
yellow		

Try again with this spinner.

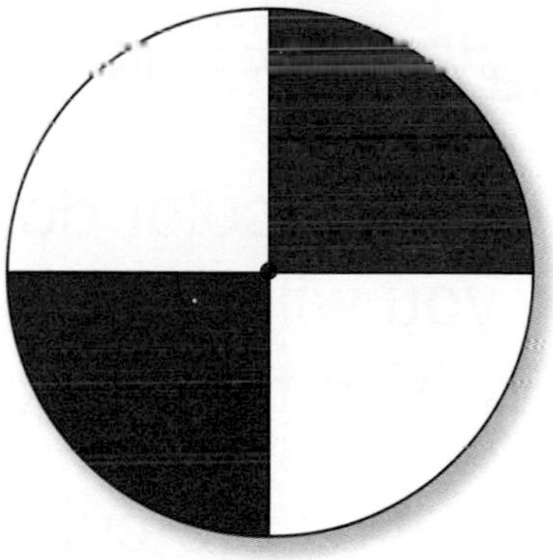

3. Which color do you predict the spinner will stop on most often?
   Circle that color.

   red   white   blue

4. Try it.
   Spin 10 times.
   Make a tally mark after each spin.
   Write the totals.

	Tally Marks	Total
red		
white		
blue		

Practice

Use a bag, 4 cubes, and 2 cubes.

1. Which color do you predict you will get more often? Circle that cube.

2. Put all the cubes into the bag. Take out a cube. Make a tally mark.

3. Put the cube back into the bag. Shake the bag. Do this 10 times. Write the totals.

	Tally Marks	Total

Use a bag, 5 cubes, and 3 cubes.

4. Which color do you predict you will get more often? Circle that cube.

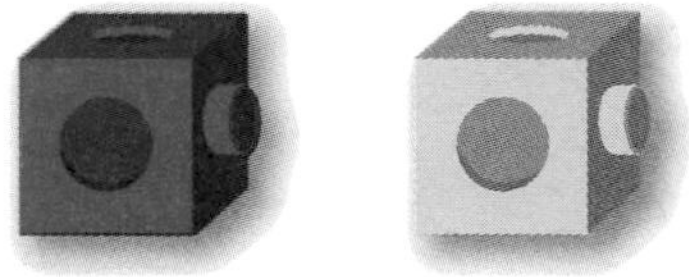

5. Put all the cubes into the bag. Shake the bag. Do this 10 times. Write the totals.

	Tally Marks	Total

Home Note **Your child made predictions and used tables.**
ACTIVITY Find 8 small objects that are alike except that 5 are one color and 3 are another color. Put them into a bag. Have your child predict which color he or she will get more often.

Name ______________________

Chapter 23

Review/Test

Concepts and Skills

1. Sort these leaves in a way you like.
Make a table.

	Tally Marks	Total

2. Look in the box. Circle the pictures that show what can come out of it.

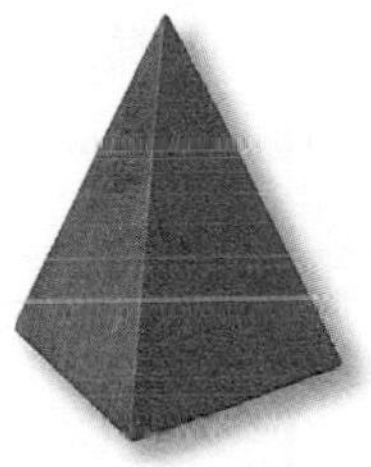

3. Look at the cubes.
Which color do you predict you will get more often if you take a cube out of the bag 10 times?
Circle that color.

red yellow

Name ______________________

Cumulative Chapters 1–23

TAAS Prep

Mark the best answer.

1 Which shows equal parts?

2 Which takes longer to do?

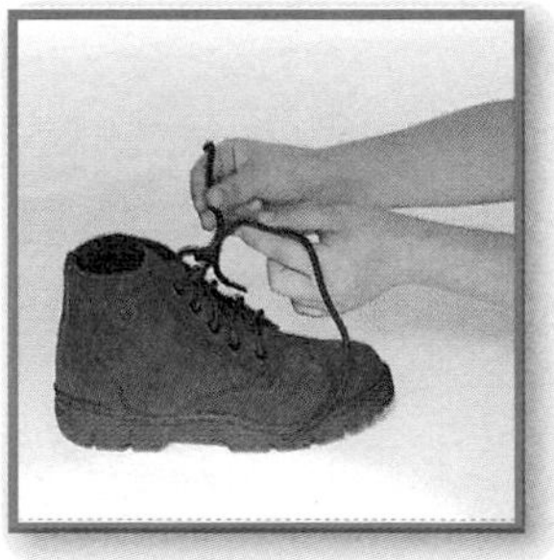

3 Which numbers are in order from least to greatest?

- 36, 61, 84
- 84, 61, 36
- 36, 84, 61
- 61, 36, 84

4 What is the time?

- 2:00
- 2:30
- 3:00
- 3:30

5 Is 13 odd or even?

- odd
- even

6 About how many centimeters long?

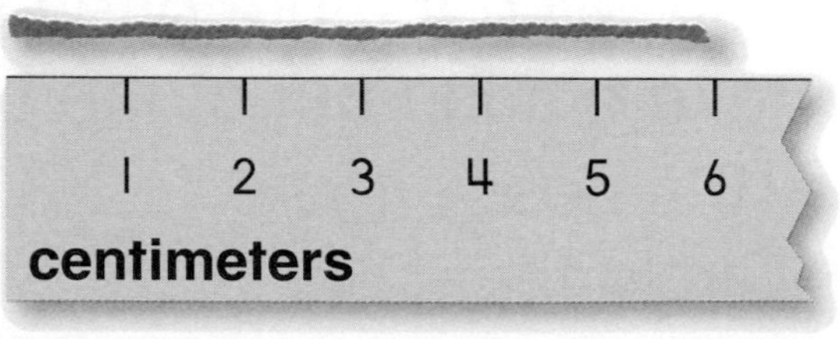

- 4 centimeters
- 5 centimeters
- 6 centimeters
- 7 centimeters

CHAPTER 24

Making Graphs

Make a graph to show what is in the picture.

Home Note In this chapter, your child will learn how to make graphs.
ACTIVITY Make a graph with your child about this picture.

SCHOOL-HOME CONNECTION

Dear Family,

Today we started Chapter 24. We will learn about picture graphs and bar graphs. Here are the new vocabulary words and an activity for us to do together.

Love,

____

Vocabulary

Picture Graph

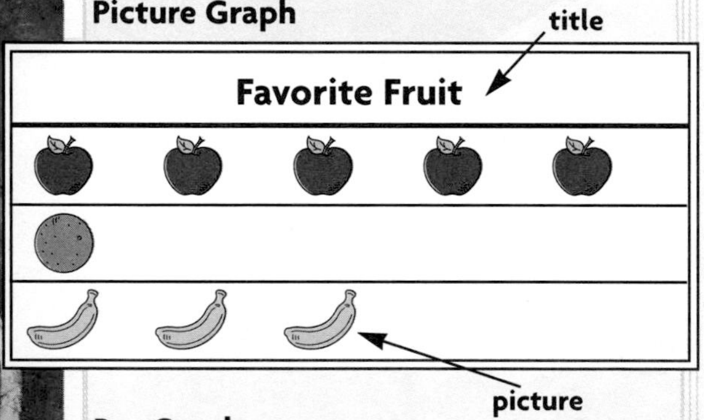

Bar Graph

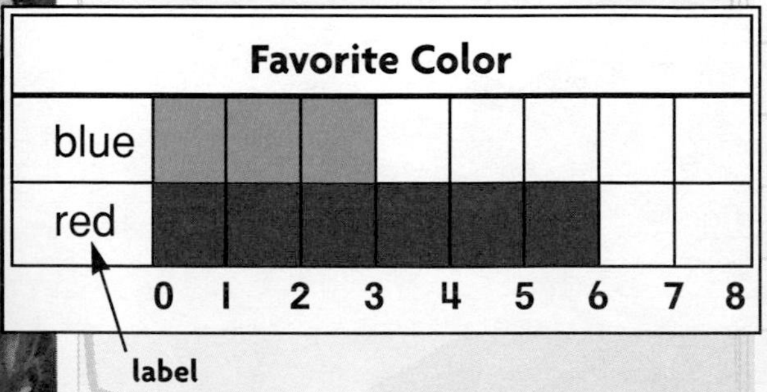

ACTIVITY

Give your child some things that go together, such as buttons, silverware, or coins, to sort into groups. Then ask him or her to count the items in each group. Work with your child to show the results on a graph.

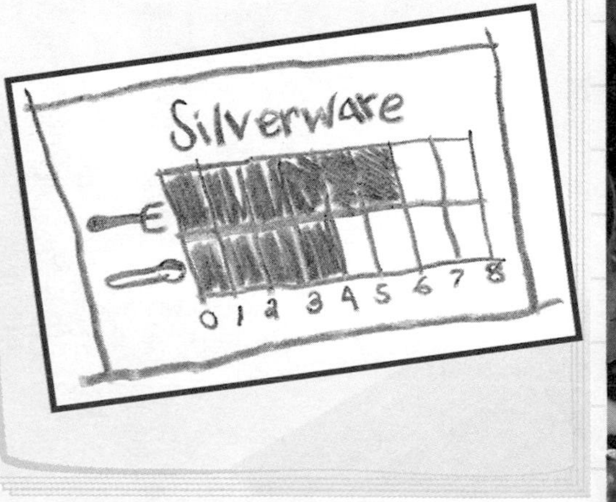

Visit our Web site for additional ideas and activities.
http://www.hbschool.com

Name ______________________________

Picture Graphs

You can make a picture graph to show when some children lost a tooth.

Count the children. Draw ☺ to fill in the graph.

	When We Lost a Tooth				
March	☺	☺	☺	☺	
April	☺	☺			
May	☺	☺	☺		

Use the graph to answer the questions.

1. How many children lost a tooth in March? ______
2. In which month did the most children lose a tooth? ______
3. In which month did the fewest children lose a tooth? ______

Practice

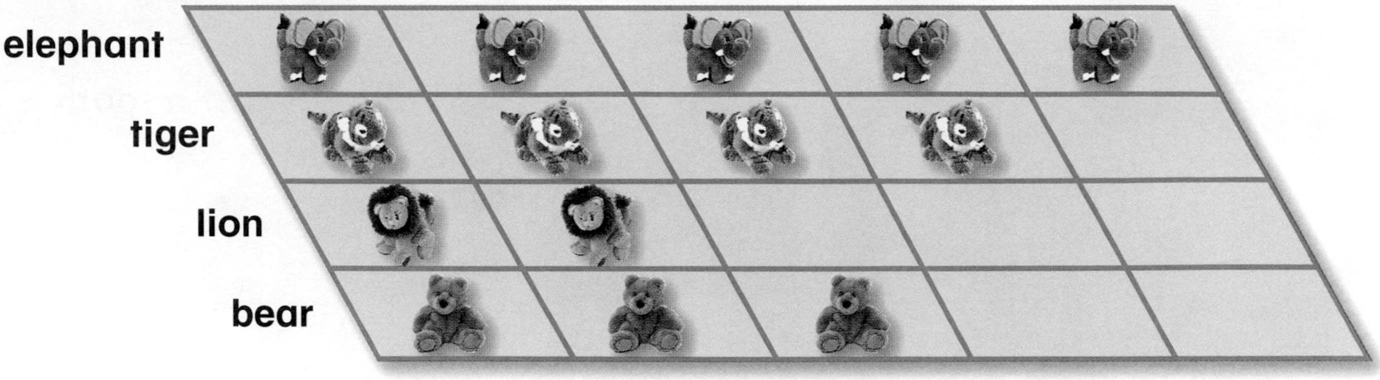

Count the toys. Draw  to fill in the graph.

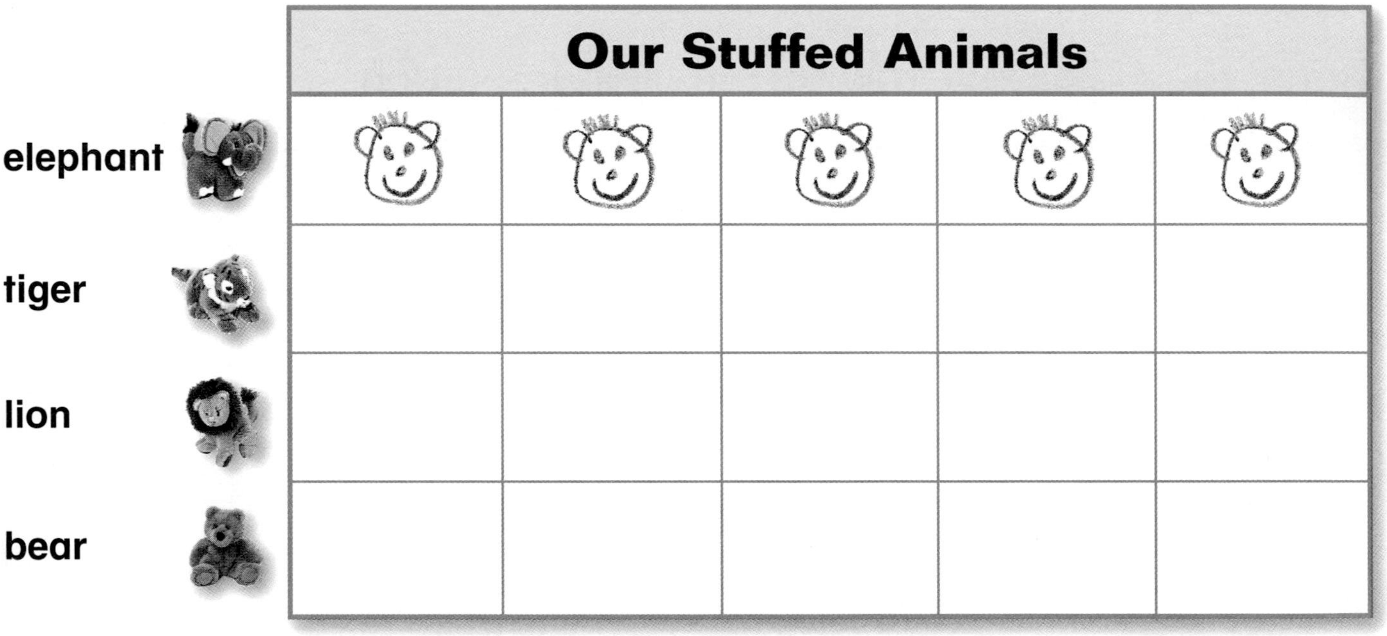

Use the graph to answer the questions.

1. How many lions are there? ______

2. How many more elephants than tigers are there? ______ more

3. Are there more elephants or bears? ______

Home Note **Your child made picture graphs.**
ACTIVITY **Have your child tell you about the graph.**

Name ______________________________

Understand • Plan • Solve • Look Back

Problem Solving
Use Data from a Graph

A group of children wants to know if more of them write with their right hand or with their left hand. They use tally marks and a bar graph to find out.

Write how many tally marks.

Our Writing Hand		Total
right hand	~~||||~~ |||	8
left hand	||	

Color the graph to match the tally marks.

Our Writing Hand										
right hand										
left hand										
	0 — 1	2	3	4	5	6	7	8	9	10

Use the graph to answer the questions.

1. How many children write with their right hand? ______
2. How many children write with their left hand? ______
3. How many more children write with their right hand than with their left hand? ______ more

Practice

A group of children wants to know if more of them are 6 or 7 years old. Write how many tally marks.

How Old We Are		Total
six	HHT II	7
seven	IIII	

Color the graph to match the tally marks.

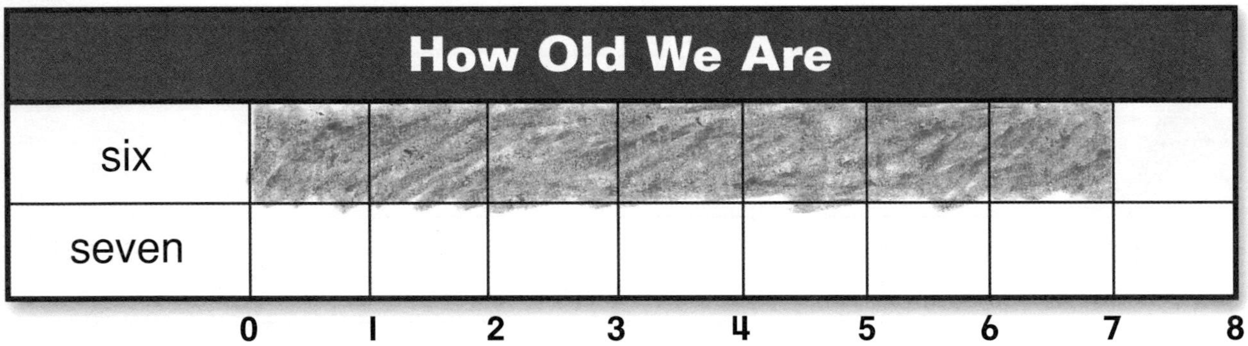

Use the graph to answer the questions.

1. How many children are 6 years old? ______

2. How many children are 7 years old? ______

3. Are more of the children 6 or 7 years old? ______

Mixed Review

Add or subtract.

4. $\begin{array}{r} 8 \\ +9 \\ \hline \end{array}$ $\begin{array}{r} 16 \\ -8 \\ \hline \end{array}$ $\begin{array}{r} 5 \\ +7 \\ \hline \end{array}$ $\begin{array}{r} 12 \\ -3 \\ \hline \end{array}$ $\begin{array}{r} 5 \\ +5 \\ \hline \end{array}$ $\begin{array}{r} 9 \\ -6 \\ \hline \end{array}$

Home Note Your child used tables to make horizontal graphs.
ACTIVITY Have your child tell you about the parts of a graph.

Name ____________________________________

Vertical Bar Graphs

Write how many tally marks.
Color the graph to match the tally marks.

When We Eat Dinner		Total
6:00	𝍸	5
6:30	𝍸 𝍸	
7:00	|||	

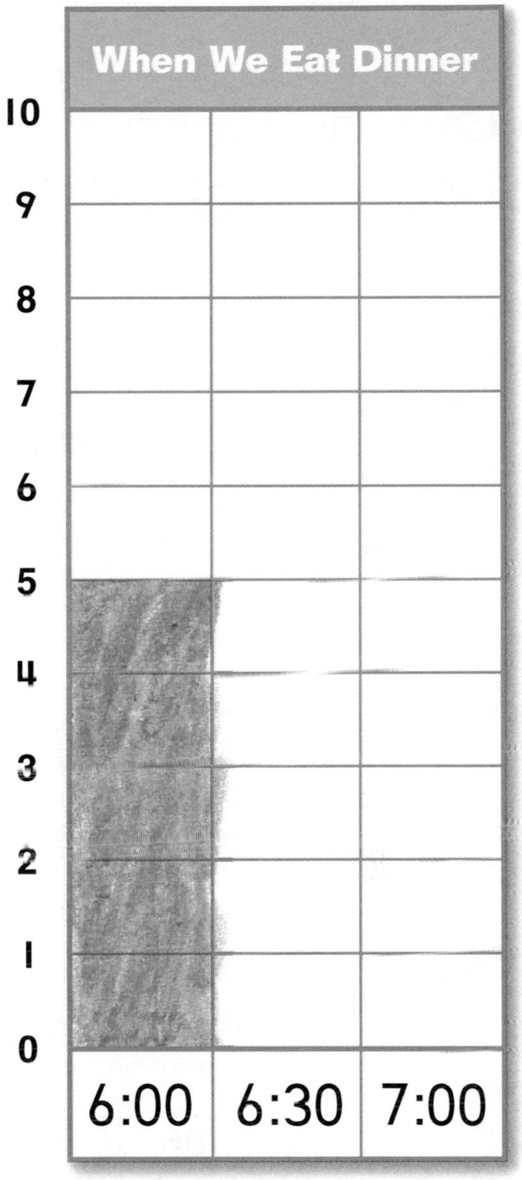

Use the graph to answer the questions.

1. How many children eat dinner at 7:00? ______

2. At what time do the most children eat dinner? ______

Talk About It • Critical Thinking

How are graphs and tally tables alike?

Practice

Write how many tally marks.
Color the graph to match the tally marks.

Our Bedtimes		Total
8:00	卌 II	7
8:30	卌 I	
9:00	IIII	

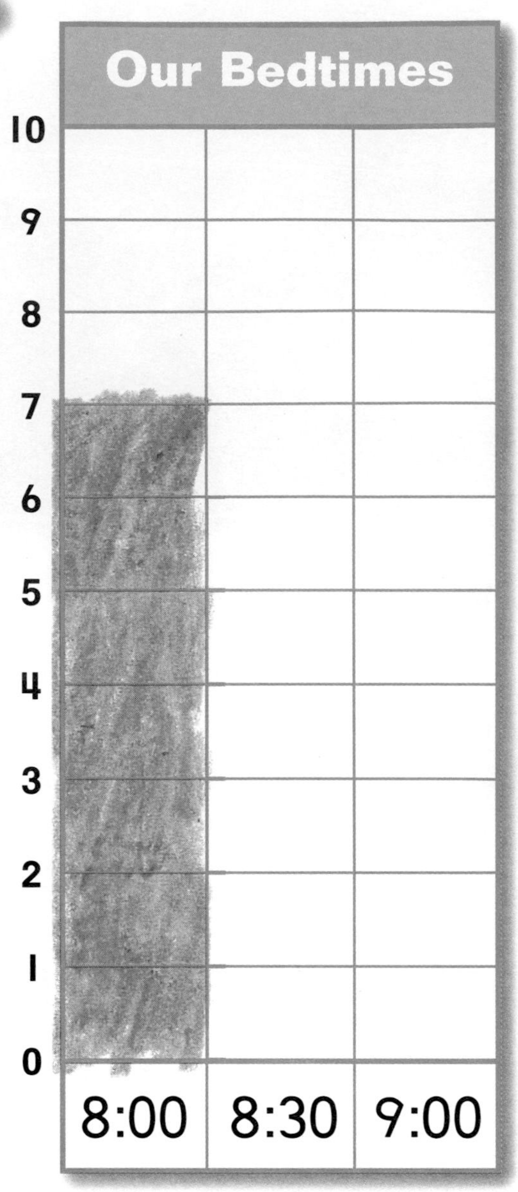

Use the graph to answer the questions.

1. How many children go to bed at 8:30? ______

2. At what time do the most children go to bed? ______

Write About It

3. Write a question someone can answer by reading this graph.

Home Note **Your child used tables to make a vertical graph.**
ACTIVITY **Have your child tell you about the graph.**

Name ______________________________

Understand • Plan • Solve • Look Back

Problem Solving
Make a Graph

Ask 10 classmates to choose their favorite season.

1. Make a tally mark for each choice. Then write how many.

2. Fill in the graph. First write the title. Then color the graph to match the tally marks.

	Our Favorite Season	Total
winter		
spring		
summer		
fall		

____

winter										
spring										
summer										
fall										

0 1 2 3 4 5 6 7 8 9 10

Use the graph to answer the questions.

3. Which season do the most children like? ______________

4. How many children like fall the best? ______

Practice

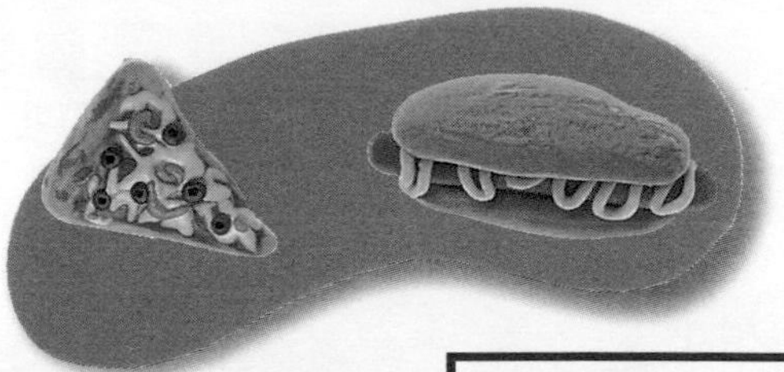

Ask 10 classmates
to choose their favorite food.

1. Make a tally mark for each choice. Then write how many.

2. Fill in the graph. First write the title. Then color the graph to match the tally marks.

Our Favorite Foods		Total
hot dog		
hamburger		
sandwich		
pizza		

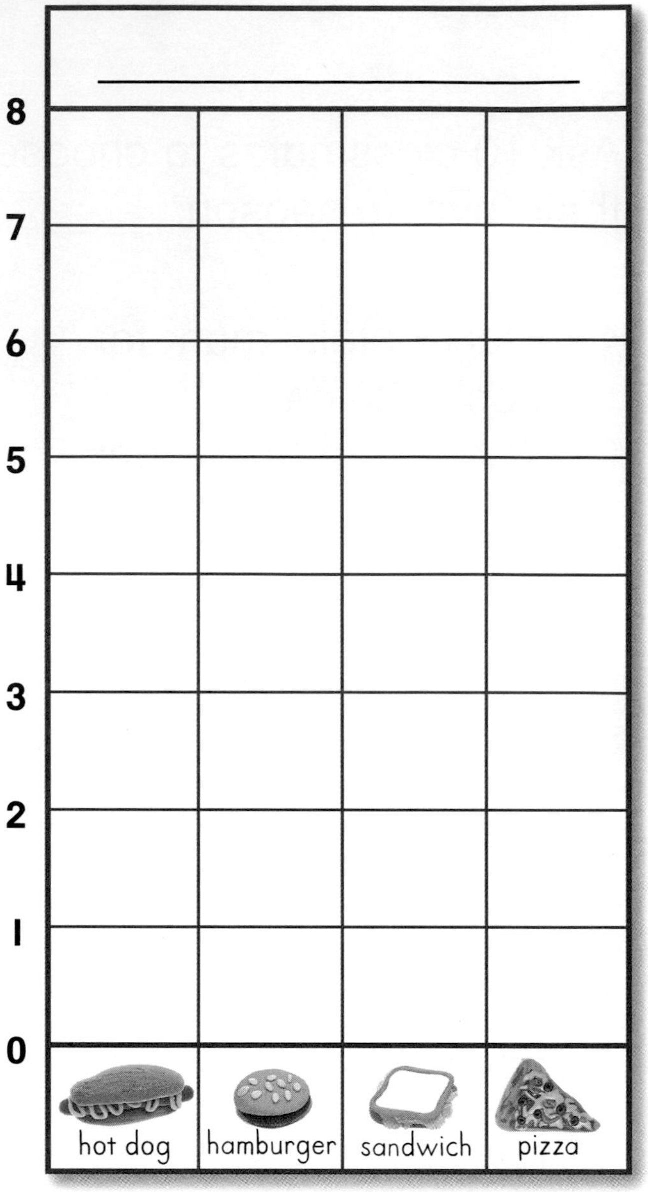

Use the graph to answer the questions.

3. Which food do the most children like? ______

4. How many children like pizza the best? ______

5. Write a question someone can answer by reading this graph.

______

Home Note Your child used the problem-solving strategy *make a graph.*
ACTIVITY Have your child tell how he or she can use graphs to help solve problems.

Name ______________________________

Chapter 24
Review/Test

Concepts and Skills

1 Count the crayons. Draw  to fill in the graph.

red

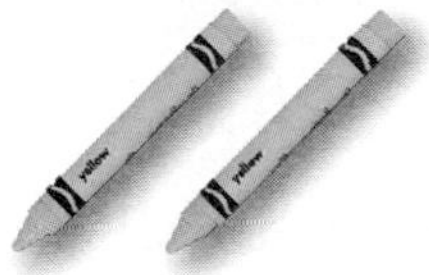
yellow

green

Crayon Colors

---

2 Count the tally marks. Color the graph to match.

Our Pets		Total
cat	III	
dog	𝍸	
bird	II	

Our Pets

cat						
dog						
bird						
	0	1	2	3	4	5	6

Use the graph to answer the questions.

3 How many dogs are there? ______

4 How many cats are there? ______

5 How many more dogs than birds are there? ______ more

Name ____________________

TAAS Prep

Mark the best answer.

1. How many children walk to school?

- 4
- 5
- 6
- 7

2. Blake has 4 green crayons and 3 red crayons. Which graph shows this?

- 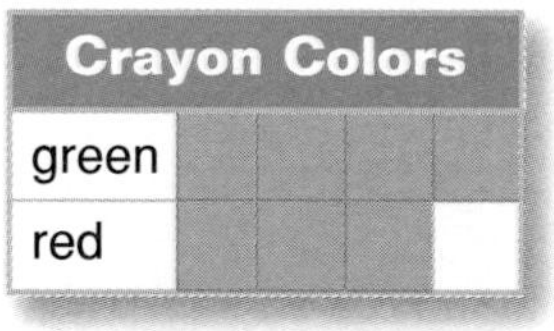

- 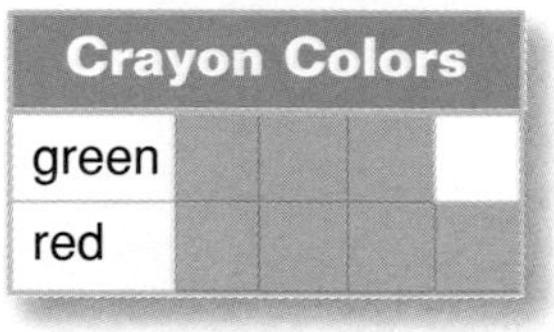

3. How could you sort the buttons?

- red and blue
- round and square
- big and little
- white and green

4. Which sentence is true?

- 3 out of the 4 glasses have milk.
- 1 out of the 3 glasses has orange juice.
- 4 out of the 4 glasses have orange juice.
- 1 out of the 4 glasses has milk.

5. Which one shows $\frac{1}{4}$ green?

- 
- 
- 
- 

6. What part is blue?

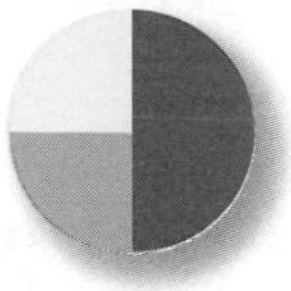

- $\frac{1}{2}$
- $\frac{1}{4}$
- $\frac{1}{3}$
- not given

Name ____________________

MATH FUN

Graphing Animals

Count the animals.
Color boxes in the graph.
Use your graph to
answer the questions.

4 legs						
2 legs						
0 legs						

0 1 2 3 4 5 6

1. How many animals have 4 legs? ____
2. How many animals have 2 legs? ____
3. How many animals have 0 legs? ____

Home Note Your child has been learning about tally tables and graphs.
ACTIVITY Work with him or her to make graphs showing objects around your home.

Name ______________________________

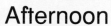

Calculator | Computer

1. What is your favorite time of day? Circle one.

Morning

Afternoon

Evening

2. Ask 10 classmates to choose their favorite time of day.
3. Make a tally mark for each choice. Then write how many.

Our Favorite Time of Day		Total
Morning		
Afternoon		
Evening		

4. How many children chose your favorite time of day? ______
5. Color the graph to match the tally marks.

Our Favorite Time of Day

Morning										
Afternoon										
Evening										
	0	1	2	3	4	5	6	7	8	9	10

6. Use a computer to make a bar graph. Print your graph. Compare it to the graph you colored.

Lunch Surprise for Rabbit

written by Nan Jackson

illustrated by Bernard Adnet

This book will help me review reading graphs.

This book belongs to ______________________.

Harcourt Brace School Publishers

Bear, Squirrel, and Rabbit
are friends.
They eat lunch at school.

What can Bear eat?

How many fish can Bear eat? ______

Count what Bear can eat.

What can Squirrel eat?

How many acorns can Squirrel eat? ______

Count what Squirrel can eat.

What can Rabbit eat?

How many beans can Rabbit eat? ______

A carrot cake for Rabbit!

Rabbit is 6 years old!

Name ______________________________

Concepts and Skills

Look at the buttons. Finish the tally table.

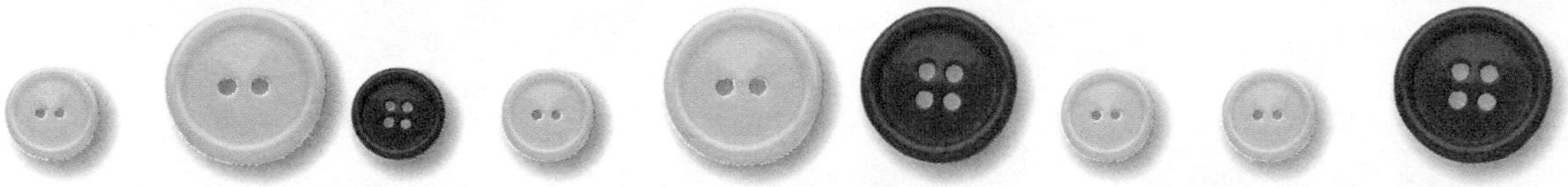

1

Buttons	
big	
little	卌

2

Buttons	
red	|||
yellow	

Look at the fruit. Circle the picture that shows what you can choose from the tray.

3

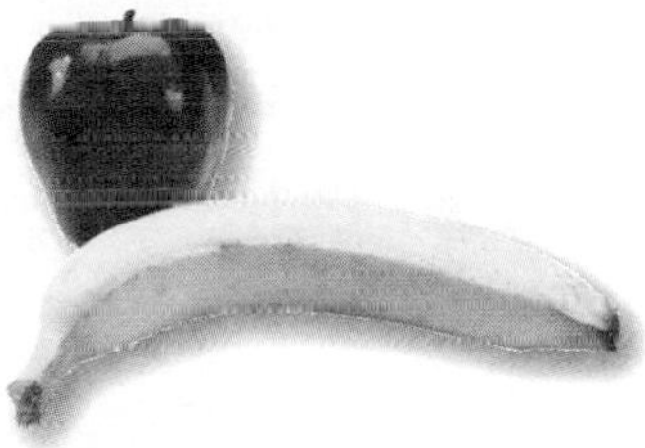

These things go in the bag.

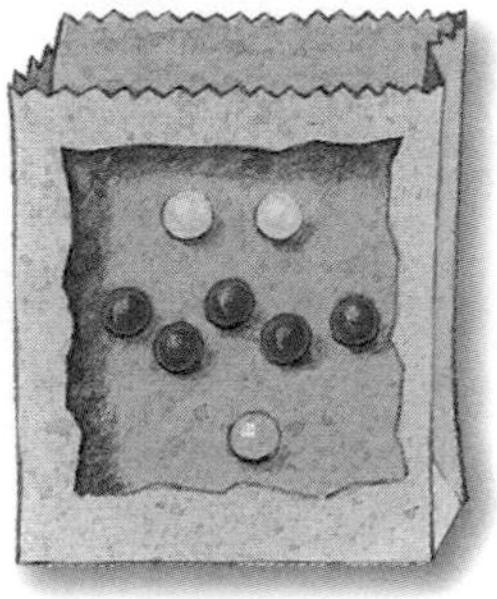

4 Which color will you get most often?

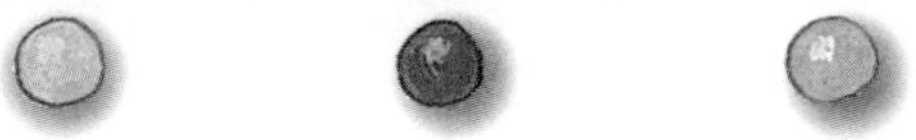

5 Which shape will you get most often?

Use the picture to finish the graphs for 6 and 7.

6

Animals We Saw			
birds			
cats			
bugs			

7

Animals We Saw

3			
2			
1			
0	birds	cats	bugs

Use the graphs to answer questions 8 and 9.

8 How many more birds than cats did we see?

_____ more birds

9 How many animals did we see in all?

_____ animals

Problem Solving

Color the graph to match the tally marks.

10

Our Favorite Animals		Total
	~~IIII~~	5
	III	3
	IIII	4

Our Favorite Animals						

0	1	2	3	4	5	6

Name ____________________

Performance Assessment

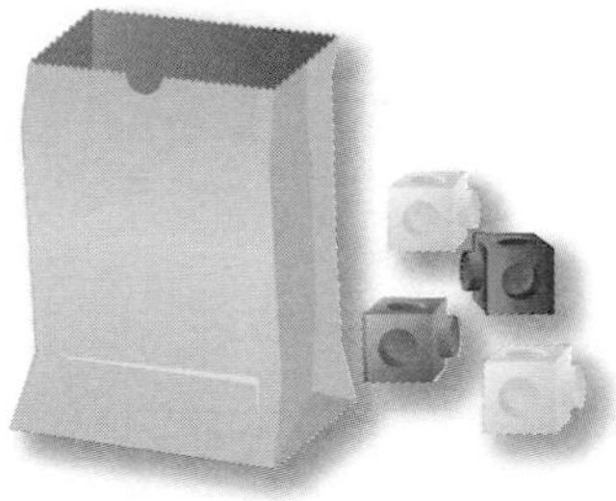

Use 1 bag, 10 , , .

1. Put some , , in the bag.
   Put 10 cubes in all.
2. Draw and color the cubes you put in the bag.

3. Take out one cube.
   Make a tally mark.
   Put the cube back in the bag.
   Shake.
   Do this 8 more times.
   Make a tally mark each time.
   Write the totals.

Cube Colors

	Tally Marks	Total

4. Color the graph to match the tally marks.

Cube Colors

	0 1	2	3	4	5	6

Write About It

5. Which color cube did you get most often? ______
6. Why do you think this happened?

Name ______________________________

Chapters 1–24 Cumulative Review

Fill in the ⬭ for the correct answer.

1.
$$\begin{array}{r} 4 \\ 6 \\ +5 \\ \hline \end{array}$$
⬭ 10
⬭ 14
⬭ 15
⬭ 16

2.
$$\begin{array}{r} 11 \\ -\ 4 \\ \hline \end{array}$$
⬭ 6
⬭ 7
⬭ 8
⬭ 9

3. Count by tens. Which number is after 30?

10, 20, 30, ______

⬭ 31
⬭ 32
⬭ 35
⬭ 40

4. Which object is heavier?

⬭  ⬭

5. Which is hot?

⬭  ⬭ 

6. There are 3 children. Each one gets an equal share. Which pizza would you use?

⬭ 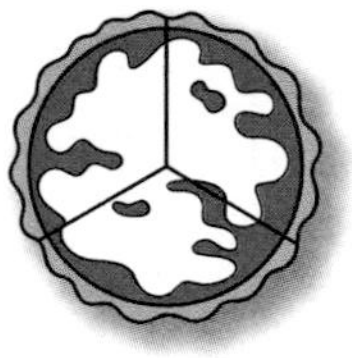 ⬭ 

7. Which shows $\frac{1}{2}$ of the apples red?

⬭ ⬭

8. How are the buttons sorted?

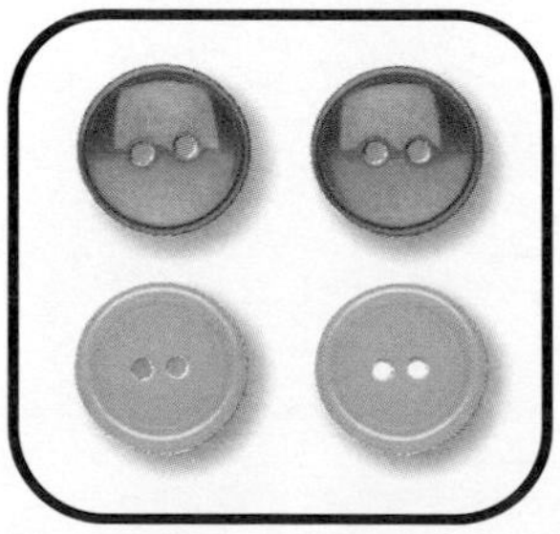 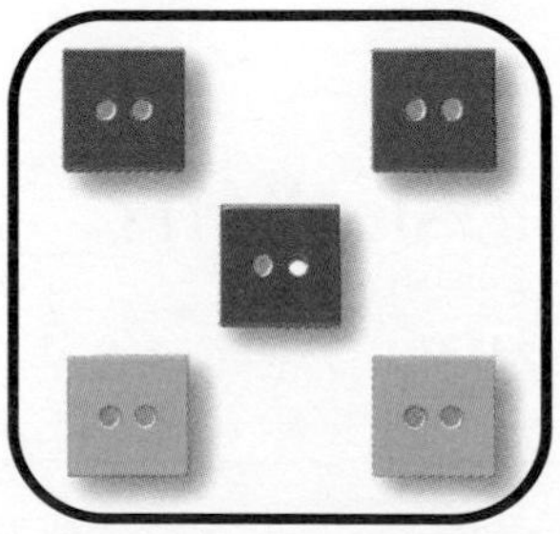

⬭ black—white ⬭ big—little
⬭ circle—square ⬭ not here

9. How many buttons are black?

Buttons	
white	IIII
black	~~IIII~~

⬭ 4 ⬭ 5
⬭ 9 ⬭ not here

CHAPTER 25

Facts to 18

Tell some stories about doubles.

Home Note In this chapter, your child will learn ways to use doubles facts to help solve addition problems.
ACTIVITY Have your child make up addition sentences about the pictures on this page.

SCHOOL-HOME CONNECTION

Dear Family,

Today we started chapter 25. We will learn some ways to make remembering the addition and subtraction facts easier. Here are the new vocabulary words and an activity for us to do together at home.

Love,

____

Vocabulary

doubles

4 + 4

doubles plus one

4 + 5

Think: 4 + 4 = 8,
so 4 + 5
is 1 more, or 9.

doubles minus one

4 + 3

Think: 4 + 4 = 8,
so 4 + 3
is 1 less, or 7.

ACTIVITY

Play "Fact of the Day" at home. Your child can tell you a fact that he or she is learning in class or finds hard to remember. Repeat the fact as many times and ways as possible throughout the day. You could sing the fact, make up dance steps to the fact, or make up a poem about the fact.

Visit our Web site for additional ideas and activities.
http://www.hbschool.com

Name ____________________

Doubles Plus One

Use ●.
Put in ● to show the doubles fact 3 + 3.
Write the sum.

$$\begin{array}{r} 3 \\ +\,3 \\ \hline 6 \end{array}$$

3 + 3 is a doubles fact.

Put in 1 more ●.
Write the sum.

$$\begin{array}{r} 3 \\ +\,4 \\ \hline 7 \end{array}$$

3 + 4 is a doubles-plus-one fact.

Use ●.
Write the sums.

1. $\begin{array}{r} 5 \\ +\,5 \\ \hline \end{array}$ $\begin{array}{r} 5 \\ +\,6 \\ \hline \end{array}$

2. $\begin{array}{r} 6 \\ +\,6 \\ \hline \end{array}$ $\begin{array}{r} 6 \\ +\,7 \\ \hline \end{array}$

3. $\begin{array}{r} 7 \\ +\,7 \\ \hline \end{array}$ $\begin{array}{r} 7 \\ +\,8 \\ \hline \end{array}$

4. $\begin{array}{r} 8 \\ +\,8 \\ \hline \end{array}$ $\begin{array}{r} 8 \\ +\,9 \\ \hline \end{array}$

5. $\begin{array}{r} 5 \\ +\,5 \\ \hline \end{array}$ $\begin{array}{r} 5 \\ +\,6 \\ \hline \end{array}$

6. $\begin{array}{r} 4 \\ +\,4 \\ \hline \end{array}$ $\begin{array}{r} 4 \\ +\,5 \\ \hline \end{array}$

Talk About It • Critical Thinking

How does knowing the sum of 4 + 4 help you know the sum of 4 + 5?

Practice

Write the sums.

1. $2 + 2 = 4$, so $2 + 3 =$ 5
2. $7 + 7 = 14$, so $7 + 8 =$ ____
3. $5 + 5 = 10$, so $5 + 6 =$ ____
4. $4 + 4 = 8$, so $4 + 5 =$ ____
5. $6 + 6 = 12$, so $6 + 7 =$ ____
6. $3 + 3 = 6$, so $3 + 4 =$ ____
7. $1 + 1 = 2$, so $1 + 2 =$ ____

Write the sums.

8. $8 + 8 =$ 16 $9 + 9 =$ ____ $5 + 5 =$ ____ $6 + 6 =$ ____ $7 + 7 =$ ____ $4 + 4 =$ ____

9. $4 + 5 =$ ____ $8 + 9 =$ ____ $6 + 7 =$ ____ $3 + 4 =$ ____ $5 + 6 =$ ____ $2 + 3 =$ ____

10. $3 + 3 =$ ____ $7 + 8 =$ ____ $8 + 9 =$ ____ $6 + 7 =$ ____ $3 + 4 =$ ____ $2 + 2 =$ ____

Home Note Your child used doubles facts to find sums of doubles-plus-one facts. ACTIVITY Have your child tell you the doubles facts and the doubles-plus-one facts for 5, 6, 7, and 8.

Name ______________________________

Doubles Minus One

Use ●.
Put in ● to show 4 + 4.
Write the sum.

$$\begin{array}{r} 4 \\ +\,4 \\ \hline 8 \end{array}$$

4 + 4 is a doubles fact.

Take away 1 ●.
Write the sum.

$$\begin{array}{r} 4 \\ +\,3 \\ \hline 7 \end{array}$$

4 + 3 is a doubles-minus-one fact.

Use ●.
Write the sums.

1. $\begin{array}{r} 6 \\ +\,6 \\ \hline \end{array}$ $\begin{array}{r} 6 \\ +\,5 \\ \hline \end{array}$	2. $\begin{array}{r} 7 \\ +\,7 \\ \hline \end{array}$ $\begin{array}{r} 7 \\ +\,6 \\ \hline \end{array}$	3. $\begin{array}{r} 8 \\ +\,8 \\ \hline \end{array}$ $\begin{array}{r} 8 \\ +\,7 \\ \hline \end{array}$
4. $\begin{array}{r} 9 \\ +\,9 \\ \hline \end{array}$ $\begin{array}{r} 9 \\ +\,8 \\ \hline \end{array}$	5. $\begin{array}{r} 5 \\ +\,5 \\ \hline \end{array}$ $\begin{array}{r} 5 \\ +\,4 \\ \hline \end{array}$	6. $\begin{array}{r} 3 \\ +\,3 \\ \hline \end{array}$ $\begin{array}{r} 3 \\ +\,2 \\ \hline \end{array}$

Talk About It • Critical Thinking

How does knowing the sum of 5 + 5 help you know the sum of 5 + 4?

Practice

Write the sums.

1. 5 + 5 = 10, so 5 + 4 = 9
2. 7 + 7 = 14, so 7 + 6 = ____
3. 4 + 4 = 8, so 4 + 3 = ____
4. 9 + 9 = 18, so 9 + 8 = ____
5. 3 + 3 = 6, so 3 + 2 = ____
6. 6 + 6 = 12, so 6 + 5 = ____
7. 8 + 8 = 16, so 8 + 7 = ____

Write the sums.

8. $\begin{array}{r} 7 \\ +6 \\ \hline \end{array}$ $\begin{array}{r} 9 \\ +8 \\ \hline \end{array}$ $\begin{array}{r} 6 \\ +5 \\ \hline \end{array}$ $\begin{array}{r} 8 \\ +7 \\ \hline \end{array}$ $\begin{array}{r} 8 \\ +9 \\ \hline \end{array}$ $\begin{array}{r} 4 \\ +5 \\ \hline \end{array}$

9. $\begin{array}{r} 5 \\ +6 \\ \hline \end{array}$ $\begin{array}{r} 5 \\ +4 \\ \hline \end{array}$ $\begin{array}{r} 6 \\ +7 \\ \hline \end{array}$ $\begin{array}{r} 8 \\ +9 \\ \hline \end{array}$ $\begin{array}{r} 4 \\ +3 \\ \hline \end{array}$ $\begin{array}{r} 7 \\ +8 \\ \hline \end{array}$

Mixed Review

Circle the figures that have all flat sides.

10. 

Home Note **Your child used doubles facts to find sums of doubles-minus-one facts.**
ACTIVITY **Say a doubles fact such as 7 + 7. Have your child name the doubles-minus-one fact.**

Name ______________________________

Doubles Patterns

Complete the chart.

1. Color the doubles.
2. Color the doubles plus one.
3. Color the doubles minus one.

+	0	1	2	3	4	5	6	7	8	9
0	0	1	2							
1										
2										
3										
4										
5										
6										
7										
8										
9										

Talk About It • Critical Thinking

How is a doubles-plus-one fact different from a doubles-minus-one fact?

Practice

Write the sums.

	doubles	doubles − 1	doubles + 1
1	6 + 6 = 12	6 + 5 = ____	6 + 7 = ____
2	7 + 7 = ____	7 + 6 = ____	7 + 8 = ____
3	4 + 4 = ____	4 + 3 = ____	4 + 5 = ____
4	9 + 9 = ____	9 + 8 = ____	9 + 10 = ____
5	3 + 3 = ____	3 + 2 = ____	3 + 4 = ____
6	8 + 8 = ____	8 + 7 = ____	8 + 9 = ____
7	5 + 5 = ____	5 + 4 = ____	5 + 6 = ____

Problem Solving

Solve. Draw a picture to show your answer.

8. I had 8 pennies. Then I found 4 more. How many pennies do I have in all?

____ pennies

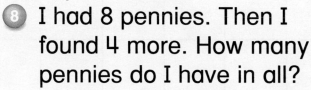

Home Note **Your child showed doubles, doubles-plus-one, and doubles-minus-one fact patterns.**
ACTIVITY **Have your child show you doubles at home, such as the two sixes of an egg carton.**

Name ______________________________

Doubles Fact Families

Add and subtract.

1

4 + 4 = 8

8 − 4 = 4

2

6 + 6 = ____

12 − 6 = ____

3

7 + 7 = ____

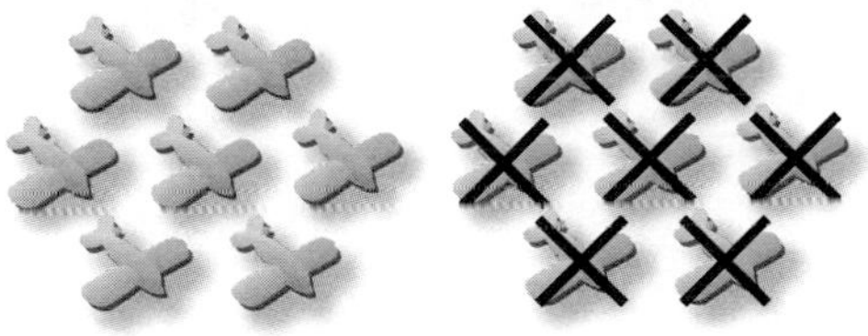

14 − 7 = ____

4

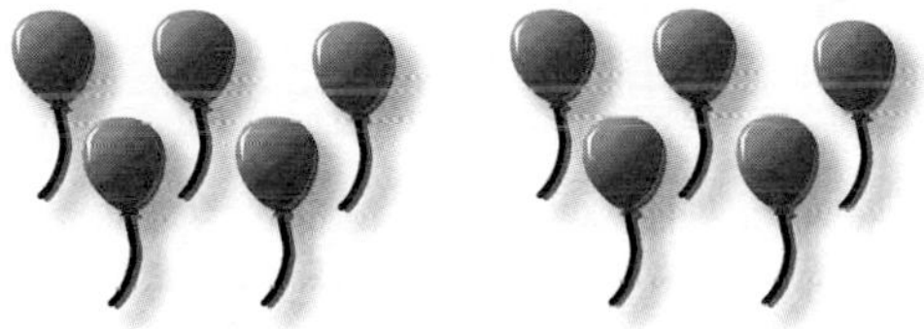

5 + 5 = ____

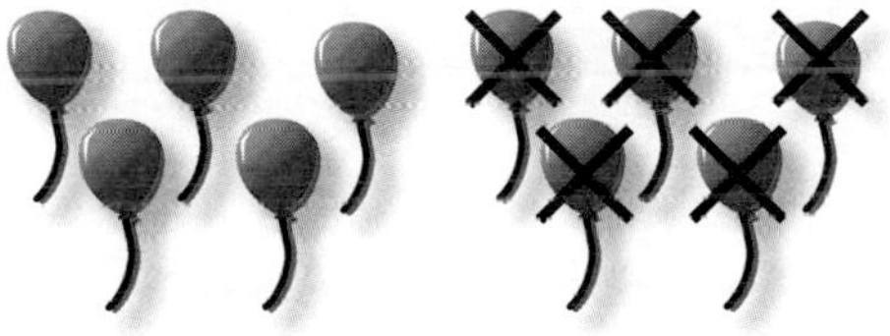

10 − 5 = ____

5

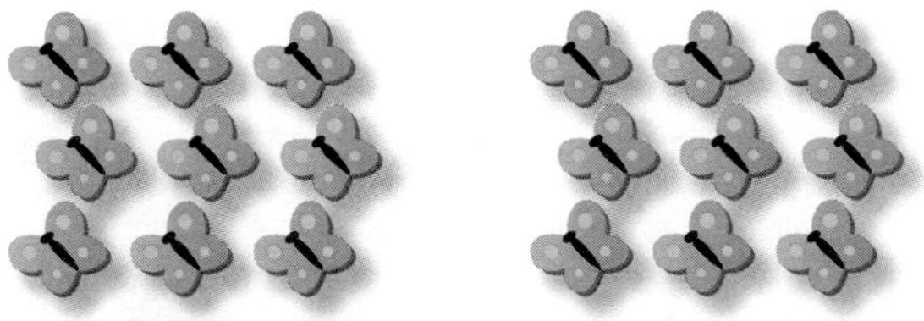

9 + 9 = ____

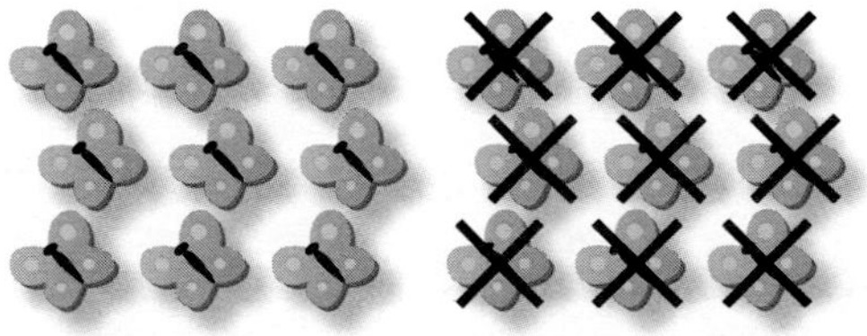

18 − 9 = ____

6

8 + 8 = ____

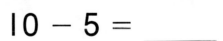

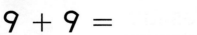

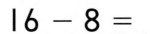

16 − 8 = ____

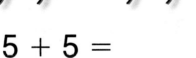

Practice

Add and subtract.

1. $\begin{array}{r} 3 \\ +3 \\ \hline 6 \end{array}$ $\quad$ $\begin{array}{r} 6 \\ -3 \\ \hline 3 \end{array}$	2. $\begin{array}{r} 1 \\ +1 \\ \hline \end{array}$ $\quad$ $\begin{array}{r} 2 \\ -1 \\ \hline \end{array}$	3. $\begin{array}{r} 2 \\ +2 \\ \hline \end{array}$ $\quad$ $\begin{array}{r} 4 \\ -2 \\ \hline \end{array}$
4. $\begin{array}{r} 9 \\ +9 \\ \hline \end{array}$ $\quad$ $\begin{array}{r} 18 \\ -9 \\ \hline \end{array}$	5. $\begin{array}{r} 8 \\ +8 \\ \hline \end{array}$ $\quad$ $\begin{array}{r} 16 \\ -8 \\ \hline \end{array}$	6. $\begin{array}{r} 7 \\ +7 \\ \hline \end{array}$ $\quad$ $\begin{array}{r} 14 \\ -7 \\ \hline \end{array}$
7. $\begin{array}{r} 6 \\ +6 \\ \hline \end{array}$ $\quad$ $\begin{array}{r} 12 \\ -6 \\ \hline \end{array}$	8. $\begin{array}{r} 5 \\ +5 \\ \hline \end{array}$ $\quad$ $\begin{array}{r} 10 \\ -5 \\ \hline \end{array}$	9. $\begin{array}{r} 4 \\ +4 \\ \hline \end{array}$ $\quad$ $\begin{array}{r} 8 \\ -4 \\ \hline \end{array}$
10. $\begin{array}{r} 12 \\ -6 \\ \hline \end{array}$ $\quad$ $\begin{array}{r} 6 \\ +6 \\ \hline \end{array}$	11. $\begin{array}{r} 18 \\ -9 \\ \hline \end{array}$ $\quad$ $\begin{array}{r} 9 \\ +9 \\ \hline \end{array}$	12. $\begin{array}{r} 14 \\ -7 \\ \hline \end{array}$ $\quad$ $\begin{array}{r} 7 \\ +7 \\ \hline \end{array}$
13. $\begin{array}{r} 10 \\ -5 \\ \hline \end{array}$ $\quad$ $\begin{array}{r} 5 \\ +5 \\ \hline \end{array}$	14. $\begin{array}{r} 16 \\ -8 \\ \hline \end{array}$ $\quad$ $\begin{array}{r} 8 \\ +8 \\ \hline \end{array}$	15. $\begin{array}{r} 8 \\ -4 \\ \hline \end{array}$ $\quad$ $\begin{array}{r} 4 \\ +4 \\ \hline \end{array}$

Write About It

Draw a picture that shows a doubles fact.
Then draw a picture that shows a doubles-plus-one fact.
Write a story about each picture.

Home Note Your child added and subtracted in doubles fact families.
ACTIVITY Say an addition doubles fact such as 6 + 6 = 12. Have your child say the matching subtraction fact, 12 − 6 = 6.

Name ______________________________

Understand • Plan • Solve • Look Back

Problem Solving
Make a Model

Use ◯.
Add or subtract.
Draw the ◯.

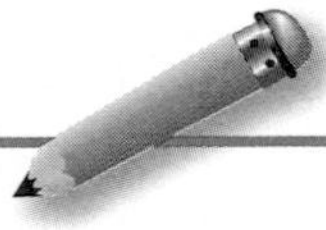

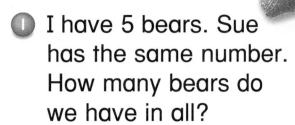

1. I have 5 bears. Sue has the same number. How many bears do we have in all?

   10 bears

2. Jan had 14 cars. She gave some away. She has 7 left. How many cars did she give away?

   ______ cars

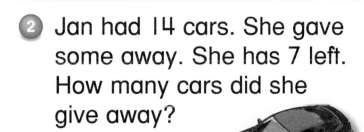

3. Sam had some trucks. Pam gave him 8 more. Now he has 17. How many trucks did he have to start?

   ______ trucks

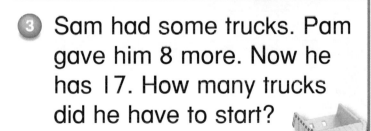

4. Rosa has 4 toy planes. Ann has two times as many. How many planes do they have in all?

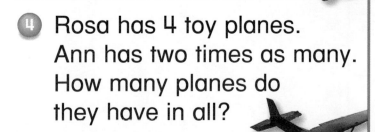

   ______ planes

Practice

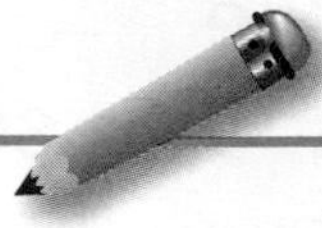

Use ●.
Add or subtract.
Draw the ●.

1. Kay has 8 bears. I have 1 more than Kay. How many bears do we have in all?

17 bears

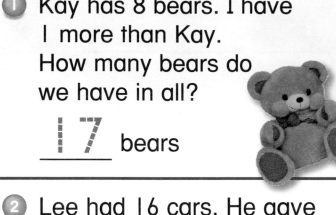

2. Lee had 16 cars. He gave some of them away. He has 8 left. How many did he give away?

______ cars

3. Pam has 3 trucks. Kim has two times as many. How many trucks do they have in all?

______ trucks

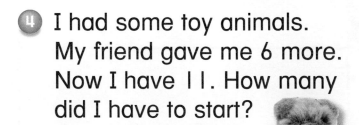

4. I had some toy animals. My friend gave me 6 more. Now I have 11. How many did I have to start?

______ animals

Home Note Your child made models to solve problems.
ACTIVITY Make up more problems. Have your child make models with small objects, such as paper clips, to solve them.

Name ______________________

Chapter 25 Review/Test

Concepts and Skills

Write the sums.

1. $\begin{array}{r} 8 \\ +8 \\ \hline \end{array}$ $\begin{array}{r} 8 \\ +9 \\ \hline \end{array}$

2. $\begin{array}{r} 6 \\ +6 \\ \hline \end{array}$ $\begin{array}{r} 6 \\ +7 \\ \hline \end{array}$

3. $\begin{array}{r} 4 \\ +4 \\ \hline \end{array}$ $\begin{array}{r} 4 \\ +5 \\ \hline \end{array}$

4. $\begin{array}{r} 9 \\ +9 \\ \hline \end{array}$ $\begin{array}{r} 9 \\ +8 \\ \hline \end{array}$

5. $\begin{array}{r} 7 \\ +7 \\ \hline \end{array}$ $\begin{array}{r} 7 \\ +6 \\ \hline \end{array}$

6. $\begin{array}{r} 8 \\ +8 \\ \hline \end{array}$ $\begin{array}{r} 8 \\ +7 \\ \hline \end{array}$

Add and subtract.

7. 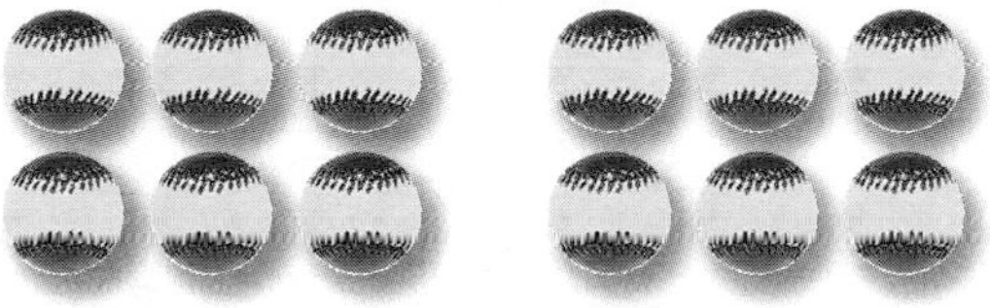

$6 + 6 =$ ____

8. 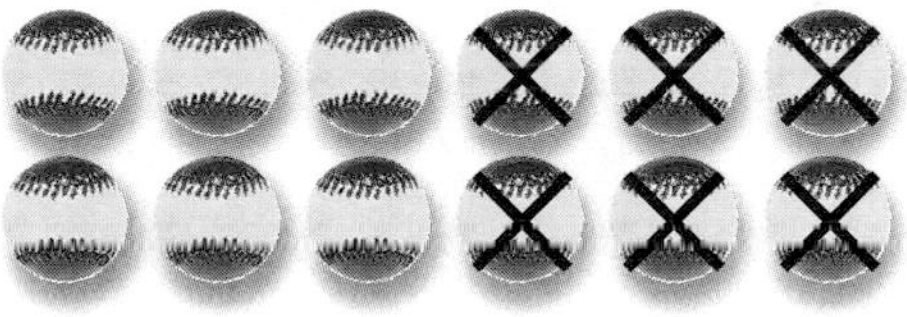

$12 - 6 =$ ____

9. $\begin{array}{r} 18 \\ -\ 9 \\ \hline \end{array}$ $\begin{array}{r} 9 \\ +9 \\ \hline \end{array}$

10. $\begin{array}{r} 6 \\ +6 \\ \hline \end{array}$ $\begin{array}{r} 12 \\ -\ 6 \\ \hline \end{array}$

11. $\begin{array}{r} 8 \\ -4 \\ \hline \end{array}$ $\begin{array}{r} 4 \\ +4 \\ \hline \end{array}$

Problem Solving

Use ○. Add or subtract.
Draw the ○.

12. I have 6 bears.
My friend has the same number.
How many do we have in all?

____ bears

Name ______________________________

Cumulative Chapters 1–25

TAAS Prep

Mark the best answer.

1. Which number comes just before 77?

- ⬭ 60
- ⬭ 67
- ⬭ 76
- ⬭ 78

2. Which number comes just after 46?

- ⬭ 40
- ⬭ 45
- ⬭ 47
- ⬭ 50

3. Which model shows 27?

- ⬭ 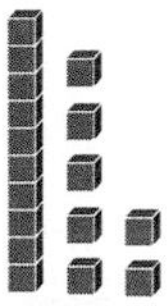
- ⬭ 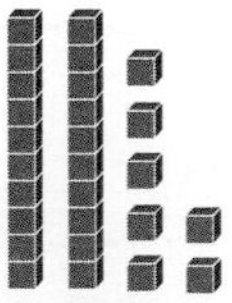
- ⬭ 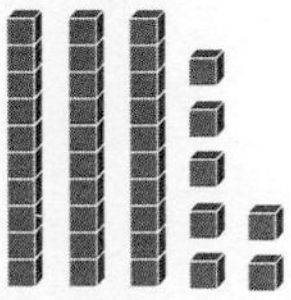
- ⬭ 

4. How many like carrots?

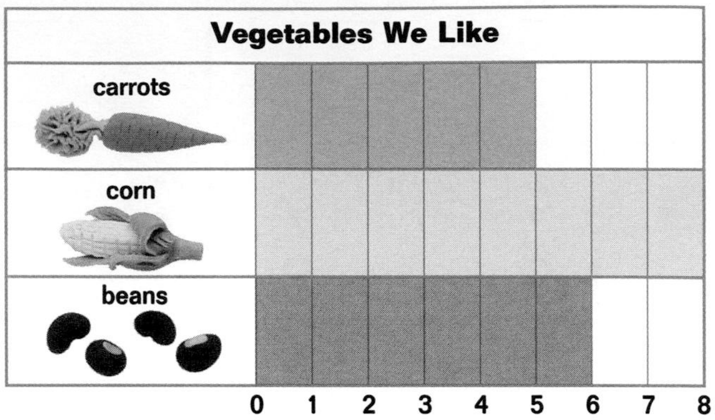

- ⬭ 4
- ⬭ 5
- ⬭ 6
- ⬭ 7

5. How many more like corn than beans?

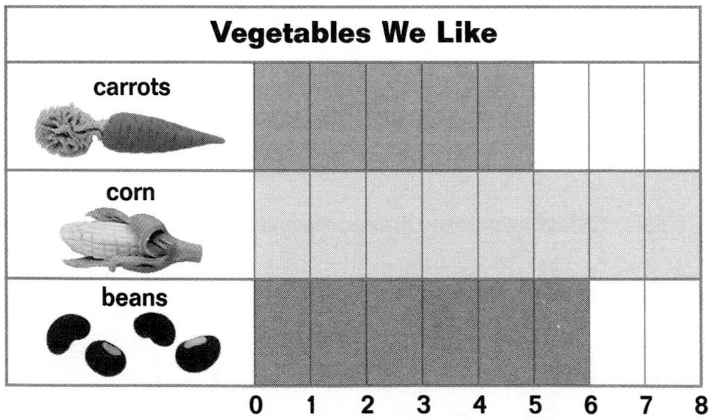

- ⬭ 2 more
- ⬭ 3 more
- ⬭ 4 more
- ⬭ 5 more

CHAPTER 26 More About Facts to 18

Tell some addition and subtraction stories about the animals.

Home Note In this chapter, your child is learning addition and subtraction facts to 18.
ACTIVITY Have your child make up addition and subtraction sentences using the pictures on this page.

SCHOOL-HOME CONNECTION

Dear Family,

Today we started Chapter 26. We will learn some ways to make adding and subtracting facts to 18 easier. Here is a vocabulary word and an activity for us to do together at home.

Love,

____

Vocabulary

Make a Ten

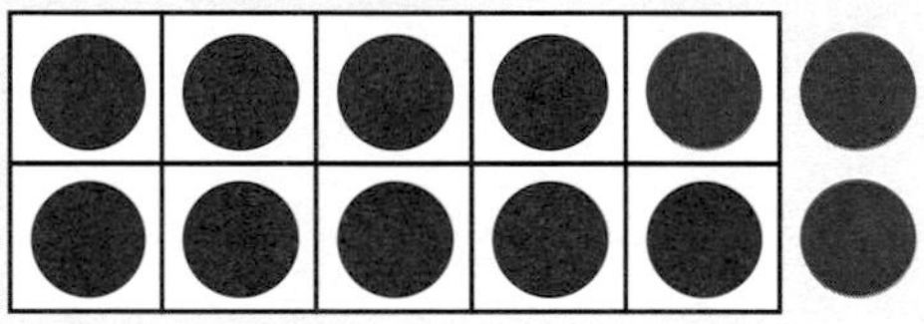

$9 + 3$

Think $10 + 2$.

$$\begin{array}{r} 9 \\ +\,3 \\ \hline 12 \end{array}$$

ACTIVITY

Give your child 2 numbers with a sum of less than 18. Have him or her choose one more number to add on and tell you the sum of the three numbers together.

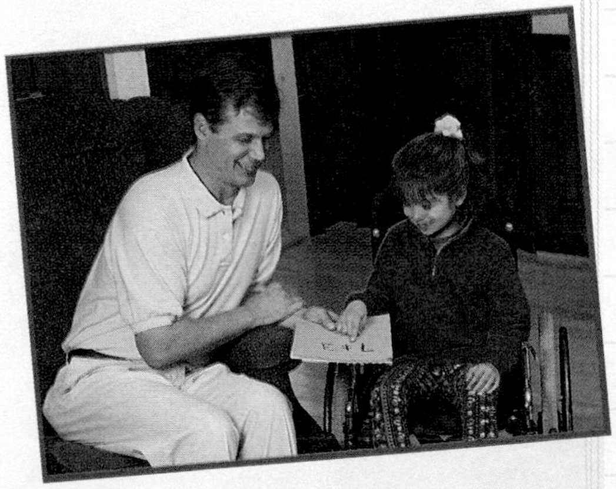

Visit our Web site for additional ideas and activities.
http://www.hbschool.com

Name ______________________________

Make a 10

Add 8 + 3.
Start with 8 ●.
Add 3 ○ .
Write the sum.

You made a 10
and have 1 extra.
10 + 1 = 11.

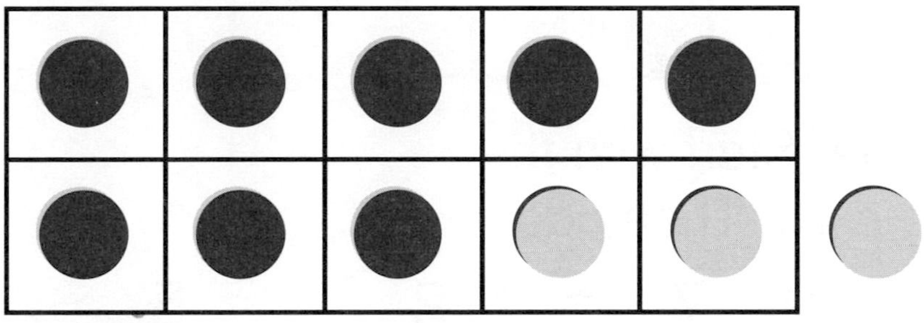

$$\begin{array}{r} 8 \\ +\,3 \\ \hline 11 \end{array}$$

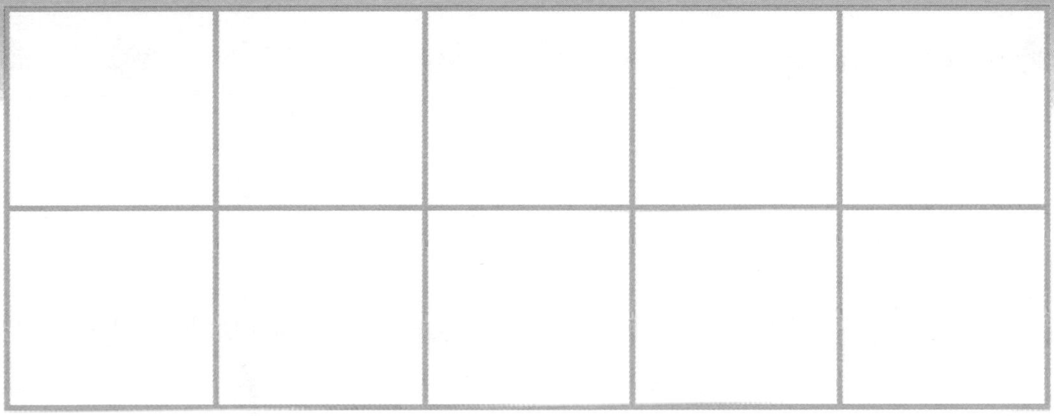

Use ○ and the 10-frame.
Start with the greater number. Make a 10. Then add.

1	$\begin{array}{r} 8 \\ +\,4 \\ \hline 12 \end{array}$	$\begin{array}{r} 9 \\ +\,3 \\ \hline \end{array}$	$\begin{array}{r} 9 \\ +\,5 \\ \hline \end{array}$	$\begin{array}{r} 7 \\ +\,4 \\ \hline \end{array}$	$\begin{array}{r} 7 \\ +\,5 \\ \hline \end{array}$	$\begin{array}{r} 8 \\ +\,6 \\ \hline \end{array}$
2	$\begin{array}{r} 3 \\ +\,9 \\ \hline \end{array}$	$\begin{array}{r} 6 \\ +\,7 \\ \hline \end{array}$	$\begin{array}{r} 8 \\ +\,7 \\ \hline \end{array}$	$\begin{array}{r} 9 \\ +\,2 \\ \hline \end{array}$	$\begin{array}{r} 8 \\ +\,5 \\ \hline \end{array}$	$\begin{array}{r} 4 \\ +\,9 \\ \hline \end{array}$

Talk About It • Critical Thinking

What are some names for 10?

Practice

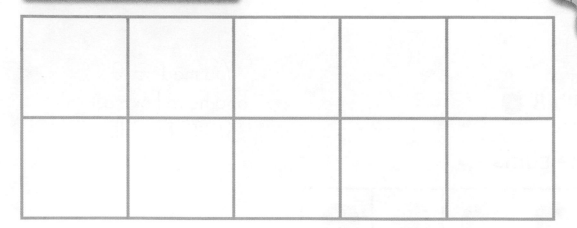

Use ● and the 10-frame. Start with the greater number. Make a 10. Then add.

1 $9 + 6 = 15$ $3 + 8 =$ ___ $7 + 5 =$ ___ $8 + 9 =$ ___ $5 + 8 =$ ___ $9 + 7 =$ ___

2 $7 + 4 =$ ___ $8 + 6 =$ ___ $3 + 9 =$ ___ $4 + 8 =$ ___ $5 + 9 =$ ___ $9 + 4 =$ ___

3 $5 + 7 =$ ___ $8 + 9 =$ ___ $4 + 7 =$ ___ $7 + 9 =$ ___ $6 + 9 =$ ___ $6 + 8 =$ ___

Problem Solving

Solve. Draw a picture.

4 One fish has 8 stripes.
The other fish has 7 stripes.
How many stripes in all?

______ stripes

Home Note Your child solved addition facts to 18.
ACTIVITY Have your child use the 10-frame to show the problems on this page.

Name ________________________________

Adding Three Numbers

You can use names for 10 when you add three numbers.
7 + 3 = 10
10 + 4 = 14

(7)
(3)
+ 4
14

You can use doubles when you add three numbers.
4 + 4 = 8
8 + 6 = 14

(4)
(4)
+ 6
14

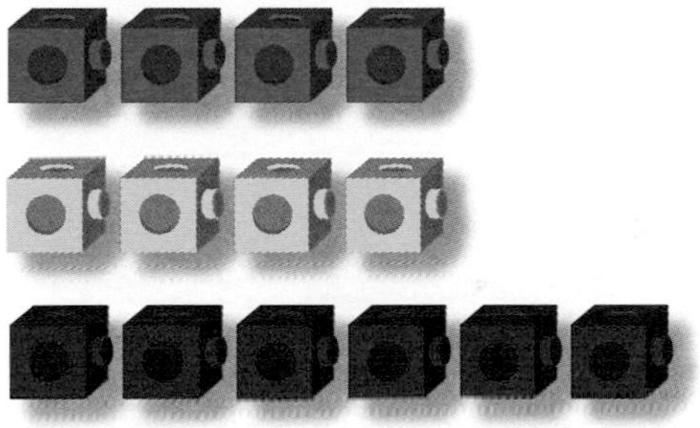

Circle names for 10 or doubles. Then add.

1	2	8	9	8	4	6
	7	2	3	2	7	5
	+ 3	+ 4	+ 1	+ 3	+ 3	+ 4
2	6	2	6	8	8	4
	6	5	2	8	2	3
	+ 3	+ 5	+ 6	+ 1	+ 7	+ 4

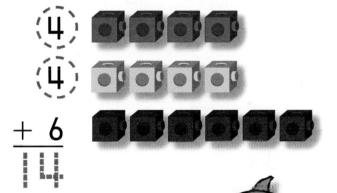

Practice

Circle names for 10 or doubles. Then add.

1

1	8	4	1	8
5	2	3	2	2
+ 5	+ 1	+ 7	+ 9	+ 6
11				

2

2	4	9	3	7
7	5	6	6	7
+ 7	+ 6	+ 1	+ 3	+ 2

3

8	2	6	8	2
8	8	3	1	7
+ 1	+ 5	+ 6	+ 9	+ 3

Mixed Review

Circle the picture that shows equal parts.

4

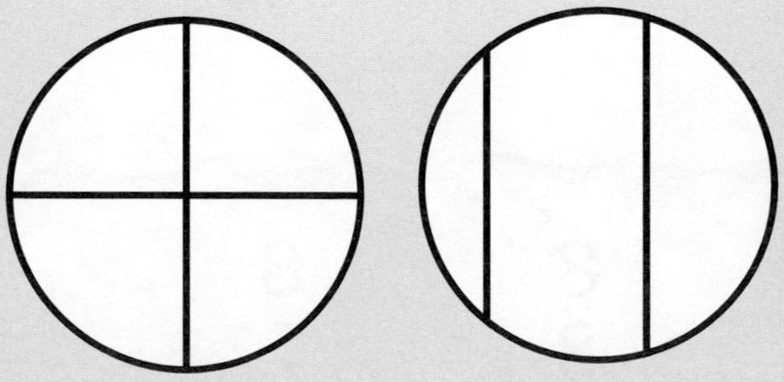

5

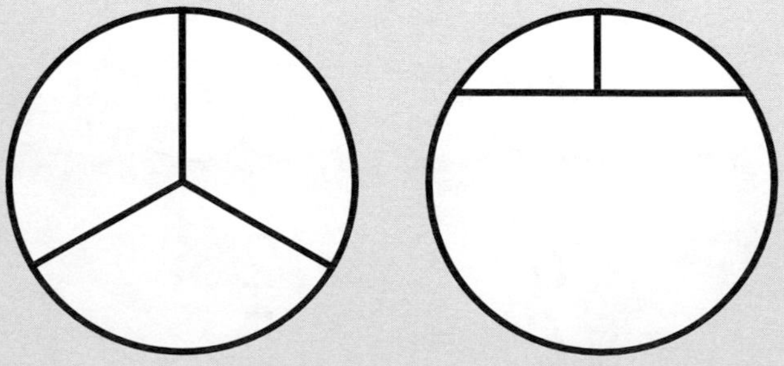

Home Note Your child found the sum of three addends by using doubles and names for ten.
ACTIVITY Have your child use pennies to show the problems on this page.

Name ____________________

Sums and Differences to 14

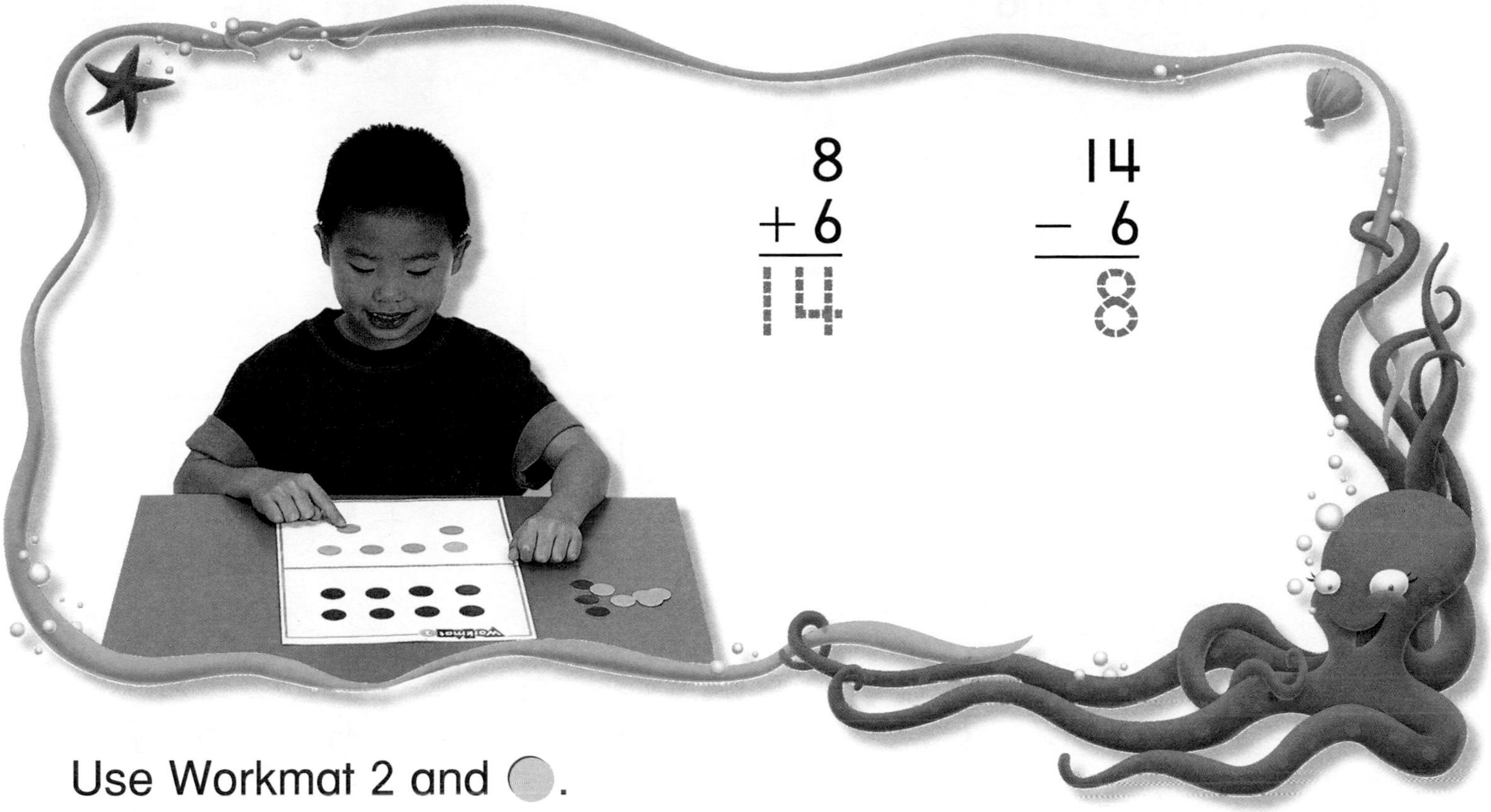

$$\begin{array}{r} 8 \\ +6 \\ \hline 14 \end{array} \qquad \begin{array}{r} 14 \\ -6 \\ \hline 8 \end{array}$$

Use Workmat 2 and ●.
Add or subtract.

1. $\begin{array}{r} 7 \\ +5 \\ \hline 12 \end{array} \qquad \begin{array}{r} 12 \\ -5 \\ \hline 7 \end{array}$

2. $\begin{array}{r} 9 \\ +2 \\ \hline \end{array} \qquad \begin{array}{r} 11 \\ -2 \\ \hline \end{array}$

3. $\begin{array}{r} 8 \\ +4 \\ \hline \end{array} \qquad \begin{array}{r} 12 \\ -4 \\ \hline \end{array}$

4. $\begin{array}{r} 9 \\ +5 \\ \hline \end{array} \qquad \begin{array}{r} 14 \\ -5 \\ \hline \end{array}$

5. $\begin{array}{r} 8 \\ +5 \\ \hline \end{array} \qquad \begin{array}{r} 13 \\ -5 \\ \hline \end{array}$

6. $\begin{array}{r} 9 \\ +3 \\ \hline \end{array} \qquad \begin{array}{r} 12 \\ -3 \\ \hline \end{array}$

7. $\begin{array}{r} 7 \\ +7 \\ \hline \end{array} \qquad \begin{array}{r} 14 \\ -7 \\ \hline \end{array}$

8. $\begin{array}{r} 7 \\ +6 \\ \hline \end{array} \qquad \begin{array}{r} 13 \\ -6 \\ \hline \end{array}$

9. $\begin{array}{r} 4 \\ +9 \\ \hline \end{array} \qquad \begin{array}{r} 13 \\ -9 \\ \hline \end{array}$

Talk About It • Critical Thinking

How does knowing 7 + 5 help you to solve 12 − 5?

Practice

Use Workmat 2 and ●.
Add or subtract.

1. $\begin{array}{r} 5 \\ +5 \\ \hline 10 \end{array}$ $\begin{array}{r} 10 \\ -\ 5 \\ \hline 5 \end{array}$

2. $\begin{array}{r} 8 \\ +3 \\ \hline \end{array}$ $\begin{array}{r} 11 \\ -\ 3 \\ \hline \end{array}$

3. $\begin{array}{r} 4 \\ +6 \\ \hline \end{array}$ $\begin{array}{r} 10 \\ -\ 6 \\ \hline \end{array}$

4. $\begin{array}{r} 9 \\ +5 \\ \hline \end{array}$ $\begin{array}{r} 14 \\ -\ 5 \\ \hline \end{array}$

5. $\begin{array}{r} 8 \\ +4 \\ \hline \end{array}$ $\begin{array}{r} 12 \\ -\ 4 \\ \hline \end{array}$

6. $\begin{array}{r} 5 \\ +9 \\ \hline \end{array}$ $\begin{array}{r} 14 \\ -\ 9 \\ \hline \end{array}$

7. $\begin{array}{r} 6 \\ +6 \\ \hline \end{array}$ $\begin{array}{r} 12 \\ -\ 6 \\ \hline \end{array}$

8. $\begin{array}{r} 5 \\ +6 \\ \hline \end{array}$ $\begin{array}{r} 11 \\ -\ 6 \\ \hline \end{array}$

9. $\begin{array}{r} 5 \\ +4 \\ \hline \end{array}$ $\begin{array}{r} 9 \\ -4 \\ \hline \end{array}$

10. $\begin{array}{r} 7 \\ +7 \\ \hline \end{array}$ $\begin{array}{r} 14 \\ -\ 7 \\ \hline \end{array}$

11. $\begin{array}{r} 9 \\ +1 \\ \hline \end{array}$ $\begin{array}{r} 10 \\ -\ 1 \\ \hline \end{array}$

12. $\begin{array}{r} 8 \\ +5 \\ \hline \end{array}$ $\begin{array}{r} 13 \\ -\ 5 \\ \hline \end{array}$

Problem Solving

Solve.
Draw a picture to show your answer.

13. 8 sea horses are swimming.
9 more join them.
How many sea horses are there in all?

_____ sea horses

Home Note Your child found sums and differences to 14.
ACTIVITY Have your child use small objects as counters and model the problems on the page.

Name ______________________________

Sums and Differences to 18

Knowing addition facts can help you subtract.
8 + 4 = 12
12 − 4 = 8

Write the sum and difference for each pair.

1. $\begin{array}{r} 9 \\ +9 \\ \hline 18 \end{array}$ $\quad$ $\begin{array}{r} 18 \\ -\ 9 \\ \hline 9 \end{array}$	2. $\begin{array}{r} 5 \\ +6 \\ \hline \end{array}$ $\quad$ $\begin{array}{r} 11 \\ -\ 6 \\ \hline \end{array}$	3. $\begin{array}{r} 6 \\ +8 \\ \hline \end{array}$ $\quad$ $\begin{array}{r} 14 \\ -\ 8 \\ \hline \end{array}$
4. $\begin{array}{r} 6 \\ +9 \\ \hline \end{array}$ $\quad$ $\begin{array}{r} 15 \\ -\ 9 \\ \hline \end{array}$	5. $\begin{array}{r} 7 \\ +7 \\ \hline \end{array}$ $\quad$ $\begin{array}{r} 14 \\ -\ 7 \\ \hline \end{array}$	6. $\begin{array}{r} 6 \\ +7 \\ \hline \end{array}$ $\quad$ $\begin{array}{r} 13 \\ -\ 7 \\ \hline \end{array}$
7. $\begin{array}{r} 8 \\ +7 \\ \hline \end{array}$ $\quad$ $\begin{array}{r} 15 \\ -\ 7 \\ \hline \end{array}$	8. $\begin{array}{r} 9 \\ +3 \\ \hline \end{array}$ $\quad$ $\begin{array}{r} 12 \\ -\ 3 \\ \hline \end{array}$	9. $\begin{array}{r} 8 \\ +8 \\ \hline \end{array}$ $\quad$ $\begin{array}{r} 16 \\ -\ 8 \\ \hline \end{array}$

Practice

Write the sum and difference for each pair.

1. $8 + 3 = 11$; $11 - 3 = 8$	2. $9 + 7$; $16 - 7$	3. $9 + 8$; $17 - 8$
4. $8 + 6$; $14 - 8$	5. $9 + 4$; $13 - 9$	6. $6 + 7$; $13 - 7$
7. $6 + 5$; $11 - 5$	8. $9 + 9$; $18 - 9$	9. $9 + 5$; $14 - 5$
10. $8 + 7$; $15 - 8$	11. $7 + 7$; $14 - 7$	12. $9 + 6$; $15 - 6$

Write About It

13. Draw a picture that shows a fact to 18.
Write an addition sentence for the picture.
Write a subtraction sentence for the picture.

Home Note **Your child found sums and differences to 18.**
ACTIVITY **Have your child tell you the two addition facts and two subtraction facts that use the numbers 9, 8, and 17.**

Name ____________________

Concepts and Skills

Use ● and the 10-frame.
Start with the greater number.
Make a 10. Then add.

1.

$$\begin{array}{r} 7 \\ +5 \\ \hline \end{array} \quad \begin{array}{r} 8 \\ +3 \\ \hline \end{array} \quad \begin{array}{r} 9 \\ +4 \\ \hline \end{array}$$

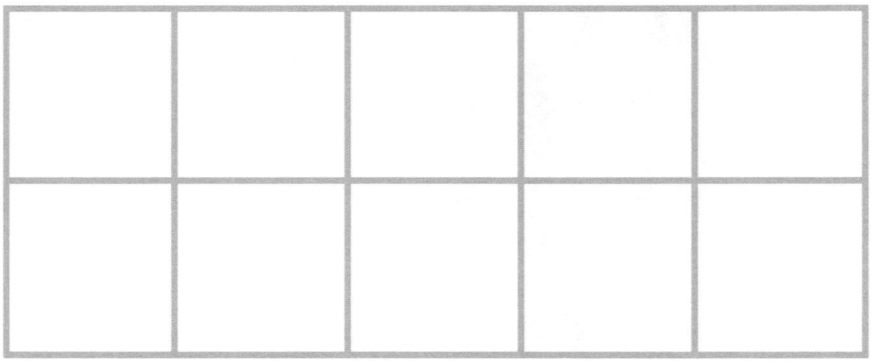

Circle names for 10 or doubles. Then add.

2.

$$\begin{array}{r} 6 \\ 5 \\ +4 \\ \hline \end{array} \quad \begin{array}{r} 2 \\ 5 \\ +5 \\ \hline \end{array} \quad \begin{array}{r} 8 \\ 2 \\ +3 \\ \hline \end{array} \quad \begin{array}{r} 4 \\ 4 \\ +5 \\ \hline \end{array} \quad \begin{array}{r} 7 \\ 5 \\ +3 \\ \hline \end{array} \quad \begin{array}{r} 3 \\ 6 \\ +3 \\ \hline \end{array}$$

Use ●. Add or subtract.

3. $\begin{array}{r} 5 \\ +9 \\ \hline \end{array} \quad \begin{array}{r} 14 \\ -9 \\ \hline \end{array}$

4. $\begin{array}{r} 8 \\ +5 \\ \hline \end{array} \quad \begin{array}{r} 13 \\ -5 \\ \hline \end{array}$

5. $\begin{array}{r} 9 \\ +4 \\ \hline \end{array} \quad \begin{array}{r} 13 \\ -4 \\ \hline \end{array}$

Write the sum and difference for each pair.

6. $\begin{array}{r} 7 \\ +9 \\ \hline \end{array} \quad \begin{array}{r} 16 \\ -9 \\ \hline \end{array}$

7. $\begin{array}{r} 8 \\ +7 \\ \hline \end{array} \quad \begin{array}{r} 15 \\ -7 \\ \hline \end{array}$

8. $\begin{array}{r} 9 \\ +9 \\ \hline \end{array} \quad \begin{array}{r} 18 \\ -9 \\ \hline \end{array}$

Name ______________________________

TAAS Prep

Cumulative Chapters 1–26

Mark the best answer.

1. How many are there?

- 4 tens and 4 ones
- 3 tens and 14 ones
- 40 tens and 4 ones
- 40 tens and 14 ones

2. Which number does the model show?

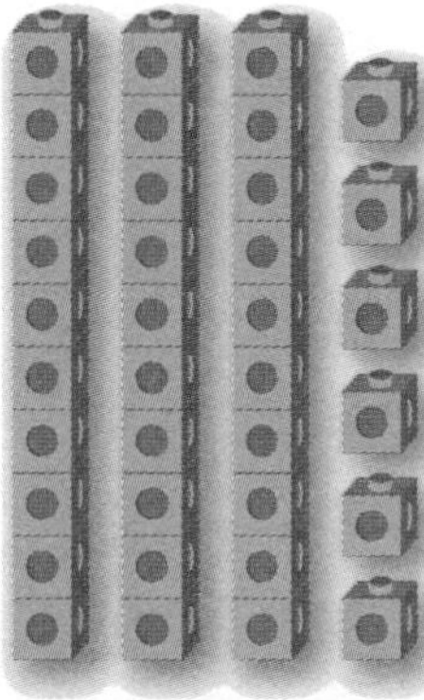

- 16
- 56
- 36
- 63

3. Which picture shows equal parts?

- 
- 
- 
- 

4. Which sentence tells about the picture?

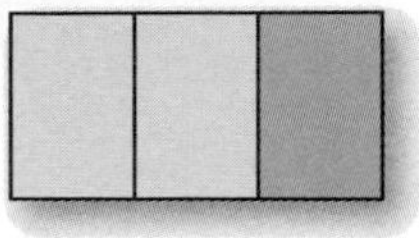

- 1 of the 3 parts is green.
- 2 of the 5 parts are green.
- 2 of the 3 parts are green
- 3 of the 3 parts are yellow.

5. Count by twos.
How many fish are there?

- 8
- 2
- 16
- 14

6. Look at the picture.
Which subtraction sentence tells how many are left?

- $7 + 5 = 12$
- $12 - 5 = 7$
- $5 + 7 = 12$
- $12 - 7 = 5$

Name ______________________________

MATH FUN

You will need:
2 [number cube],
2 [game piece]

Buried Treasure

1. Roll the [number cube]. Add the numbers.
2. Move your [game piece] the number of spaces.
3. Take turns.
4. The first person to reach the treasure is the winner.

Home Note **Your child has been learning addition facts to 18.**
ACTIVITY **Play this game to help your child practice addition.**

Name ______________________________

Calculator | Computer

Use a [calculator].
Add the numbers in the magic square
all the ways the arrows show.
Write the sums.
Then write the magic number.

3 + 4 + 5 = 12

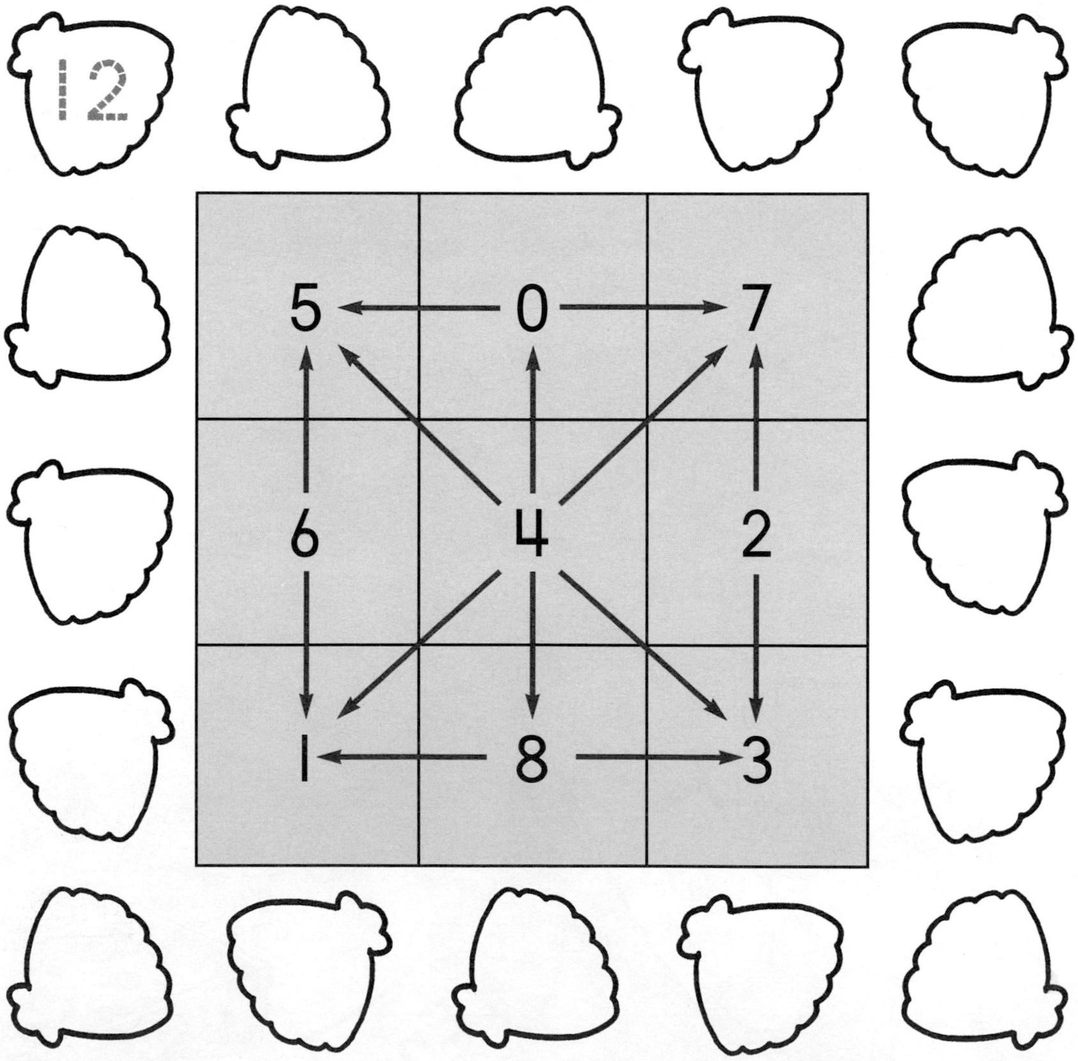

The magic number is ______.

The Hungry King

written by
Lucy Floyd

illustrated by
Deborah Haley Melmon

This book will help me review doubles plus one.

This book belongs to ____________________.

A

Once there was a very hungry king.
"I have only 6 muffins," said the king.
"I need MORE!"

So the cook gave him 6 more.

"Now I have 6 + 6 = ______ muffins," said the hungry king. "I still need more."

B

So the cook gave him 1 more muffin.
“Goody!” said the hungry king.

“Now I have 6 + 7 = _____ muffins!”

And he ate every one of them.

"I am still hungry," said the king.
So the cook gave him 7 muffins.
"I need MORE!" said the hungry king.
So the cook gave him 7 more muffins.
"Goody!" said the hungry king.

"Now I have 7 + 7 = ____ muffins!
I still need more."

D

So the cook gave him 1 more muffin.
"Goody!" said the hungry king.

"Now I have $7 + 8 =$ ______ muffins!"

And he ate every one of them.

E

Then the cook gave him 8 muffins.
"I need MORE!" said the hungry king.

So the cook gave him 8 more muffins.
"Goody!" said the hungry king.

"Now I have 8 + 8 = ______ muffins.
I still need more."

F

So the cook gave him 1 more muffin.
"Goody!" said the hungry king.

"Now I have $8 + 9 =$ ____ muffins!
Should I eat them all?"

WHAT DID THE KING DO?

G

The king ate all the muffins!
He was a VERY, VERY sick king!

Name ____________________

Concepts and Skills

Write the sums.

1. $\begin{array}{r} 5 \\ +5 \\ \hline \end{array}$ $\begin{array}{r} 5 \\ +6 \\ \hline \end{array}$

2. $\begin{array}{r} 4 \\ +4 \\ \hline \end{array}$ $\begin{array}{r} 4 \\ +5 \\ \hline \end{array}$

3. $\begin{array}{r} 6 \\ +6 \\ \hline \end{array}$ $\begin{array}{r} 6 \\ +7 \\ \hline \end{array}$

4. $\begin{array}{r} 7 \\ +7 \\ \hline \end{array}$ $\begin{array}{r} 7 \\ +6 \\ \hline \end{array}$

5. $\begin{array}{r} 9 \\ +9 \\ \hline \end{array}$ $\begin{array}{r} 9 \\ +8 \\ \hline \end{array}$

6. $\begin{array}{r} 8 \\ +8 \\ \hline \end{array}$ $\begin{array}{r} 8 \\ +7 \\ \hline \end{array}$

Add or subtract.

7.

$4 + 4 =$ ____

$8 - 4 =$ ____

8.

$6 + 6 =$ ____

$12 - 6 =$ ____

9.
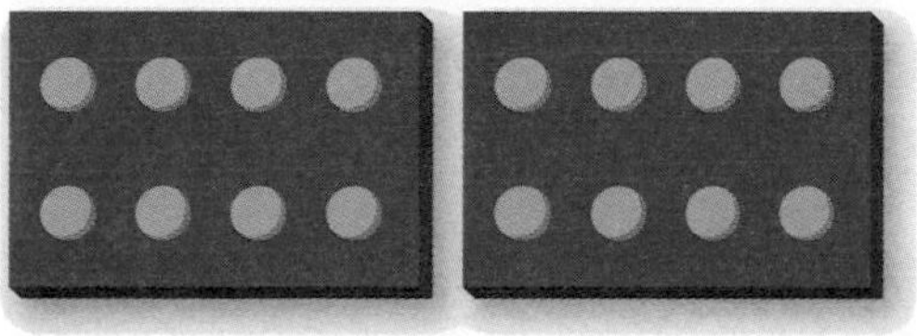

$8 + 8 =$ ____

$16 - 8 =$ ____

Use ● and the 10-frame. Add.

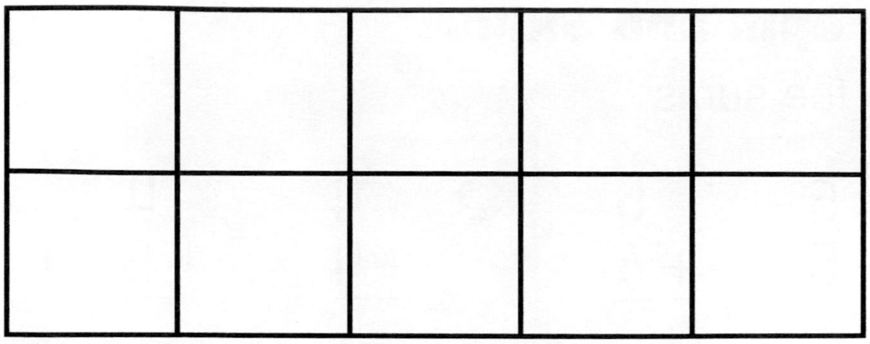

(10) $\begin{array}{r} 7 \\ +5 \\ \hline \end{array}$

(11) $\begin{array}{r} 9 \\ +4 \\ \hline \end{array}$

(12) $\begin{array}{r} 8 \\ +3 \\ \hline \end{array}$

(13) $\begin{array}{r} 6 \\ +7 \\ \hline \end{array}$

Circle names for 10 or doubles. Then add.

(14) $\begin{array}{r} 5 \\ 4 \\ +6 \\ \hline \end{array}$

(15) $\begin{array}{r} 6 \\ 2 \\ +6 \\ \hline \end{array}$

(16) $\begin{array}{r} 7 \\ 3 \\ +2 \\ \hline \end{array}$

(17) $\begin{array}{r} 4 \\ 4 \\ +5 \\ \hline \end{array}$

Write the sum and difference for each pair.

(18) $\begin{array}{r} 4 \\ +9 \\ \hline \end{array}$ $\begin{array}{r} 13 \\ -\ 9 \\ \hline \end{array}$

(19) $\begin{array}{r} 8 \\ +6 \\ \hline \end{array}$ $\begin{array}{r} 14 \\ -\ 6 \\ \hline \end{array}$

(20) $\begin{array}{r} 7 \\ +9 \\ \hline \end{array}$ $\begin{array}{r} 16 \\ -\ 9 \\ \hline \end{array}$

Problem Solving

Use ● to solve. Draw them.

(21) I have 5 dogs. My friend has the same number. How many dogs do we have in all?

_____ dogs

Name ________________

Performance Assessment

Use Workmat I and ●.

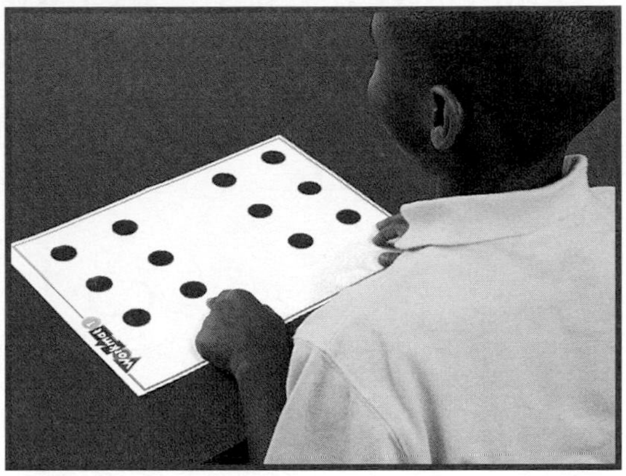

1. Show a doubles fact with a sum of 12, 14, or 16.

Show the **doubles-plus-one** fact.
Show the **doubles-minus-one** fact.

2. Draw a picture to show each fact.
Write the addition fact.
Write a related subtraction fact.

doubles	doubles plus one	doubles minus one
___ + ___ = ___	___ + ___ = ___	___ + ___ = ___
___ − ___ = ___	___ − ___ = ___	___ − ___ = ___

Write About It

3. Find the sum two different ways.
Draw a picture to show each way.

$$\begin{array}{r} 7 \\ 3 \\ +\,7 \\ \hline \end{array} \qquad \begin{array}{r} 7 \\ 3 \\ +\,7 \\ \hline \end{array}$$

Name ______________________________

Fill in the ⬭ for the correct answer.

Chapters 1–26 Cumulative Review

1. Which figure is a triangle?

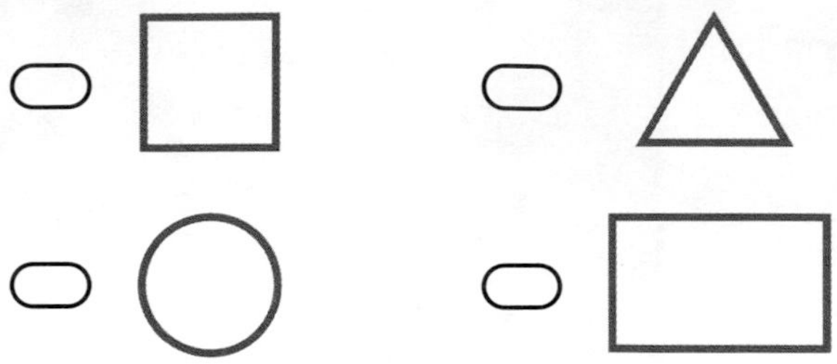

2. How many?

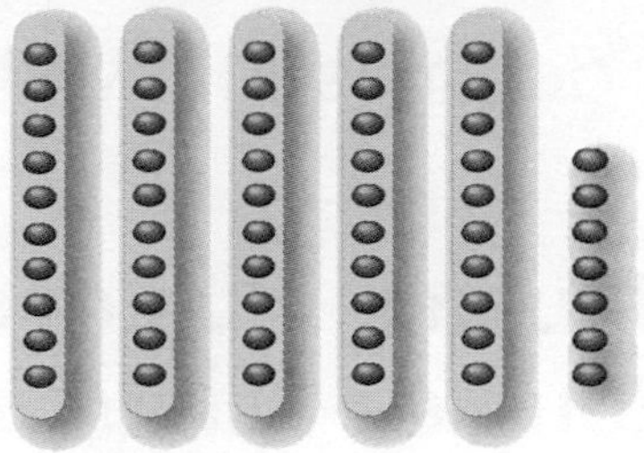

- ⬭ 12
- ⬭ 50
- ⬭ 57
- ⬭ 75

3. Which coins equal a  ?

- ⬭ 5 pennies
- ⬭ 5 nickels
- ⬭ 5 dimes
- ⬭ not here

4. Which takes more than a minute to do?

- ⬭ paint a picture
- ⬭ fill a glass

5.

$$\begin{array}{r} 14 \\ -\ 7 \\ \hline \end{array}$$

- ⬭ 5
- ⬭ 6
- ⬭ 7
- ⬭ 8

6.

$$\begin{array}{r} 7 \\ 3 \\ +\,2 \\ \hline \end{array}$$

- ⬭ 12
- ⬭ 13
- ⬭ 14
- ⬭ 15

Use the graph for questions 7 and 8.

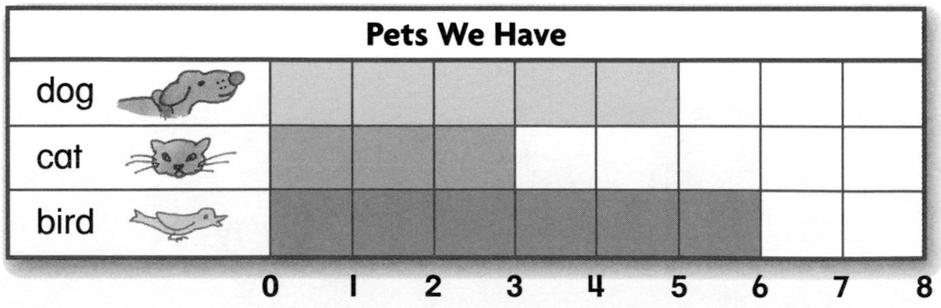

7. How many pets do we have in all?

- ⬭ 5
- ⬭ 14
- ⬭ 6
- ⬭ not here

8. How many more dogs than cats do we have?

- ⬭ 1 more
- ⬭ 3 more
- ⬭ 2 more
- ⬭ 4 more

CHAPTER 27

Multiply and Divide

Tell some stories about joining and separating equal groups.

Home Note In this chapter, your child is learning about multiplication and division situations.
ACTIVITY Have your child find the equal groups in the picture.

SCHOOL-HOME CONNECTION

Dear Family,

Today we started Chapter 27. In this chapter we will begin to learn about multiplication and division. Here are the new vocabulary words and an activity for us to do together at home.

Love,

____

Vocabulary

equal groups

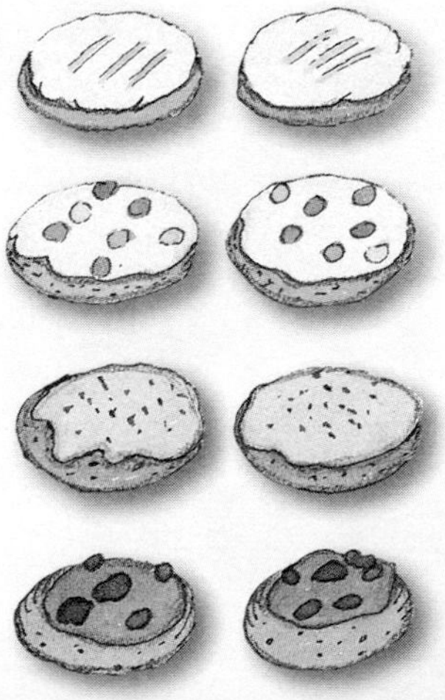

4 groups of 2

ACTIVITY

Have your child pretend that he or she is working in a store, setting up a display of cans. Every shelf in the display must have the same number of cans. Then give your child different numbers of cans, and ask him or her to put them on cabinet or closet shelves.

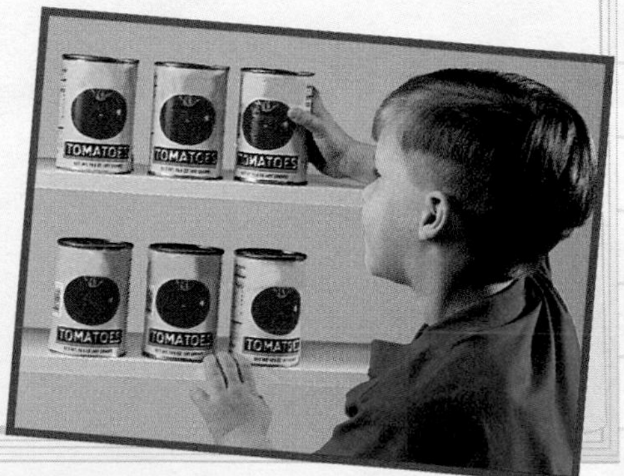

Visit our Web site for additional ideas and activities.
http://www.hbschool.com

Name ______________________________

Counting Equal Groups

Use ■, ■, ■, ■. Draw them.
Write how many in all.

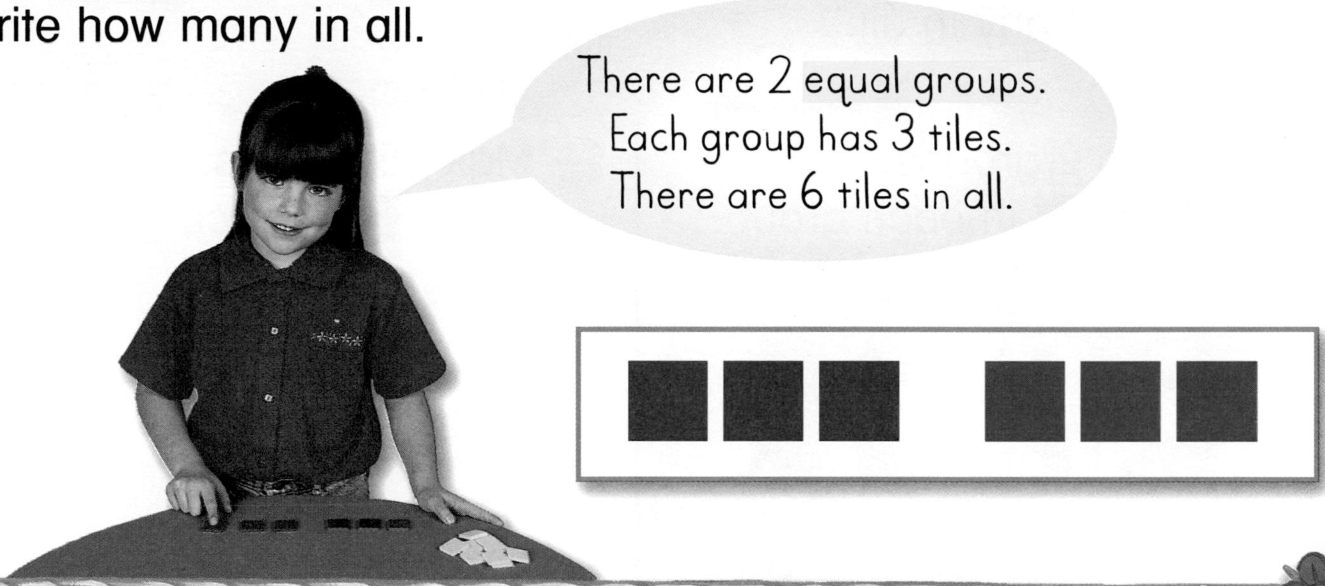
There are 2 equal groups.
Each group has 3 tiles.
There are 6 tiles in all.

1. Make 4 groups.
   Put 2 ■ in each group.

How many in all? 8

2. Make 3 groups.
   Put 2 ■ in each group.

How many in all? ______

3. Make 4 groups.
   Put 4 ■ in each group.

How many in all? ______

4. Make 3 groups.
   Put 3 ■ in each group.

How many in all? ______

Practice

Use ■, ■, ■, ■. Draw them.
Write how many in all.

1. Make 2 groups.
   Put 4 ■ in each group.

How many in all? 8

2. Make 5 groups.
   Put 2 ■ in each group.

How many in all? ______

3. Make 2 groups.
   Put 5 ■ in each group.

How many in all? ______

4. Make 3 groups.
   Put 2 ■ in each group.

How many in all? ______

Mixed Review

5.

7	10	8	11	9	12
+ 7	− 8	+ 5	− 6	+ 9	− 4
☐	☐	☐	☐	☐	☐

Home Note **Your child made equal groups.**
ACTIVITY **Have your child point out equal groups at home, such as the legs on a set of chairs, and then find how many in all.**

Name ______________________________

How Many in Each Group?

Use ●. Draw them.
Write how many in each group.

1. Use 10 ●.
Make 2 equal groups.

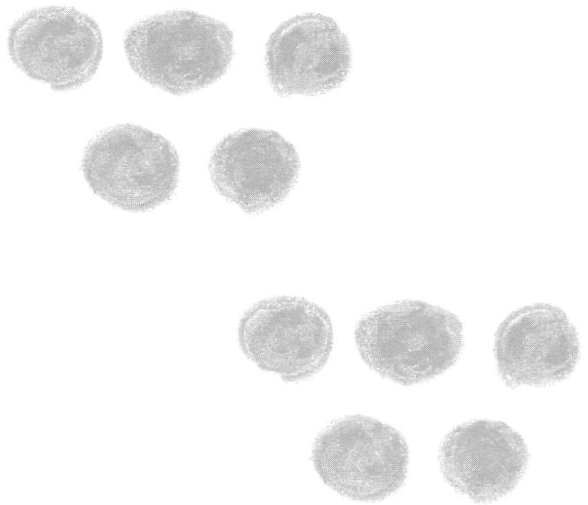

How many in each group? 5

2. Use 6 ●.
Make 3 equal groups.

How many in each group? ______

3. Use 8 ●.
Make 2 equal groups.

How many in each group? ______

4. Use 8 ●.
Make 4 equal groups.

How many in each group? ______

Practice

Use ◯. Draw them.
Write how many in each group.

1 Use 12 ◯.
Make 3 equal groups.

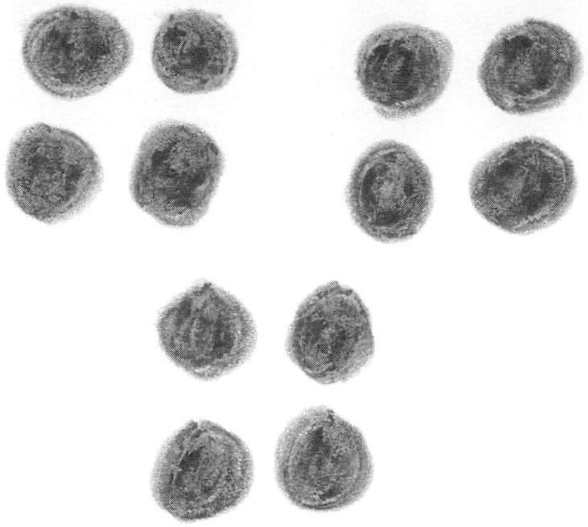

How many in each group? 4

2 Use 4 ◯.
Make 2 equal groups.

How many in each group? ______

3 Use 6 ◯.
Make 2 equal groups.

How many in each group? ______

4 Use 12 ◯.
Make 4 equal groups.

How many in each group? ______

Home Note **Your child made equal groups.**
ACTIVITY **Have your child separate a group of 10 small objects first into groups of 5 and then into groups of 2. Each time, ask him or her how many are in each group.**

Name ______________________________

How Many Groups?

Use ●. Draw them.
Write how many groups.

1. Use 10 ●.
Put 2 in each group.

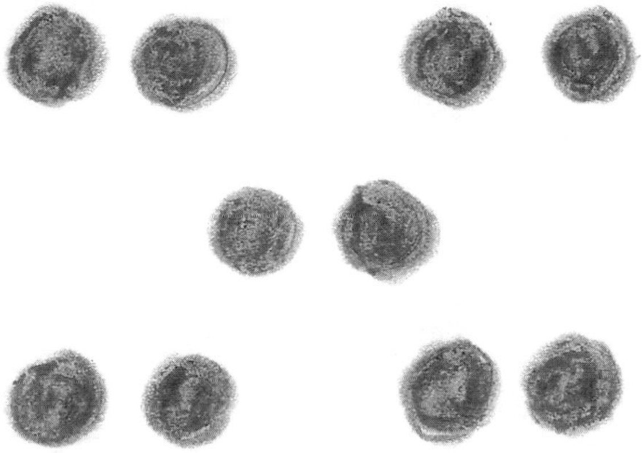

How many groups? 5

2. Use 12 ●.
Put 4 in each group.

How many groups? ______

3. Use 8 ●.
Put 2 in each group.

How many groups? ______

4. Use 10 ●.
Put 5 in each group.

How many groups? ______

Talk About It • Critical Thinking

How do you know the groups are equal?

Practice

Use ◯. Draw them.
Write how many groups.

1 Use 6 ◯.
Put 2 in each group.

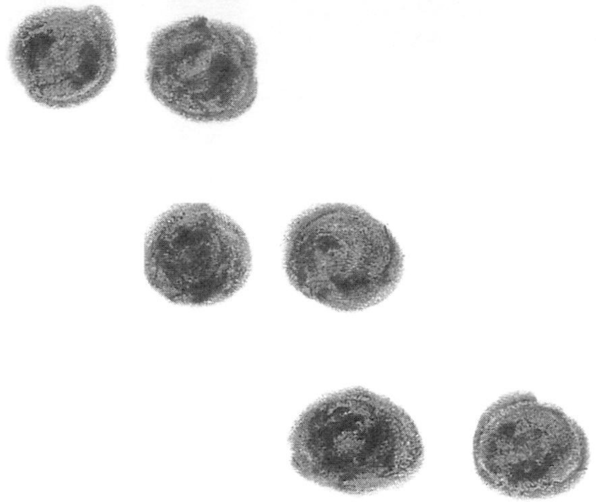

How many groups? 3

2 Use 8 ◯.
Put 4 in each group.

How many groups? ______

3 Use 12 ◯.
Put 6 in each group.

How many groups? ______

4 Use 6 ◯.
Put 3 in each group.

How many groups? ______

Home Note Your child made equal groups.
ACTIVITY Have your child separate 12 objects into groups of 2, 3, 4, and 6. Each time, ask him or her how many groups this makes.

Name ______________________

Understand • Plan • Solve • Look Back

Problem Solving
Draw a Picture

Draw a picture to solve the problem.

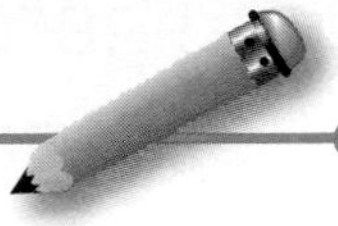

1. There are 3 hot dogs on each plate.
There are 3 plates.
How many hot dogs are there?

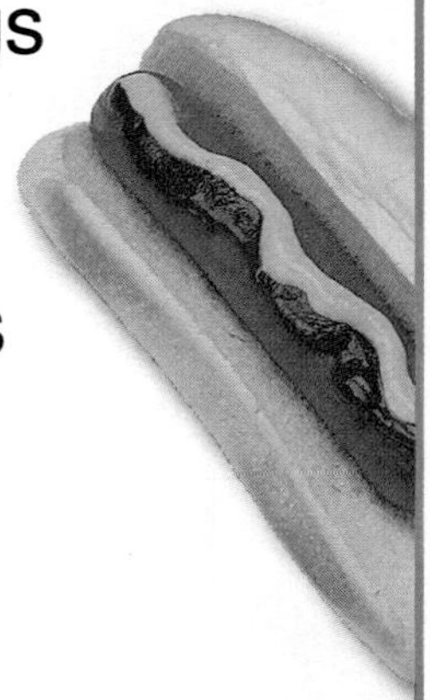

9 hot dogs

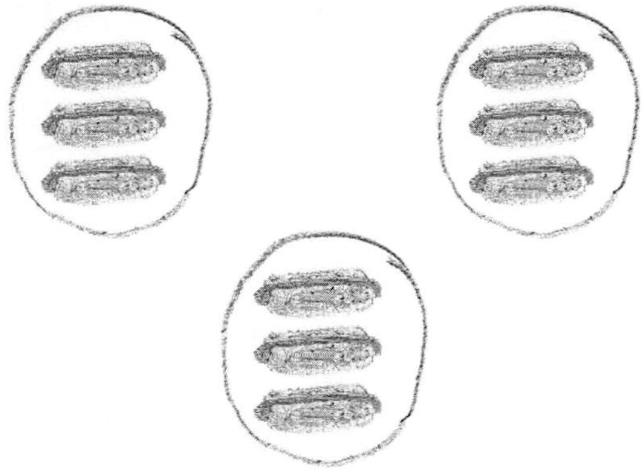

2. There are 4 carrots.
Two children want the same number of carrots.
How many can each child have?

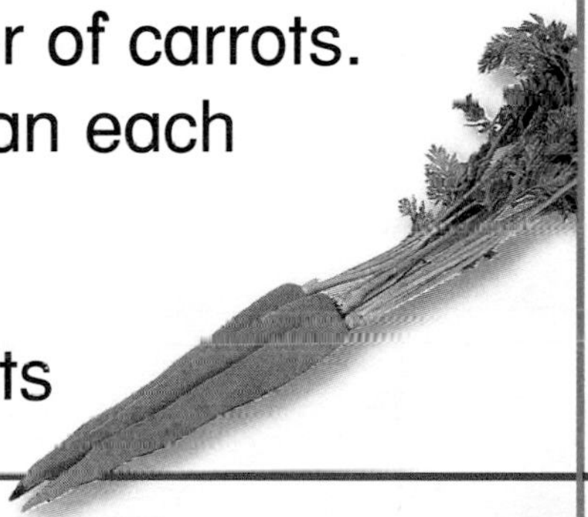

______ carrots

3. There are 3 children.
Each child needs 2 napkins.
How many napkins will be needed?

______ napkins

4. There are 7 apples.
We eat 4.
How many apples are left?

______ apples

Practice

Draw a picture to solve the problem.

1. There are 4 slices of bread. We need 4 more. How many slices do we need?

8 slices

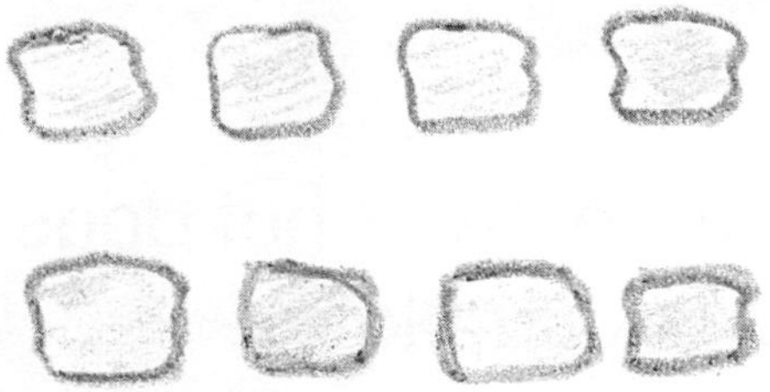

2. There are 6 bowls of fruit. Each bowl has 2 cherries. How many cherries are there?

______ cherries

3. There are 6 crackers. We give 2 to each child. How many children will get crackers?

______ children

4. There are 10 pretzels. Every child gets 2 pretzels. How many children will get pretzels?

______ children

Home Note Your child drew pictures to solve problems.
ACTIVITY Give your child simple addition, subtraction, multiplication, and division problems. Have him or her draw pictures to solve them.

Name ______________________

Chapter 27

Review/Test

Concepts and Skills

Use ■. Draw them. Write how many.

1. Make 5 groups. Put 2 ■ in each group. How many in all? _____	2. Make 2 groups. Put 3 ■ in each group. How many in all? _____
3. Use 6 ■. Make 2 equal groups. How many in each group? _____	4. Use 9 ■. Put 3 in each group. How many groups? _____

Problem Solving

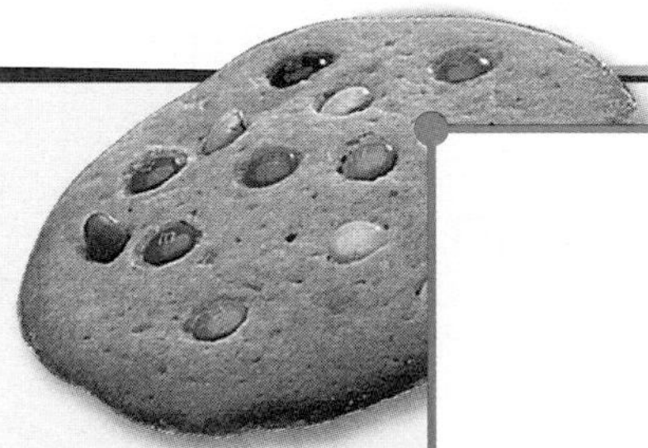

Draw a picture to solve.

5. There are 4 cookies on each plate. There are 2 plates. How many cookies are there?

_____ cookies

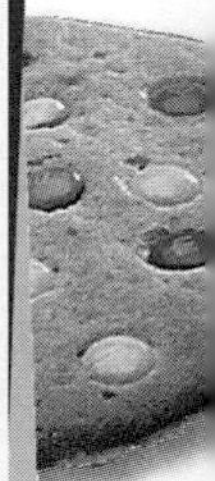

Name ______________________________

Cumulative Chapters 1–27

TAAS Prep

Mark the best answer.

1. Which coins add up to the amount on the tag?

2. About how many 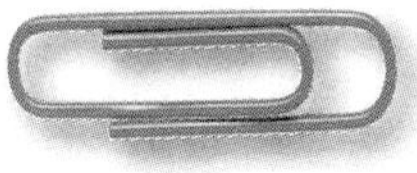 long?

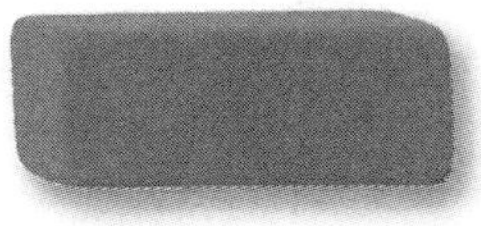

- 1
- 6
- 3
- 8

3. What numbers are in this fact family?

$5 + 8 = 13$ $13 - 5 = 8$
$8 + 5 = 13$ $13 - 8 = 5$

- 5, 3, 8
- 15, 3, 5
- 5, 8, 13
- 8, 6, 5

4. How can you complete the sentence?

__________ buttons are red

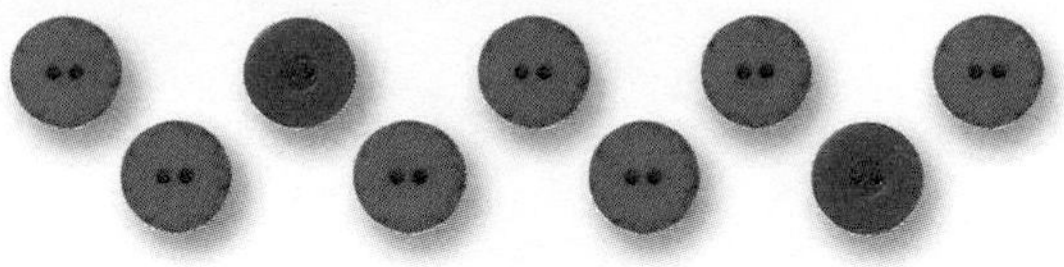

- Two of the seven
- Seven of the nine
- Five of the seven
- Two of the nine

5. What time is it?

- 12:30
- 12:00
- 3:00
- 3:30

6. Look at the picture. What kind of day is it?

- a hot day
- a cold day

CHAPTER 28

Two-Digit Addition and Subtraction

What addition and subtraction questions can you ask about the picture?

Home Note In this chapter, your child is learning to add and subtract tens and ones.
ACTIVITY A group of seals is called a pod. Have your child tell an addition or subtraction story about a pod of seals.

SCHOOL-HOME CONNECTION

Dear Family,

Today we started Chapter 28. We will learn about adding and subtracting two-digit numbers. Here are the vocabulary words and an activity we can do together at home.

Love,

____

Vocabulary

In the number 23, the 2 stands for the number of tens, and the 3 stands for the number of ones. 23 is 2 tens and 3 ones.

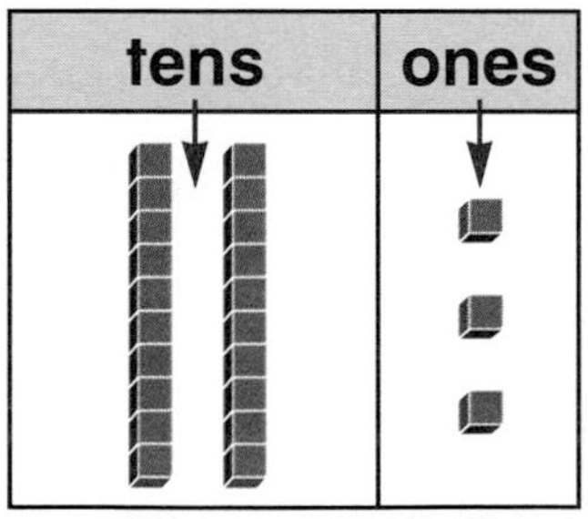

2 tens   3 ones

23

ACTIVITY

Have your child find pictures in magazines or ads of items such as a teddy bear. Then write two prices for each picture, one reasonable and one unreasonable. Have your child tell you something about each item. Then have him or her look at the two prices, decide which price is more reasonable, and tell you why.

Visit our Web site for additional ideas and activities.
http://www.hbschool.com

Name ______________________

Adding and Subtracting Tens

Tony counts 30 penguins. Then he sees 10 more. How many penguins are there in all?

Put in 3 tens to show 30.

$$\begin{array}{r} 30 \\ +\ 10 \\ \hline \end{array}$$

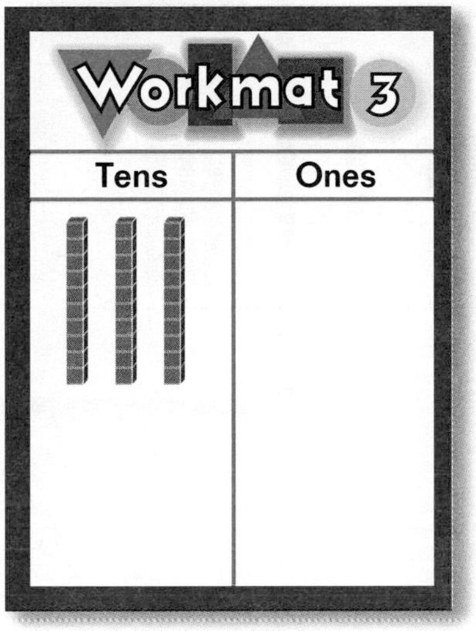

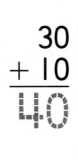

Put in 1 ten to show 10 more. Write the sum.

$$\begin{array}{r} 30 \\ +\ 10 \\ \hline 40 \end{array}$$

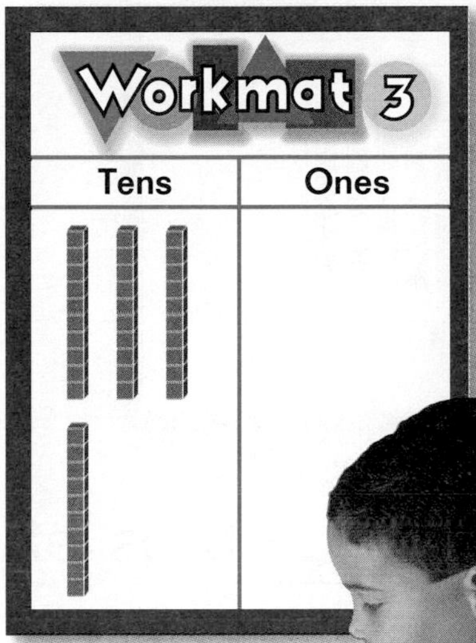

There are __40__ penguins in all.

Use Workmat 3 and .
Add.

1.

$$\begin{array}{r} 20 \\ +\ 10 \\ \hline \end{array} \quad \begin{array}{r} 50 \\ +\ 20 \\ \hline \end{array} \quad \begin{array}{r} 30 \\ +\ 50 \\ \hline \end{array} \quad \begin{array}{r} 10 \\ +\ 10 \\ \hline \end{array} \quad \begin{array}{r} 60 \\ +\ 20 \\ \hline \end{array}$$

2.

$$\begin{array}{r} 70 \\ +\ 20 \\ \hline \end{array} \quad \begin{array}{r} 40 \\ +\ 40 \\ \hline \end{array} \quad \begin{array}{r} 20 \\ +\ 30 \\ \hline \end{array} \quad \begin{array}{r} 80 \\ +\ 10 \\ \hline \end{array} \quad \begin{array}{r} 10 \\ +\ 40 \\ \hline \end{array}$$

Talk About It • Critical Thinking

How does knowing 7 + 2 help you to find 70 + 20?

Practice

Patty sees 50 birds.
10 fly away.
How many birds are left?

Put in 5 tens.

$$\begin{array}{r} 50 \\ -\ 10 \\ \hline \end{array}$$

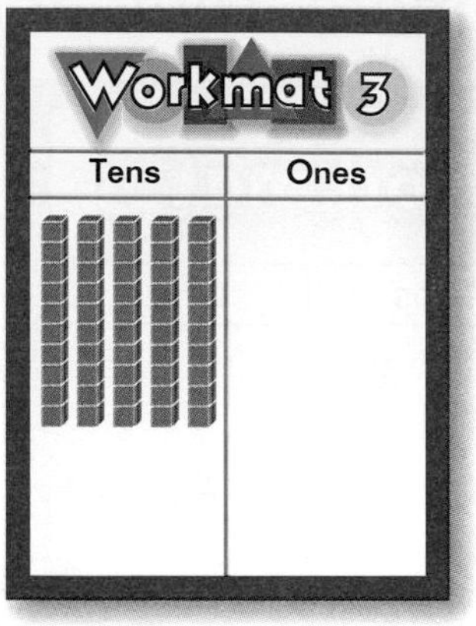

Take away 1 ten.
Write the difference.

$$\begin{array}{r} 50 \\ -\ 10 \\ \hline 40 \end{array}$$

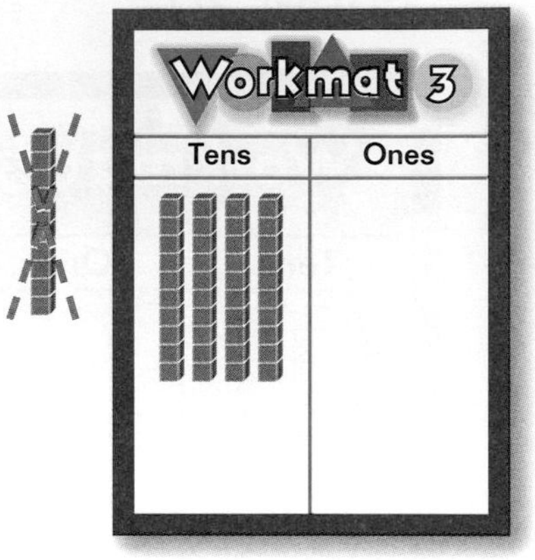

There are 40 birds left.

Use Workmat 3 and tens. Subtract.

1. $\begin{array}{r} 70 \\ -\ 30 \\ \hline \end{array}$ $\begin{array}{r} 50 \\ -\ 10 \\ \hline \end{array}$ $\begin{array}{r} 60 \\ -\ 20 \\ \hline \end{array}$ $\begin{array}{r} 40 \\ -\ 30 \\ \hline \end{array}$ $\begin{array}{r} 90 \\ -\ 10 \\ \hline \end{array}$

---

2. $\begin{array}{r} 30 \\ -\ 20 \\ \hline \end{array}$ $\begin{array}{r} 80 \\ -\ 50 \\ \hline \end{array}$ $\begin{array}{r} 20 \\ -\ 10 \\ \hline \end{array}$ $\begin{array}{r} 60 \\ -\ 50 \\ \hline \end{array}$ $\begin{array}{r} 80 \\ -\ 30 \\ \hline \end{array}$

Mixed Review

Write the time.

3. 

___ : ___

4. 

___ : ___

5. 

___ : ___

Home Note Your child added and subtracted tens.
ACTIVITY Have your child use objects to practice adding and subtracting tens.

Name ________________________________

Adding Tens and Ones

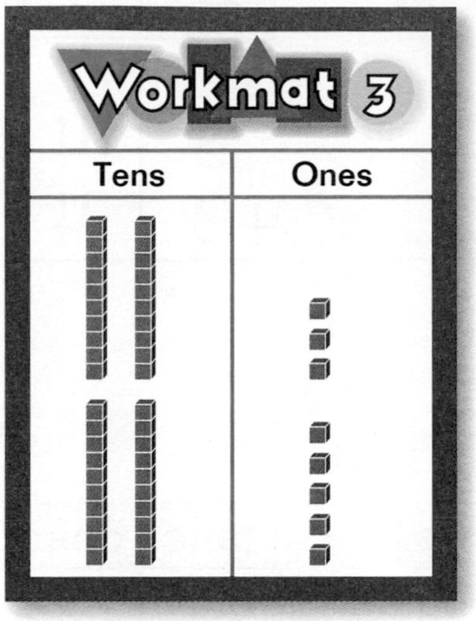

tens	ones
2	3
+2	5

Add the ones.

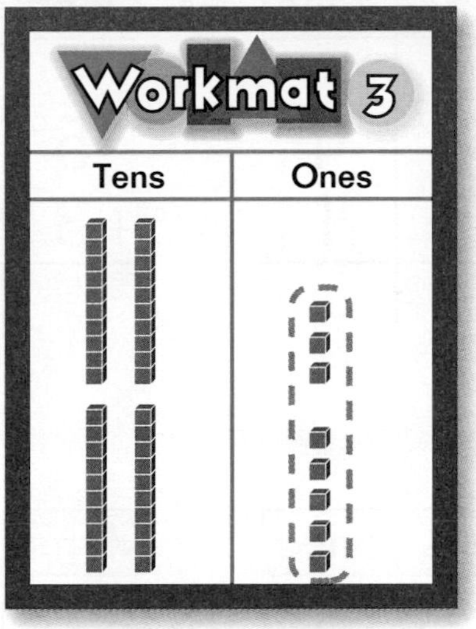

tens	ones
2	3
+2	5
	8

Add the tens.

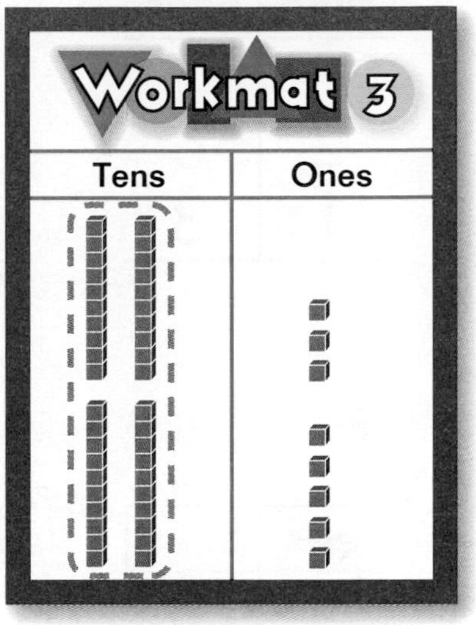

tens	ones
2	3
+2	5
4	8

Add.

1.

tens	ones
2	3
+2	2

tens	ones
3	1
+1	4

tens	ones
4	3
+3	3

tens	ones
6	2
+2	6

2.

tens	ones
5	3
+1	1

tens	ones
2	7
+4	2

tens	ones
3	6
+2	3

tens	ones
7	1
+1	6

Practice

Add.

1

tens	ones
2	6
+1	3

tens	ones
2	4
+3	1

tens	ones
5	7
+3	2

tens	ones
3	4
+3	4

2

tens	ones
3	2
+1	5

tens	ones
6	2
+1	5

tens	ones
3	3
+2	4

tens	ones
6	7
+1	1

3

tens	ones
4	3
+3	1

tens	ones
5	4
+2	4

tens	ones
8	1
+1	0

tens	ones
7	3
+2	5

Write About It

4 Write a story about some animals at the North Pole. Use numbers between 11 and 49 in your story.

Home Note **Your child added two-digit numbers.**
ACTIVITY **Choose some problems from pages 443 and 444. Have your child explain how he or she found the answers.**

Name ____________________

Subtracting Tens and Ones

Show 38.

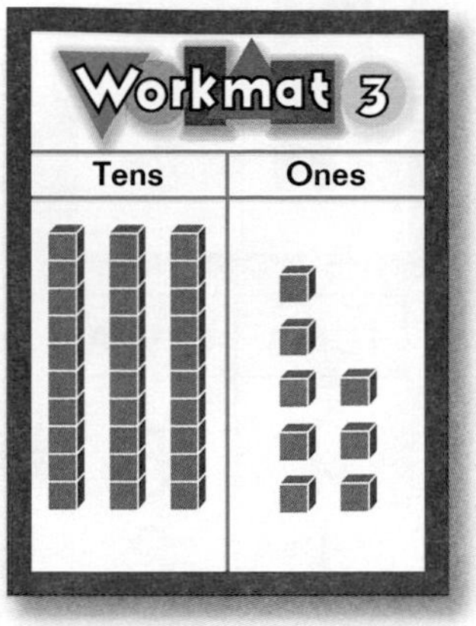

tens	ones
3	8
−1	2

Subtract the ones.

tens	ones
3	8
−1	2
	6

Subtract the tens.

tens	ones
3	8
−1	2
2	6

Use Workmat 3 and tens and ones blocks.
Subtract.

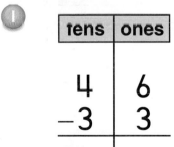

1.

tens	ones
4	6
−3	3

tens	ones
6	8
−2	4

tens	ones
8	3
−5	2

tens	ones
7	7
−4	3

2.

tens	ones
5	8
−4	4

tens	ones
9	4
−1	2

tens	ones
5	1
−3	0

tens	ones
6	6
−2	6

Practice

Show 45.

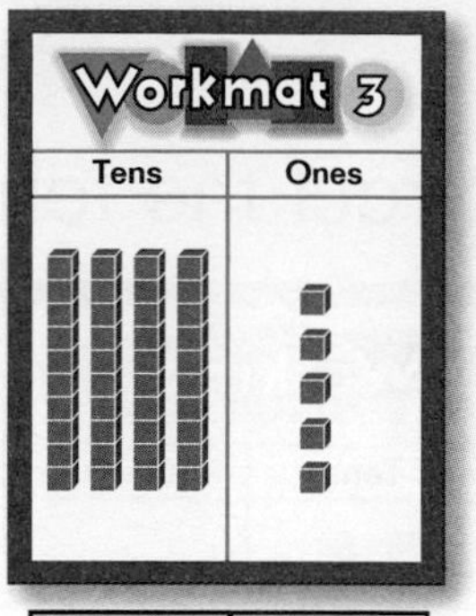

tens	ones
4	5
−1	3

Subtract the ones.

tens	ones
4	5
−1	3
	2

Subtract the tens.

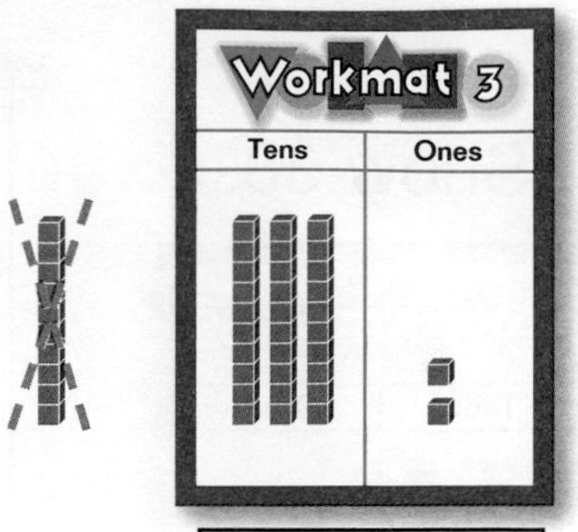

tens	ones
4	5
−1	3
3	2

Use Workmat 3 and ▭ ▫. Subtract.

1

tens	ones
8	2
−5	1

tens	ones
4	6
−3	5

tens	ones
9	8
−8	2

tens	ones
7	5
−6	0

2

tens	ones
9	6
−3	2

tens	ones
4	4
−3	4

tens	ones
6	5
−2	0

tens	ones
7	3
−3	1

Problem Solving

3 Kay saved 35¢. She spent 15¢ at the store. How much money does she have left?

_____ ¢

Home Note Your child subtracted two-digit numbers.
ACTIVITY Choose some problems from pages 445 and 446. Have your child explain how he or she found the answers.

Name ______________________________

Understand • Plan • Solve • Look Back

Problem Solving

Logical Reasoning

Circle the reasonable answer.

1. Jenny had 12 toy animals.
She got 14 more.
How many does she have now?

26 animals | 4 animals | 260 animals

2. Mary read 10 books.
Then she read 10 more.
How many books did she read in all?

10 books

20 books

100 books

3. Liz had 44 pennies.
She gave away 12.
How many does she have now?

56 pennies

32 pennies

320 pennies

Practice

Circle the reasonable answer.

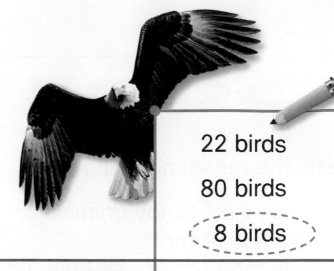

1. Jane saw 15 birds. Then 7 flew away. How many birds are left?	22 birds 80 birds 8 birds
2. Sam had 26 pennies. He spent 22 pennies. How many does he have now?	4 pennies 444 pennies 44 pennies
3. Luis had 38 cards. He gave away 10. How many cards does he have now?	48 cards 28 cards 280 cards
4. There are 15 boys and 14 girls in Paolo's class. How many children are there in all?	290 children 300 children 29 children
5. Sue has 35 stickers. She got 12 more. How many does she have now?	47 stickers 470 stickers 23 stickers
6. Mark collected 48 stamps. He gave 10 away. How many does he have now?	58 stamps 480 stamps 38 stamps

Home Note Your child used logical reasoning to solve story problems.
ACTIVITY Have your child tell you how he or she knew which answers made sense.

Name ______________________________

Chapter 28
Review/Test

Concepts and Skills

Add or subtract.

1.

40	30	70	40	50	60
+ 10	+ 20	+ 10	− 30	− 20	− 30

2.

tens	ones
3	3
+1	2

tens	ones
6	4
+1	4

tens	ones
5	2
+2	1

tens	ones
7	5
+1	3

3.

tens	ones
6	7
−2	3

tens	ones
8	2
−4	1

tens	ones
9	7
−5	2

tens	ones
7	4
−3	3

Problem Solving

Circle the answer that makes sense.

4. Carol got 12 flowers. Then she got 12 more. How many flowers did she have in all?

240 flowers

24 flowers

42 flowers

5. Ann had 25 pennies. She spent 22 pennies. How many does she have now?

3 pennies

333 pennies

23 pennies

Name ____________________

TAAS Prep

Mark the best answer.

1. Which numbers come next?

   25, 30, 35, 40, ___, ___

   - ○ 41, 42
   - ○ 45, 46
   - ○ 45, 50
   - ○ 50, 60

2. How many tens and ones?

   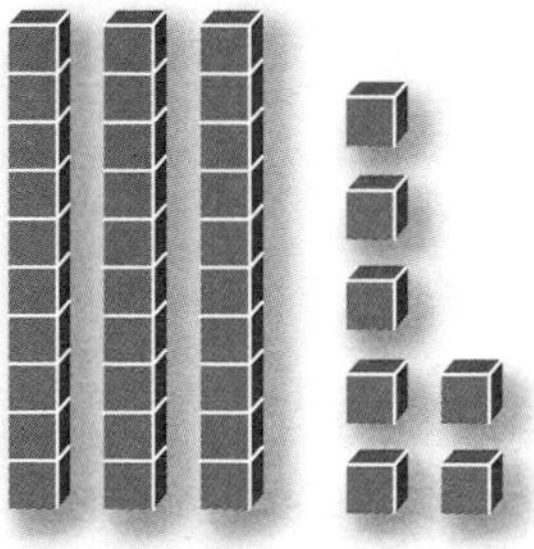

   - ○ 2 tens, 5 ones
   - ○ 3 tens, 7 ones
   - ○ 4 tens, 8 ones
   - ○ 7 tens, 3 ones

3. Sam had 13 books. He got 4 more. How many books does Sam have in all?

   - ○ 14
   - ○ 15
   - ○ 17
   - ○ 19

4. Which is true?

   - ○ 4 crayons are red.
   - ○ 4 crayons are blue.
   - ○ 1 out of 4 crayons are blue.
   - ○ 1 out of 4 crayons are red.

5. Which figure is a square?

   - ○ 
   - ○ 
   - ○ 
   - ○  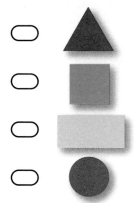

6. Which coins equal  ?

   - ○ 
   - ○ 
   - ○ 
   - ○ 

Name ______________________________

MATH FUN

You will need:
2 pennies,
▭ ▫.

Addition Toss

1. Drop 2 pennies on the game board.
2. Find the sum of the two numbers the pennies land on.
3. Have your partner use ▭ ▫ to check the sum. Keep trying until you have the correct answer.
4. Play until you and your partner each have 5 turns.

13	24	20	41	32
43	10	33	12	23
20	42	14	30	25
34	40	21	44	10
30	22	31	50	11

Home Note **Your child has been learning about two-digit addition.**
ACTIVITY **Play this game at home to help your child practice this new skill.**

Name ______________________________

Calculator | Computer

Use a  .
Find the sums and differences.
Write which keys you press.
Write what you see.

1. 25 + 33 = ____

ON/C	2	5	+	3	3	=	58

2. 43 − 21 = ____

ON/C			−			=	

3. 61 + 24 = ____

ON/C			+			=	

4. 56 − 25 = ____

ON/C			−			=	

5. 72 + 22= ____

ON/C			+			=	

6. 59 − 36 = ____

ON/C			−			=	

7. 35 + 34 = ____

ON/C			+			=	

Can I Play?

written by Shirley Frederick
illustrated by Ed Martinez

This book will help me review equal groups.

This book belongs to ____________________.

A

Rebecca has 12 marbles.

"Here are ____ marbles for you

and ____ marbles for me," said Rebecca.

"Now can we play?" asked John.

Jeff asked, "Can I play?"

"Yes, you can play," said Rebecca.

"Here are ____ marbles for you,

and ____ marbles for you,

and ____ marbles for me," said Rebecca.

"Now can we play?" asked John.

D

Melissa asked, "Can I play?"

"Yes, you can play," said Rebecca.

"Here are _____ marbles for you,

and _____ marbles for you,

and _____ marbles for you,

and _____ marbles for me," said Rebecca.

"I had 4 marbles. But now I only have 3 marbles!" said Jeff.

"More of us are playing, so now you get 3," said Rebecca.

"Now can we play?" asked John.

“Can I play?” asked Tom.

“Can I play?” asked Susan.

“Yes, you can play,” said Rebecca.

“Now how many marbles will I get?” asked Jeff.

“Each of us will get _____ marbles,” said Rebecca.

“Let’s play!” said John.

Name ______________________

Concepts and Skills

Use ◯. Draw them.
Write how many in all.

1. Make 2 groups.
Put 2 ◯ in each group.

How many in all? ______

2. Make 2 groups.
Put 3 ◯ in each group.

How many in all? ______

Use ◯. Draw them.
Write how many in each group.

3. Use 6 ◯.
Make 3 equal groups.

How many in each group? ______

4. Use 10 ◯.
Make 2 equal groups.

How many in each group? ______

Use ◯. Draw them.
Write how many groups.

5. Use 9 ◯.
Put 3 in each group.

How many groups ? ______

6. Use 8 ◯.
Put 4 in each group.

How many groups? ______

Add or subtract.

7. $\begin{array}{r} 70 \\ +10 \\ \hline \end{array}$

8. $\begin{array}{r} 40 \\ +20 \\ \hline \end{array}$

9. $\begin{array}{r} 60 \\ -10 \\ \hline \end{array}$

10. $\begin{array}{r} 30 \\ -20 \\ \hline \end{array}$

11.

tens	ones
2	2
+1	3

12.

tens	ones
3	8
+1	1

13.

tens	ones
4	4
+1	4

14.

tens	ones
5	6
+2	1

15.

tens	ones
8	3
−3	2

16.

tens	ones
5	4
−3	1

17.

tens	ones
8	7
−4	2

18.

tens	ones
9	6
−2	2

Problem Solving

Circle the reasonable answer.

19. Bo ate 20 peanuts.
Then he ate 10 more.
How many peanuts
did he eat in all?

10 peanuts
30 peanuts
300 peanuts

20. Sue had 88 pennies.
She spent 14 pennies.
How many pennies
does she have now?

38 pennies
74 pennies
112 pennies

Name ____________________

Performance Assessment

Use Workmat 1 and ◯.
Draw and color to show your work.

1. Make 2 groups.
Put the same number of ◯ in each group.

How many in each group? ___

How many in all? ___

2. Make 4 groups.
Put the same number of ◯ in each group.

How many in each group? ___

How many in all? ___

Write About It

Use Workmat 3 and .

3. Find two numbers that have a sum of 46.

Write the numbers.
Draw base-ten blocks to show how you know.

_____ _____

Name ______________________________

Chapters 1–28 Cumulative Review

Fill in the ⬭ for the correct answer.

1. Which is the same shape as this cylinder? 

⬭ 
⬭ 
⬭ 
⬭ 

2. Which numbers are in order from least to greatest?

⬭ 37, 57, 87, 17
⬭ 87, 57, 17, 37
⬭ 17, 37, 87, 57
⬭ 17, 37, 57, 87

3. Which tells the amount?

⬭ 4¢
⬭ 11¢
⬭ 21¢
⬭ 30¢

4. Which tells the time?

⬭ 6:00
⬭ 6:30
⬭ 8:30
⬭ 9:30

5. Which are you likely to take out of the bag most often?

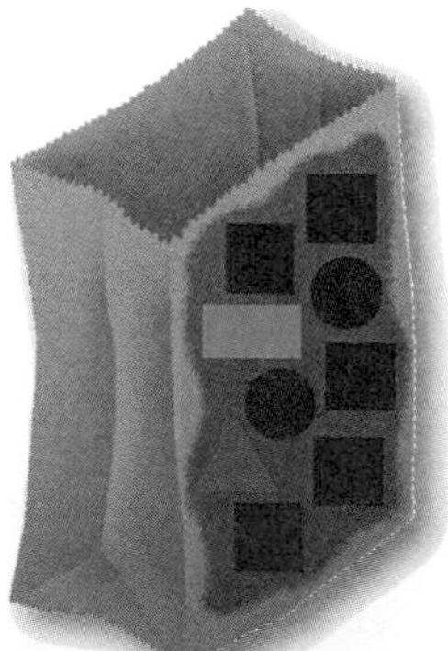

⬭ 
⬭ 
⬭ 
⬭ 

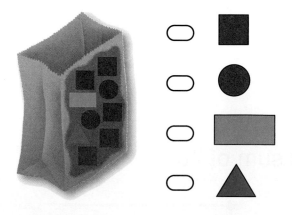

6. Which group shows $\frac{1}{3}$ of the pears yellow?

⬭ 
⬭ 

7. $\begin{array}{r} 11 \\ -\ 2 \\ \hline \end{array}$

⬭ 7
⬭ 9
⬭ 11
⬭ 13

8. $\begin{array}{r} 7 \\ +8 \\ \hline \end{array}$

⬭ 13
⬭ 14
⬭ 15
⬭ 16

9. $\begin{array}{r} 14 \\ -\ 7 \\ \hline \end{array}$

⬭ 5
⬭ 7
⬭ 8
⬭ 9

PICTURE GLOSSARY

add (page 27)

3 + 2 = 5

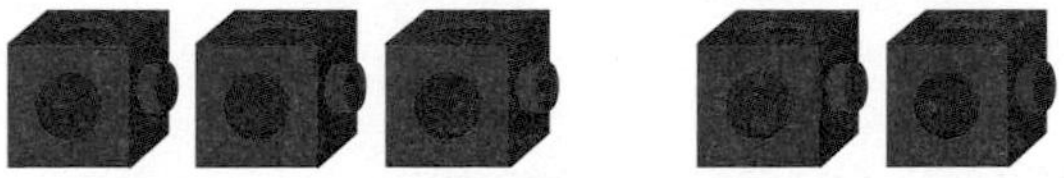

addition sentence (page 31)

4 + 1 = 5

after (page 233)

19, **20**

balance (page 331)

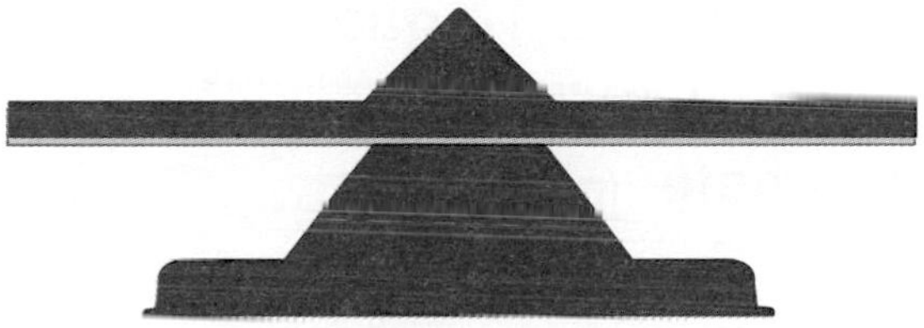

bar graph (page 381)

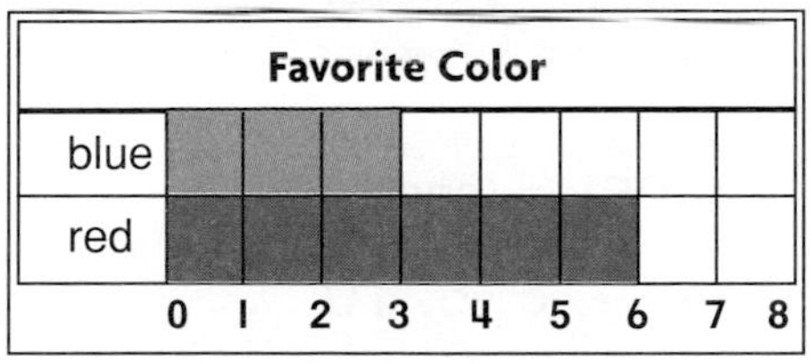

before (page 233)

39, 40

between (page 233)

29, **30,** 31

centimeter (page 325)

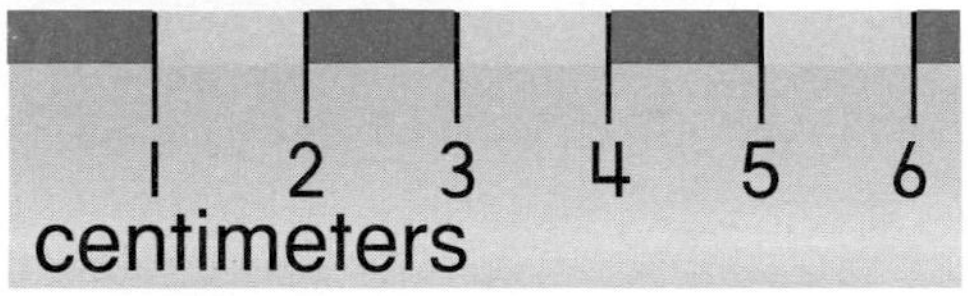

circle (page 135)

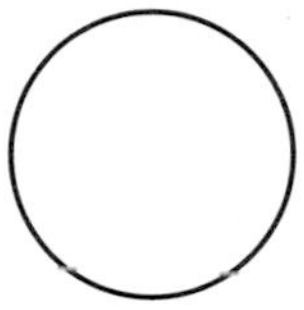

closed figure (page 147)

cone (page 121)

corner (page 137)

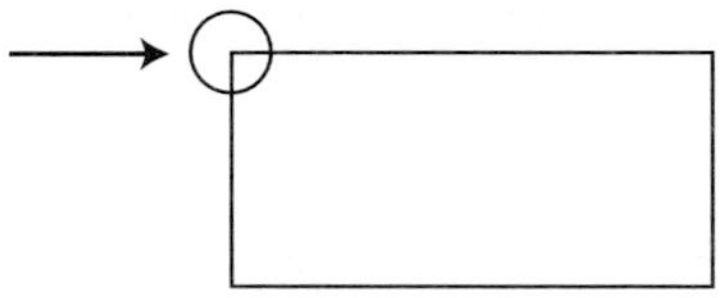

count back (page 101)

9, 8

10 − 2 = 8

count on (page 73)

8 + 2 = 10

cube (page 123)

cylinder (page 123)

difference (page 43)

9 − 3 = **6**

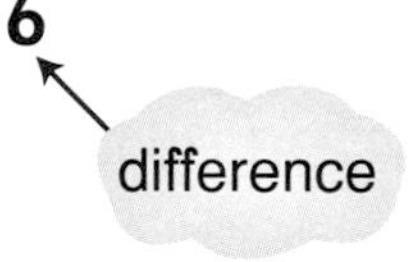

dime (page 261)

10¢
10 cents

doubles (page 77)

4 + 4, 5 + 5, 6 + 6

doubles minus one (page 399)

6 + 6 = 12, so 6 + 5 = 11

doubles plus one (page 397)

4 + 4 = 8, so 4 + 5 = 9

equal groups (page 429)

equal parts (page 345)

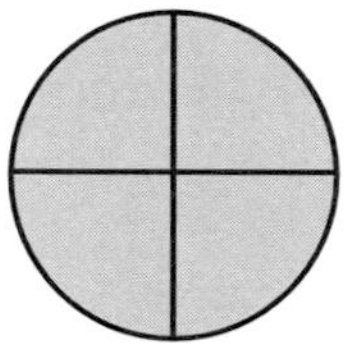

equals = (page 27)

the same as

4 + 1 = 5

4 plus 1 **equals** 5.

estimate (page 221)

Estimate—about 4

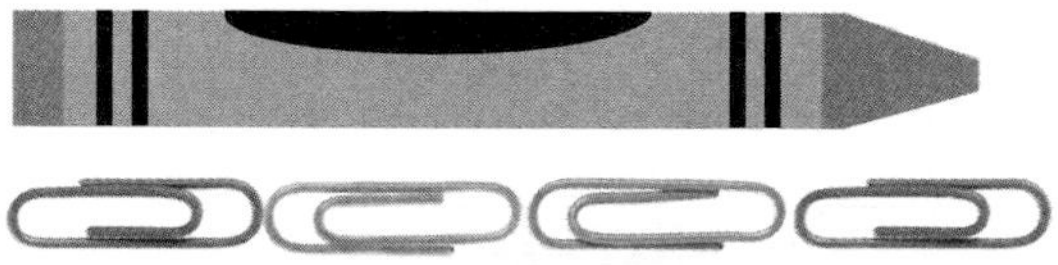

even numbers (page 247)

0, 2, 4, 6, 8, 10 . . .

face (page 127)

fact family (page 93)

$5 + 3 = 8 \quad 3 + 5 = 8$

$8 - 3 = 5 \quad 8 - 5 = 3$

fewer (page 5)

4 balls are **fewer** than 8 balls.

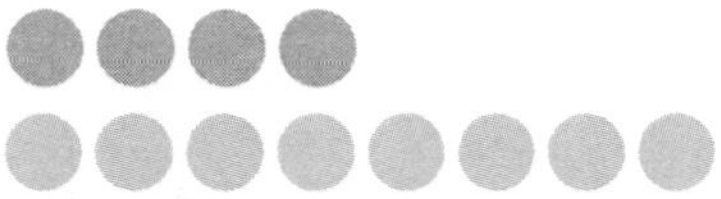

fractions (pages 347–352)

halves thirds fourths

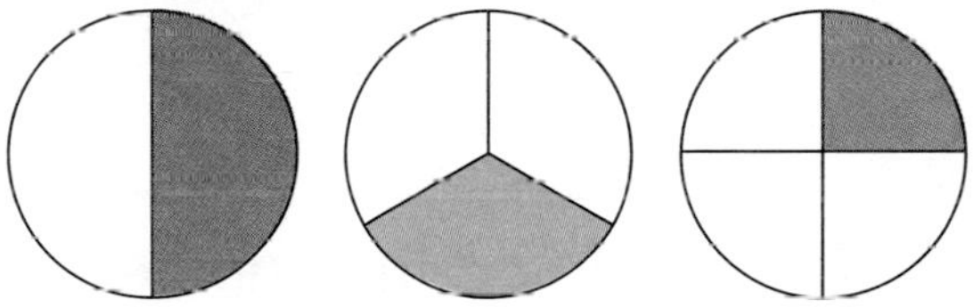

$\frac{1}{2}$ $\frac{1}{3}$ $\frac{1}{4}$

greater than (page 229)

5 is **greater than** 1.

half-hour (page 305)

30 minutes

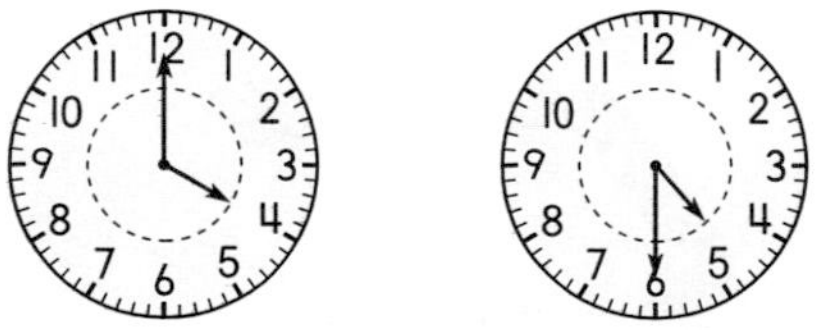

4:00 4:30

heavier (page 333)

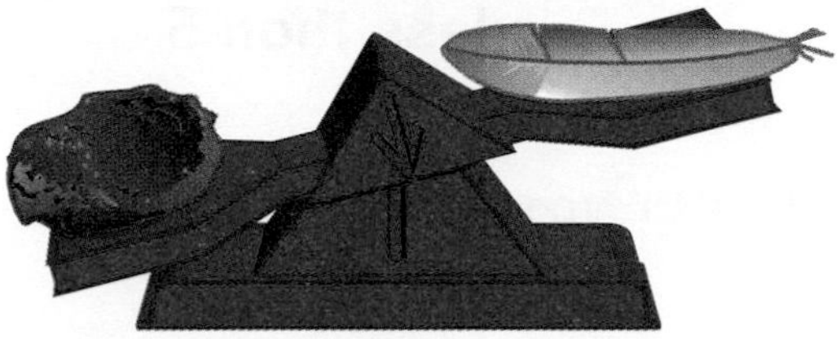

A rock is **heavier** than a feather.

hour (page 301)

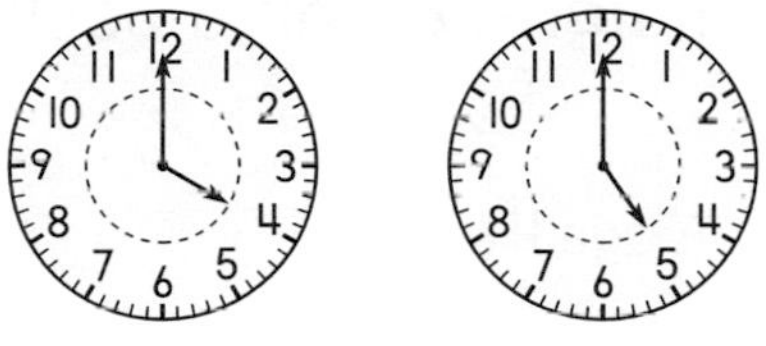

4:00 5:00

hour hand (page 299)

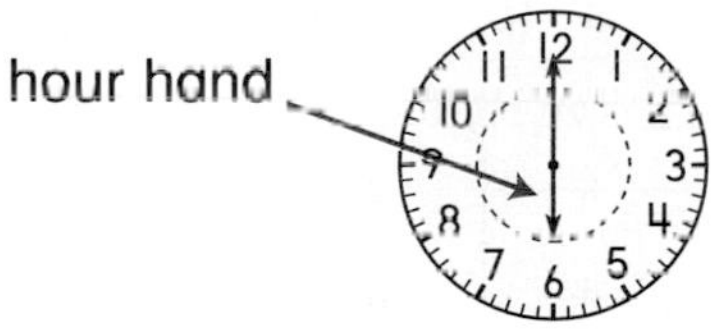

inch (page 321)

inside (page 149)

The star is **inside** the square.

less than (page 231)

3 is **less than** 5.

lighter (page 333)

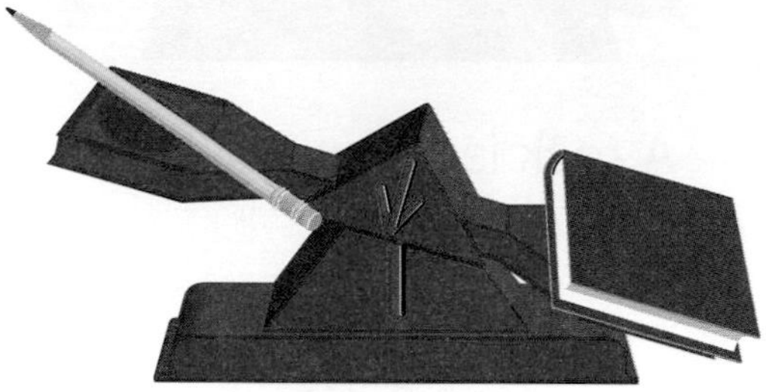

A pencil is **lighter** than a book.

minus – (page 41)

3 – 2 = 1

3 **minus** 2 equals 1.

minute hand (page 299)

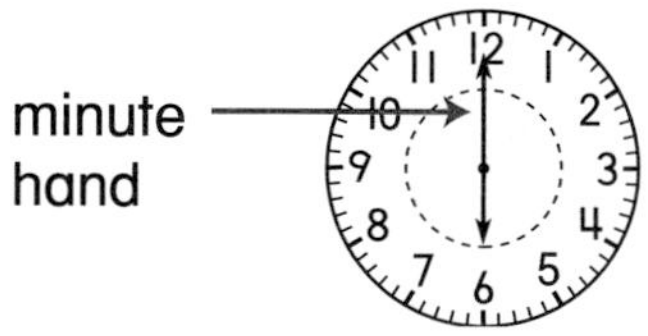

more (page 5)

8 stars are **more** than 4 stars.

nickel (page 259)

5¢
5 cents

number sentence (page 185)

4 + 2 = 6 or 6 – 3 = 3

o'clock (page 299)

The clock shows 1 **o'clock**.

odd numbers (page 247)

1, 3, 5, 7, 9 . . .

on (page 149)

The star is **on** the square.

ones (page 213)

open figure (page 147)

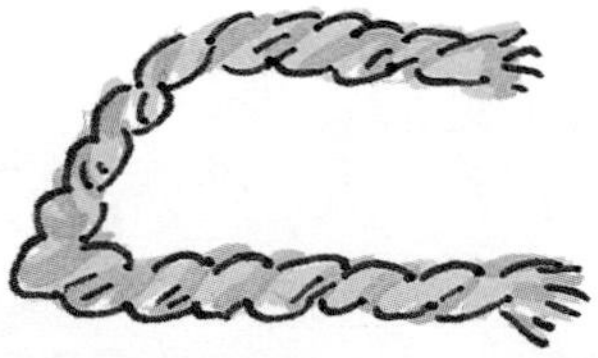

Order Property (page 59)

2 + 3 = 5   3 + 2 = 5

ordinal numbers (page 19)

first second third

outside (page 149)

The star is **outside** the square.

pattern (page 159)

penny (pennies) (page 67)

1 ¢
1 cent

picture graph (page 379)

plus + (page 27)

4 + 3 = 7

4 **plus** 3 equals 7.

pyramid (page 123)

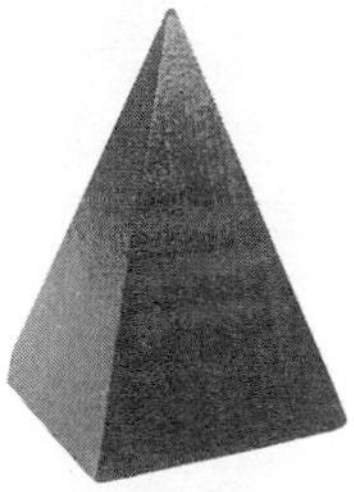

quarter (page 279)

25¢
25 cents

rectangle (page 135)

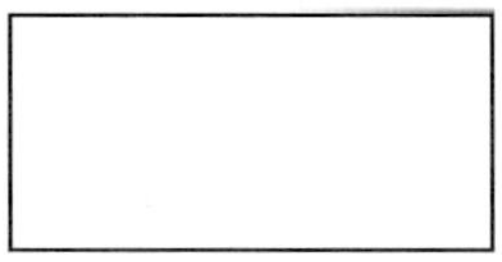

rectangular prism (page 121)

ruler (pages 323, 327)

inch **ruler**

centimeter **ruler**

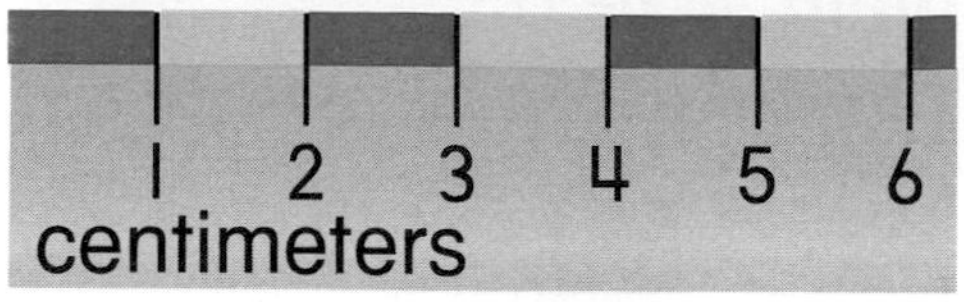

side (page 137)

sphere (page 121)

square (page 135)

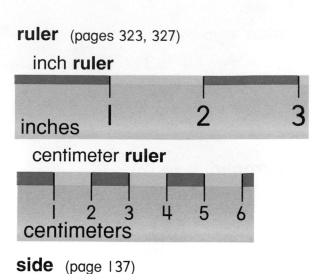

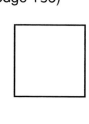

subtract (page 41)

6 − 2 = 4

subtraction sentence (page 45)

9 − 2 = 7

sum (page 27)

9 + 1 = **10**

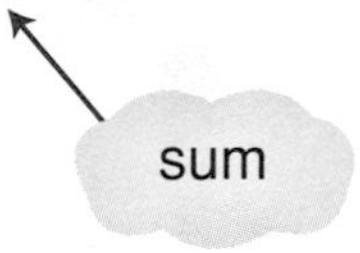

symmetry (page 141)

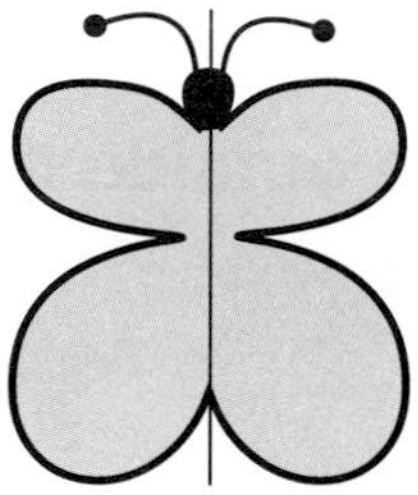

tally marks (page 367)

tens (page 211)

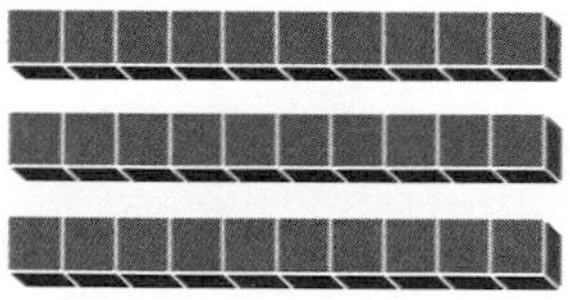

triangle (page 135)

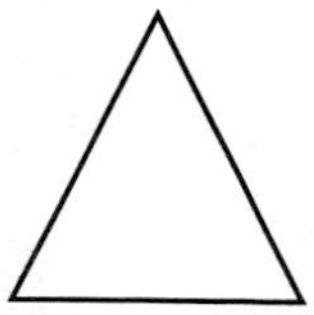

Photo Credits
Page Placement Key: (t)-top (c)-center (b)-bottom (l)-left (r)-right (fg)-foreground (bg)-background (i)-inset

All photos Harcourt Brace & Company except:
Cover: Mark J. Thomas/Dembinsky Photo Associates
Page: 47 (ct) Larry West/Bruce Coleman, Inc.; 47 (cb) Phil Savoie/Bruce Coleman, Inc.; 47 (b) Arthur C. Smith/Grant Heilman Photography; 48 (t) Larry West/Bruce Coleman, Inc.; 48 (ct) Chris Luneski/Natural Selection Stock Photography; 48 (cb) Larry Allan/Bruce Coleman, Inc.; 48 (bl), 48 (br) J.C. Carton/Bruce Coleman, Inc.; 49 Runk/Schoenberger/Grant Heilman; 50 G K & Vikki Hart/The Image Bank; 54 (t) Roy Morsch/The Stock Market; 54 (b) Charles Krebs/The Stock Market; 56 (br) Larry Allen/Bruce Coleman Inc.; 81 (t) G K & Vikki Hart/The Image Bank; 81 (cb) Stephen J. Krasemann/DRK Photo; 94 (t) S. Nielsen/Bruce Coleman, Inc.; 94 (ct) Phil Degginger/Bruce Coleman, Inc.; 94 (cb) Sullivan & Rogers/Bruce Coleman, Inc.; 94 (b), 97 (c) Robert Falls/Bruce Coleman, Inc.; 182 (b) Larry Allan/Bruce Coleman, Inc.; 186 (t) G K & Vikki Hart/The Image Bank; 186 (ct) Rosemary Calvert/Tony Stone Images; 186 (cb) David W. Hamilton/The Image Bank; 186 (b) Zefa Germany/The Stock Market; 286 (t) Lin Miller; 412 George Disario/The Stock Market; 420 (cr) Craig Cook/EarthWater; 420 (cr) George Disario/The Stock Market; 448 (t) Jim Simmen/Tony Stone Images.

Art Credits:
Page: 1, 2, 3, 4, 5, 7, 8, 14, 17, 19, 21, 189, 190, 191, 192, 271, 272, 274, 441, 442, 445, 446, 451 Dennis Greenlaw; 6, 52, 122, 305, 311 Simon Galkin; 6, 30, 67, 82, 127, 151, 152, 243, 249, 273, 274, 275, 276, 355, 356, 431 Silver Editions; 8, 11, 12, 21, 22, 159, 166, 339, 340, 342 Barbara Friedman; 15, 101, 102, 106, 107, 108, 113, 220, 321, 322, 325, 326, 329, 379, 385, 386 Obadinah Heavner; 20, 121, 122, 123, 131 Jenny Campbell; 23, 24 Jackie Snider; 24, 32, 33, 35 Jane Caminos; 25, 26, 135, 136, 141, 164, 263, 264, 266, 267, 268, 411, 412, 415, 416, 439, 440, 443, 444, 456 Bernard Adnet; 28, 35, 171, 174, 316, 392, 426 Leslie Watkins; 31 Jim Paillot; 34, 36, 384, 389, 392, 393 Kenneth Batelman; 37, 38, 51 Toby Williams; 38 Rachel Geswaldo; 39, 40, 429, 430, 431, 432 Kenneth Spengler; 41, 43, 45, 46, 147, 148, 153, 154, 155 Jennifer Bolten; 51 Stacey Schuett; 57, 58, 61, 62, 69, 70, 151, 152, 157, 158 Cynthia Jabar; 59, 60, 63, 64, 244, 251 Eileen Hine; 65, 66 Matt Straub; 71, 72, 75, 76, 83 C.D. Hullinger; 73, 74, 77, 78 Victoria Raymond; 79, 80, 84, 120, 121, 123, 319, 320, 329 Claude Martinot; 85, 86, 91, 92, 97, 98 Franklin Ayers; 87, 88, 95, 96, 97, 98, 116, 420 Georgia Shola; 89, 90, 93, 94, 179, 180, 183, 184, 187, 188, 227, 228, 232, 236, 237, 253, 349, 350, 355, 356 Jennifer Haywood; 99, 100, 104, 108, 111, 118, 299, 314 Deborah Melmon; 103, 109, 110 John Berg; 119, 120 Tracy Sabin; 382, 390 Daniel Moreton; 133, 134, 139, 140, 289, 290 Seth Larson; 137, 138, 193, 194, 195, 259, 260, 261, 265, 266, 287, 288, 291, 292, 323, 324, 343, 344, 345, 346, 353, 354, 357, 369, 370, 373, 374, 386, 401, 402, 408 Terry Taylor; 145, 146, 149, 150, 155, 156 Scott Scheidly; 146, 172, 254, 256, 378 Rita Lascaro; 158 Lane Yerkes; 162, 169 Valeria Petrone; 175, 176 Joy Allen; 177, 178, 181, 182 Annette Cable; 209, 210, 213, 214, 216 Tim Haggerty; 211, 212, 218 Nathan Jarvis; 225, 226, 230, 233, 234 Don Tate; 235, 236, 332, 347, 348, 351, 352, 358, 359, 361, 362, 364, 397, 398, 399, 400, 403 Nancy Tobin; 239, 240, 241, 242, 245, 246, 248, 249 Mike Dammer; 242, 243, 244, 247, 248 Susan Swan; 257, 258, 297, 298, 302, 306, 308 Liisa Chauncy Guida; 273, 278 Jenny Vainisi; 275, 276, 279, 280, 367, 371, 372, 374, 417, 418, 421, 426 Gary Colby; 285, 286, 295 Luisa D'Augusta; 286 Stan Tusan; 317, 413, 414 Darin Johnston; 318 Wendy Wallenberg; 327, 328, 329 Nancy Freeman; 331, 332, 335, 336 David Slonim; 333, 334, 337, 338 Dave Jonason; 365, 366 Sally Jo Vitsky; 385 Lisa Berrett; 395, 396 Bob Holt; 409, 410 Deborah Spina Dixon; 427, 428, 433, 434, 438 Melissa Iwai; 428 Marsha Winborn; 447 Joe Bartos; Book 1: A–H Nathan Jarvis; Book 2: A–H Karen Strelecki; Book 3: A–H Tim Bowers; Book 4: A–H David McPhail; Book 5: A–H Anna Rich; Book 6: A–H Lindy Burnett; Book 7: A–H Gerald McDermot; Book 8: A–H Bernard Adnet; Book 9: A–H Deborah Haley Melmon; Book 10: A–H Ed Martinez